Fundamental Theories of Physics

Volume 199

AF852605

The international monograph series "Fundamental Theories of Physics" aims to stretch the boundaries of mainstream physics by clarifying and developing the theoretical and conceptual framework of physics and by applying it to a wide range of interdisciplinary scientific fields. Original contributions in well-established fields such as Quantum Physics, Relativity Theory, Cosmology, Quantum Field Theory, Statistical Mechanics and Nonlinear Dynamics are welcome. The series also provides a forum for non-conventional approaches to these fields. Publications should present new and promising ideas, with prospects for their further development, and carefully show how they connect to conventional views of the topic. Although the aim of this series is to go beyond established mainstream physics, a high profile and open-minded Editorial Board will evaluate all contributions carefully to ensure a high scientific standard.

More information about this series at http://www.springer.com/series/6001

Silvia De Bianchi · Claus Kiefer
Editors

One Hundred Years of Gauge Theory

Past, Present and Future Perspectives

Springer

Editors
Silvia De Bianchi
Department of Philosophy
Universitat Autonoma de Barcelona
Bellaterra, Spain

Claus Kiefer
Institute for Theoretical Physics
University of Cologne
Cologne, Nordrhein-Westfalen, Germany

ISSN 0168-1222 ISSN 2365-6425 (electronic)
Fundamental Theories of Physics
ISBN 978-3-030-51199-9 ISBN 978-3-030-51197-5 (eBook)
https://doi.org/10.1007/978-3-030-51197-5

This Springer imprint is published by the registered company Springer Nature Switzerland AG
The registered company address is: Gewerbestrasse 11, 6330 Cham, Switzerland

Preface

The origin of gauge theory has been studied by scientists and historians of science in recent decades, but a complex outlook taking into account the historical and philosophical implications is still missing. The aim of this volume is to celebrate hundred years of gauge theory, by considering as seminal starting point of its history the publication of Hermann Weyl's *Raum-Zeit-Materie*. In 1918 Hermann Weyl published the first edition of his masterpiece in which he draws the conceptual underpinnings of gauge invariance later reframed within the context of relativistic quantum mechanics in 1929. This volume aims at stimulating the reflection upon the origin and development of gauge theory and its scientific and philosophical importance. Taking into account one of the central concepts of Weyl's work, symmetry, this volume sheds light on several aspects of Weyl's work and gauge theory and connects theoretical physics with other fields, including mathematics, history and philosophy. The multidisciplinary approach proposed by the volume makes it a unique in the landscape of previous books on the history of gauge theory. Indeed, our scope is to discuss not only the historical and philosophical underpinnings of gauge theory, but also to put forward a discussion about future perspectives of gauge theory taking into account the state of art in both theoretical and experimental physics.

Before resuming the content of the contributors, it is worth mentioning that our aim is to stimulate the interaction and future collaborations among philosophers, physicists and historians in order to grasp from a fresh perspective both Weyl's work and the development and rationale behind gauge theory. This is pretty much in the spirit of Weyl's thought. As it emerges in the contributions, Weyl strongly supported the interaction between philosophical reflection and scientific research, especially in the light of the great revolutions introduced by relativity theory and Quantum Mechanics. For this reason, we decided to group the contributions in this volume to constitute three parts focused on the historical and philosophical underpinnings of gauge theory inspired by Weyl's work, those devoted to Weyl's *Raum-Zeit-Materie* and the philosophical underpinning of his approach, and finally those exploring the theoretical and mathematical physics of gauge theory.

The first part of the volume is introduced by Norbert Straumann's contribution titled "Hermann Weyl's Space-Time Geometry and the Origin of Gauge Theory 100 Years Ago". It focuses on the historical roots of gauge theory by describing the gradual recognition that a common feature of the known fundamental interactions is their gauge structure. Central to his reconstruction is the work of Hermann Weyl and Wolfgang Pauli's early construction in 1953 of a non-Abelian Kaluza-Klein theory. In "Gauging the Spacetime Metric—Looking Back and Forth a Century Later", Erhard Scholz reviews Weyl's 1918 proposal for generalizing Riemannian geometry by local scale gauge, its mathematical foundations, as well as his philosophical and physical implications. Scholz reviews in detail Weyl's disillusion with this research programme and the rise of a convincing alternative for the gauge idea by translating it to the phase of wave functions and spinor fields in quantum mechanics. In mid-20th century years the question of conformal and/or local scale gauge transformation were reconsidered in high energy physics (Bopp, Wess, et al.) and, independently, in gravitation theory (Jordan, Fierz, Brans, Dicke). As Scholz underlines, it is in this context that Weyl geometry attracted new interest among different groups of physicists (Omote-Utiyama-Kugo, Dirac-Canuto-Maeder, Ehlers-Pirani-Schild). The merit of Scholz's contribution is to show that, albeit modified, Weyl's first proposal of his basic geometrical structure finds new interest in present day studies of elementary particle physics, cosmology and philosophy of physics. On the philosophical aspects that Weyl's 1918 proposal implies, Sebastian De Haro proposes an analysis regarding empirical equivalence and duality. In "On Empirical Equivalence and Duality", he argues that theories can be taken to be empirically equivalent on the ground of the judicious reading: very different-looking theories can have equivalent empirical content. The last two contributions regarding this first part of our collection both mark the relevance of gauge symmetry and the necessity of not taking it as mathematical redundancy. This topic is briefly exposed in Carlo Rovelli's contribution "Gauge Is More Than Mathematical Redundancy" and largely debated from a conceptual standpoint by Gabriel Catren in "Homotopic Identities and the Limits of the Interpretation of Gauge Symmetries as 'surplus structure'".

In the second part of the volume, we grouped contributions that can fall under the approach of integrating the history and philosophy of science. They are devoted to Weyl's *Raum-Zeit-Materie*, its conceptual roots and implications, as well as the reconstruction of the debates surrounding philosophical debates. In Dennis Dieks' contribution titled "Reichenbach, Weyl, Philosophy and Gauge", Weyl's approach and phenomenologist stance is compared and contrasted with Reichenbach's logical empiricism. By following the guideline of the reflection upon the nature of space and time and the revolution introduced by relativity, Dieks assesses the nature of Weyl's phenomenological stance, mostly influenced by Husserl's philosophy. In Dieks' view, Weyl's use of phenomenology should be seen as a case of personal heuristics rather than as a systematic modern philosophy of physics. Also in Thomas Ryckman's contribution, Weyl's philosophical views are taken into account. In "Hermann Weyl, the Gauge Principle, and Symbolic Construction from the 'Purely Infinitesimal'", Ryckman reconstructs the history of the development of

Weyl's 1918 formal unification of Einstein's theory and electromagnetism. Then he focuses on its consequences and Weyl's purely mathematical turn in 1925–6 to Lie theory and of course Lie groups and Lie algebras that played prominent roles in the subsequent development of the gauge principle leading up to the Standard Model of particle physics. In Ryckman's view, Weyl's predominant interest in Lie theory stems from two complementary philosophical interests, phenomenology and an epistemologically driven assumption of the "*Nahewirkungsphysik*". Both inform Weyl's notion of symbolic construction, a pillar in his works from 1927 onward. In Silvia De Bianchi's "Weyl's *Raum-Zeit-Materie* and the Philosophy of Science" the philosophical underpinning of Weyl's interpretation of Relativity as emerging from the pages of *Raum-Zeit-Materie* is analysed in detail. In particular, the distinction between the philosophical and the mathematical methods is underlined. De Bianchi underscores the dichotomy and relationship between time and consciousness that is identified by Weyl as the conceptual engine moving the whole history of Western philosophy, and the revolutionary relevance of relativity for its representation is investigated together with the conceptual underpinning of Weyl's philosophy of science. In identifying the main traits of Weyl's philosophy of science in 1918, this paper also offers a philosophical analysis of some underlying concepts of unified field theory.

In the third part of our collection, the reader will find a number of contributions exploring past and current perspectives of gauge theory in different branches of physics, including cosmology, quantum gravity and high energy physics.

Claus Kiefer in "Space, Time, Matter in Quantum Gravity" investigates the role that the three central concepts of Weyl's book play in a quantum theory of the gravitational field. He focuses on quantum geometrodynamics where the key concept is a wave functional on the configuration space of all three geometries and matter fields (Wheeler's superspace). At the most fundamental level, time is absent; the standard notion of time (and spacetime) only emerges in an appropriate semiclassical limit. He reviews ideas about the origin of matter from topology and from a unified quantum theory of interactions—problems which so far remain unsolved.

Friedrich Hehl and Yuri Obukhov in "Conservation of Energy-Momentum of Matter as the Basis for the Gauge Theory of Gravitation" give a concise overview of gauge theories of gravity. These are constructed by starting from a rigid symmetry that is made local. Of great relevance is the Poincaré gauge theory of gravity for which the global Poincaré symmetry of special relativity is employed. Therefore, they emphasize the role that Gauge theories of gravity may play in the construction of a unified field theory.

Christian Steinwachs in "Higgs Field in Cosmology" investigates features of the Standard Model when applied to cosmology. A central role in this is played by the Higgs field, and Steinwachs entertains the idea that this field could lead to the inflationary expansion of the early universe. This is, in fact, a promising idea because no new speculative field is needed in this case. Steinwachs also elaborates on the role of Higgs inflation in quantum cosmology and the quantum equivalence (or non-equivalence) of different field parametrizations.

In “The Gauge Theoretical Underpinnings of General Relativity”, Thomas Schücker compares various structural approaches to general relativity: the field theoretic approach, chrono- and geometric approaches and, in more detail, the gauge theoretic approach. The latter approach exhibits many similarities with the gauge theory underlying the Standard Model, although important differences remain.

Finally, we decided to close our volume with a contribution by Gerard ’t Hooft, titled “Past and Future of Gauge Theory”. ‘t Hooft is himself one of the key figures in the historic development of gauge theories. In his contribution, he gives a colourful and personal account of this development and of the main scientists who were involved in it. He makes a strong case for the importance of gauge theories in the future and speculates in particular about the fundamental role that conformal symmetry might play in the unification of the Standard Model with gravity. Whatever the future will bring, gauge theories will continue being of interest for another hundred years.

Barcelona, Spain — Silvia De Bianchi
Cologne, Germany — Claus Kiefer
April 2020

Acknowledgements We would like to acknowledge the Wilhelm and Else Heraeus Stiftung for their most generous and efficient support of our conference “Hundred Years of Gauge Theory” that took place from July 30 to August 3, 2018, at Physikzentrum Bad Honnef. For this research Silvia De Bianchi benefited from the Ramón y Cajal program (RYC-2015-17289) and the publication has been made possible by receiving funding from the European Union’s Horizon 2020 research and innovation programme under grant agreement No 758145.

Contents

Theoretical and Mathematical Physics of Gauge Theory

Contributors

Gabriel Catren Laboratoire SPHERE (UMR 7219), Université de Paris - CNRS, Paris Cedex 13, France

Silvia De Bianchi Department of Philosophy, Autonomous University of Barcelona, Bellaterra, Spain

Sebastian De Haro Trinity College, Cambridge, UK;
Department of History and Philosophy of Science, University of Cambridge, Cambridge, UK;
Vossius Center for History of Humanities and Sciences, University of Amsterdam, Amsterdam, Netherlands

Dennis Dieks History and Philosophy of Science, Utrecht University, Utrecht, Netherlands

Friedrich W. Hehl Institute of Theoretical Physics, University of Cologne, Köln, Germany

Gerard 't Hooft Institute for Theoretical Physics Utrecht University, TB, Utrecht, The Netherlands

Claus Kiefer Institute for Theoretical Physics, University of Cologne, Köln, Germany

Yuri N. Obukhov Nuclear Safety Institute, Russian Academy of Sciences, Moscow, Russia

Carlo Rovelli Aix Marseille Université CNRS CPT, Marseille, France;
Université de Toulon CNRS CPT, La Garde, France

Thomas Ryckman Stanford University, Stanford, CA, USA

Erhard Scholz Faculty of Mathematics/Natural Sciences, Interdisciplinary Centre for History and Philosophy of Science, University of Wuppertal, Wuppertal, Germany

Thomas Schücker Aix Marseille Univ, Université de Toulon, CNRS, CPT, Marseille, France

Christian F. Steinwachs Department of Physics, University of Freiburg, Freiburg, Germany

Norbert Straumann Physik-Institut University of Zürich, Zürich, Switzerland

History and Philosophy of Gauge Theory: Weyl's *Raum-Zeit-Materie* and Its Legacy

Hermann Weyl's Space-Time Geometry and the Origin of Gauge Theory 100 Years Ago

Norbert Straumann

Abstract One of the major developments of twentieth century physics has been the gradual recognition that a common feature of the known fundamental interactions is their gauge structure. In this lecture the early history of gauge theory is reviewed, emphasizing especially Hermann Weyl's seminal contributions of 1918 and 1929. Wolfgang Pauli's early construction in 1953 of a non-Abelian Kaluza-Klein theory is described in some detail.

1 Introduction

The history of gauge theories begins with General Relativity, which can be regarded as a non-Abelian gauge theory of a special type. To a large extent the other gauge theories emerged in a slow and complicated process gradually from General Relativity. Their common geometrical structure—best expressed in terms of connections of fiber bundles—is now widely recognized.

It all began with Weyl [1], who made in 1918 the first attempt to extend General Relativity in order to describe gravitation and electromagnetism within a unifying geometrical framework. This brilliant proposal contains the germs of all *mathematical* aspects of non-Abelian gauge theory. For what was later called by Weyl 'gauge' (German: 'Eich-') invariance he used in this paper the word scale-invariance ('Maßstab-Invarianz'), meaning invariance under change of length or change of calibration.

Einstein admired Weyl's theory as[1] "*a coup of genius of the first rate*" but immediately realized that it was physically untenable. After a long discussion Weyl finally admitted that his attempt was a failure as a physical theory. (For a discussion of the intense Einstein-Weyl correspondence, see Ref. [2].) It paved, however, the way for

[1]German original: "Es ist ein Genie-Streich ersten Ranges".

N. Straumann (✉)
Physik-Institut University of Zürich, Zürich, Switzerland
e-mail: norbert.straumann@gmail.com

S. De Bianchi and C. Kiefer (eds.), *One Hundred Years of Gauge Theory*, Fundamental Theories of Physics 199, https://doi.org/10.1007/978-3-030-51197-5_1

Fig. 1 Hermann Weyl. *Source* ETH-Bibliothek Zürich, Bildarchiv. Licensed under the Creative Commons Attribution-Share Alike 3.0 Unported license

the correct understanding of gauge invariance. Weyl himself reinterpreted in 1929 his original theory after the advent of quantum theory in a grand paper [3]. Weyl's reinterpretation of his earlier speculative proposal had actually been suggested before by London [4]. Fock [5], Klein [6], and others arrived at the principle of gauge invariance in the framework of wave mechanics along a completely different line. It was, however, Weyl who emphasized the role of gauge invariance as a *constructive principle* from which electromagnetism can be derived. This point of view became very fruitful for our present understanding of fundamental interactions. (For a more extensive discussion, see [7]) (Fig. 1).

2 Weyl's Attempt to Unify Gravitation and Electromagnetism

On the 1st of March 1918 Weyl writes in a letter to Einstein ([8], Vol. 8B, Doc.472)[2] *"These days I succeeded, as I believe, to derive electricity and gravitation from a common source ...".* Einstein's prompt reaction by postcard indicates already a physical objection which he explained in detail shortly afterwards. Before we come to this we have to describe Weyl's theory of 1918.

Weyl's starting point was purely mathematical. He felt a certain uneasiness about Riemannian geometry, as is clearly expressed by the following sentences early in his paper:

> But in Riemannian geometry described above there is contained a last element of geometry "at a distance" (ferngeometrisches Element)—with no good reason, as far as I can see; it is due only to the accidental development of Riemannian geometry from Euclidean geometry. The metric allows the two magnitudes of two vectors to be compared, not only at the same point, but at any arbitrarily separated points. A true infinitesimal geometry should, however, recognize only a principle for transferring the magnitude of a vector to an infinitesimally close point and then, on transfer to an arbitrary distant point, the integrability of the magnitude of a vector is no more to be expected than the integrability of its direction.

After these remarks Weyl turns to physical speculation and continues as follows:

> On the removal of this inconsistency there appears a geometry that, surprisingly, when applied to the world, explains not only the gravitational phenomena but also the electrical. According to the resultant theory both spring from the same source, indeed in general one cannot separate gravitation and electromagnetism in a unique manner. In this theory all physical quantities have a world geometrical meaning; the action appears from the beginning as a pure number. It leads to an essentially unique universal law; it even allows us to understand in a certain sense why the world is four-dimensional.

3 Weyl's Generalization of Riemannian Geometry

In this section we describe in some detail Weyl's geometry in a bundle theoretical language. I prefer this, because it is common with that of non-abelian gauge theories (on the classical level).

In Weyl's geometry the spacetime manifold M is equipped with a *conformal structure*, i.e., with a class $[g]$ of conformally equivalent Lorentz metrics g (and not a definite metric as in General Relativity). For such a conformal manifold $(M, [g])$ we can introduce the bundle of *conformal frames*, which are linear frames (X_0, X_1, X_2, X_3) for which $g_p(X_\mu, X_\nu) = \exp(2\lambda(p))\eta_{\mu\nu}$, where $\eta = (\eta_{\mu\nu}) = \mathrm{diag}(-1, 1, 1, 1)$, for any (and thus all) $g \in [g]$. The set $W(M)$ of conformal frames on M can be regarded in an obvious manner as the total space of a principle fibre bundle, whose structure group

[2] German original: "Dieser Tage ist es mir, wie ich glaube, gelungen, Elektrizität und Gravitation aus einer gemeinsamen Quelle herzuleiten ...".

G is the group consisting of all positive multiples of homogeneous Lorentz transformations, i.e., $G \cong O(1,3) \times \mathbb{R}_+$. This *conformal (Weyl) bundle* is a reduction of the bundle of linear frames $L(M)$ (and an extension of the bundle of orthonormal frames for every $g \in [g]$). A *Weyl connection* is a torsion-free connection on $W(M)$, defined by a connection form ω. (As such it has a unique extension to $L(M)$.) The canonical 1-form θ on $W(M)$, i.e., the restriction of the soldering form on $L(M)$, satisfies $D^\omega\theta = 0$, where D^ω is the exterior covariant derivative belonging to ω, expressing the vanishing torsion. Since the connection form has values in the Lie algebra $\mathcal{G}$ of G, i.e., in $o(1,3) \oplus \mathbb{R}$, we can split ω uniquely

$$\omega = \hat{\omega} + \phi \cdot 1, \tag{1}$$

where $\hat{\omega}$ has values in $o(1,3)$ and ϕ is an $\mathbb{R}$-valued 1-form on $W(M)$. Thus in matrix notation

$$\hat{\omega}^T\eta + \eta\hat{\omega} = 0, \quad \omega^T\eta + \eta\omega = 2\phi\eta. \tag{2}$$

A Weyl connection can be considered as a torsion free linear connection, which is reducible to a connection in $W(M)$. The restriction of $\hat{\omega}$ to any orthonormal frame bundle $O_g(M) \subset W(M)$, $g \in [g]$, defines a metric connection in $O_g(M)$ with torsion. Since the torsion of the Weyl connection vanishes, the first structure equation reads

$$d\theta + \omega \wedge \theta = 0. \tag{3}$$

The curvature $\Omega = D^\omega\omega$ is determined by the second structure equation

$$\Omega = d\omega + \omega \wedge \omega, \tag{4}$$

which can be written as

$$\Omega = (d\hat{\omega} + \hat{\omega} \wedge \hat{\omega}) + d\phi \cdot 1. \tag{5}$$

A *Weyl space* is a conformal manifold together with a Weyl connection.

The frames $\sigma(x) = \{e_\mu(x)\}$ of a local section $\sigma : U \to W(M)$ are dual to to the components θ^μ of $\sigma^*\theta$,

$$\theta^\mu(e_\nu) = \delta^\mu_\nu. \tag{6}$$

For any metric $g \in [g]$ we can choose local sections such that the frames $\{e_\mu(x)\}$ are orthonormal with respect to g,

$$g = \eta_{\mu\nu}\theta^\mu \otimes \theta^\nu. \tag{7}$$

The exterior covariant derivative of g has relative to the dual basis $\{\theta^\mu\}$ the components[3]

---

[3]In the local equations ω^α_β denotes the pull-back $\sigma^*(\omega^\alpha_\beta)$.

$$(Dg)_{\mu\nu} = d\eta_{\mu\nu} - \omega^{\lambda}{}_{\mu}\eta_{\lambda\nu} - \omega^{\lambda}{}_{\nu}\eta_{\lambda\mu} \overbrace{=}^{(2)} -2A\eta_{\mu\nu}, \tag{8}$$

with $A := \sigma^*\phi$. Thus

$$Dg = -2A \otimes g. \tag{9}$$

If g is replaced by $\tilde{g} = e^{2\lambda}g \in [g]$ then $D\tilde{g} = -2\tilde{A} \otimes \tilde{g}$, where $\tilde{A} = A - d\lambda$.

This leads us to the concept of a covariant Weyl derivative on a conformal manifold $(M, [g])$: A *covariant Weyl derivative* ∇ on a conformal manifold $(M, [g])$ is a covariant torsionless derivative on the spacetime manifold M which satisfies the condition

$$\nabla g = -2A \otimes g, \tag{10}$$

where the map $A : [g] \to \Lambda^1(M)$ satisfies

$$A(e^{2\lambda}g) = A(g) - d\lambda. \tag{11}$$

$A(g)$ is the *gauge potential* belonging to g, and (11) is what Weyl called a *gauge transformation.*

It is not difficult to show that there is a bijective relation between the set of covariant Weyl derivatives on a conformal manifold $(M, [g])$ and the set of Weyl connection forms on the corresponding conformal bundle.

Existence of covariant Weyl derivatives For the existence and explicit formulae of covariant Weyl derivatives we generalize the well-known Koszul treatment of the Levi-Civita connection. In particular we generalize the Koszul formula (see, e.g., [9], Eq. (15.42)) for the covariant Levi-Civita derivative ∇^{LC} to

$$g(\nabla_Z Y, X) = g(\nabla_Z^{LC} Y, X) + [-A(X)g(Y, Z) + A(Y)g(Z, X) + A(Z)g(X, Y)]. \tag{12}$$

This equation defines ∇_X in terms of g and A.

Derivation of (12). Equation (10) reads explicitly

$$(\nabla_X g)(Y, Z) = Xg(Y, Z) - g(\nabla_X Y, Z) - g(Y, \nabla_X Z) = -2A(X)g(Y, Z). \tag{13}$$

Since the torsion vanishes, i.e., $\nabla_X Y - \nabla_Y X - [X, Y] = 0$, we can write this as

$$Xg(Y, Z) = g(\nabla_Y X, Z) + g([X, Y], Z) + g(Y, \nabla_X Z) + 2A(X)g(Y, Z). \tag{14}$$

After cyclic permutations, we obtain as in the derivation of the standard Koszul formula, Eq. (12).

With routine calculations one verifies that the generalized Koszul formula (12) defines a covariant derivative with vanishing torsion, and moreover it satisfies the defining property (10). (In these calculations one uses that the Levi-Civita derivative has vanishing torsion and that the metricity of ∇^{LC} is equivalent to the Ricci identity [9], Eq. (15.39)).

Local formula. Choose in (12) $X = \partial_\mu,\ Y = \partial_\nu,\ ,Z = \partial_\lambda$ of local coordinates. Then we obtain

$$\langle \nabla_{\partial_\mu} \partial_\nu, \partial_\lambda \rangle = \frac{1}{2}(-g_{\nu\mu,\lambda} + g_{\mu\lambda,\nu} + g_{\lambda\nu,\mu)}) + (-A_\mu g_{jk} + A_\nu g_{\nu\lambda} + A_\lambda g_{\mu\nu}). \quad (15)$$

In other words, one has to perform in the Christoffel symbols of the Levi-Civita connection the substitution

$$g_{\mu\nu,\lambda} \to g_{\mu\nu,\lambda} - 2A_\lambda g_{\mu\nu}. \quad (16)$$

Consider now a curve $\gamma : [0, 1] \to M$ and a parallel-transported vector field X along γ. If $l(t)$ is the length of $X(t)$, measured with a representative $g \in [g]$, we obtain from (10)

$$\frac{\dot{l}}{l} = \frac{1}{2l^2}(\nabla_{\dot\gamma} g)(X(t), X(t)) = -A(\dot\gamma), \quad (17)$$

and thus the following relation between $l(p)$ for the initial point $p = \gamma(0)$ and $l(q)$ for the end point $q = \gamma(1)$:

$$l(q) = \exp\left(-\int_\gamma A\right) l(p). \quad (18)$$

Equation (11) implies that this relation holds for all $g \in [g]$. Therefore, the ratio of lengths in q and p (measured with $g \in [g]$) *depends in general on the connecting path* γ (see Fig. 2). The length is only independent of γ if the exterior differential of A,

$$F = dA \quad (F_{\mu\nu} = \partial_\mu A_\nu - \partial_\nu A_\mu), \quad (19)$$

vanishes.

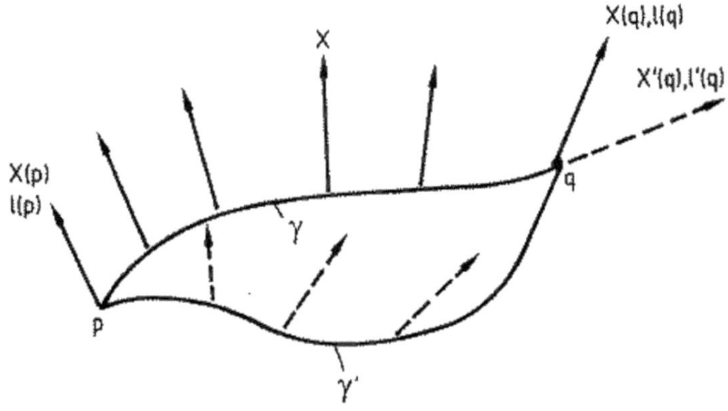

Fig. 2 Path dependence of parallel displacement and transport of length in Weyl spacetime

Note that (18) holds in particular for a geodesic ($\nabla_{\dot{\gamma}}\dot{\gamma} = 0$) and $X = \dot{\gamma}$. So the length ot the tangent vector $\dot{\gamma}$ does not remain constant as in the pseudo-Riemannian case.

4 Electromagnetism and Gravitation

Turning to physics, Weyl assumes that his "purely infinitesimal geometry" describes the structure of spacetime and consequently he requires that physical laws should satisfy a double-invariance: 1. They must be invariant with respect to arbitrary smooth coordinate transformations. 2. They must be *gauge invariant*, i.e., invariant with respect to substitutions

$$g \to e^{2\lambda} g, \quad A \to A - d\lambda, \tag{20}$$

for an arbitrary smooth function λ.

Nothing is more natural to Weyl, than identifying A_μ with the vector potential and $F_{\mu\nu}$ in Eq. (19) with the field strength of electromagnetism. In the absence of electromagnetic fields ($F_{\mu\nu} = 0$) the scale factor $\exp(-\int_\gamma A)$ in (18) for length transport becomes path independent (integrable) and one can find a gauge such that A_μ vanishes for simply connected spacetime regions. In this special case one is in the same situation as in General Relativity.

Weyl proceeds to find an action which is generally invariant as well as gauge invariant and which would give the coupled field equations for g and A. We do not want to enter into this, except for the following remark. In his first paper [1] Weyl proposes what we call nowadays the Yang-Mills action

$$S(g, A) = -\frac{1}{4} \int Tr(\Omega \wedge *\Omega). \tag{21}$$

Here Ω denotes the curvature form and $*\Omega$ its Hodge dual. Note that the latter is gauge invariant, i.e., independent of the choice of $g \in [g]$. In Weyl's geometry the curvature form splits as $\Omega = \hat{\Omega} + F$, where $\hat{\Omega}$ is the metric piece [10]. Correspondingly, the action also splits,

$$S(g, A) = -\frac{1}{4} \int Tr(\hat{\Omega} \wedge *\hat{\Omega}) - \frac{1}{4} \int F \wedge *F. \tag{22}$$

The second term is just the Maxwell action. Weyl's theory thus contains formally all aspects of a non-Abelian gauge theory.[4]

Weyl emphasizes, of course, that the Einstein-Hilbert action is not gauge invariant. Later work by Pauli [12] and by Weyl himself [1, 11] led soon to the conclusion that

[4]The integrand in Eq. (21) is in local coordinates indeed identical to the scalar density $R_{\alpha\beta\gamma\delta}R^{\alpha\beta\gamma\delta}\sqrt{-g}dx^0 \wedge \ldots \wedge dx^3$ which is used by Weyl ($R_{\alpha\beta\gamma\delta}$= the curvature tensor of the Weyl connection).

the action (21) could not be the correct one, and other possibilities were investigated (see the later editions of Weyl's classic treatise [11]).

Independent of the precise form of the action Weyl shows that in his theory gauge invariance implies the *conservation of electric charge* in much the same way as general coordinate invariance leads to the conservation of energy and momentum.[5] This beautiful connection pleased him particularly [6] "*...[it] seems to me to be the strongest general argument in favour of the present theory—insofar as it is permissible to talk of justification in the context of pure speculation.*" The invariance principles imply five 'Bianchi type' identities. Correspondingly, the five conservation laws follow in two independent ways from the coupled field equations and may be "termed the eliminants" of the latter. These structural connections hold also in modern gauge theories.

4.1 Einstein's Objection and Reactions of Other Physicists

After this sketch of Weyl's theory we come to Einstein's striking counterargument which he first communicated to Weyl by postcard. The problem is that if the idea of a nonintegrable length connection (scale factor) is correct, then the behavior of clocks would depend on their history. Consider two identical atomic clocks in adjacent world points and bring them along different world trajectories which meet again in adjacent world points. According to (21) their frequencies would then generally differ. This is in clear contradiction with empirical evidence, in particular with the existence of stable atomic spectra. Einstein therefore concludes (see [8], Vol. 8B, Doc. 507) [7]

> ...(if) one drops the connection of the ds to the measurement of distance and time, then relativity loses all its empirical basis.

Nernst shared Einstein's objection and demanded on behalf of the Berlin Academy that it should be printed in a short amendment to Weyl's article. Weyl had to accept this. We have described the intense and instructive subsequent correspondence between Weyl and Einstein elsewhere [2] (see also Vol. 8B of [8]). As an example, let us quote from one of the last letters of Weyl to Einstein ([8], Vol. 8B, Doc. 669):

> This [insistence] irritates me of course, because experience has proven that one can rely on your intuition; so unconvincing as your counterarguments seem to me, as I have to admit ...

[5]We adopt here the somewhat naive interpretation of energy-momentum conservation for generally invariant theories of the older literature.

[6]German original:"...[dies] erscheint mir als eines der stärksten Argumente zugunsten der hier vorgetragenen Theorie—soweit im rein Spekulativen überhaupt von einer Bestätigung die Rede sein kann".

[7]German original:"Lässt man den Zusammenhang des ds mit Massstab- und Uhr-Messungen fallen, so verliert die Relativitätstheorie ihre empirische Basis".

> By the way, you should not believe that I was driven to introduce the linear differential form in addition to the quadratic one by physical reasons. I wanted, just to the contrary, to get rid of this 'methodological inconsistency (Inkonsequenz)' which has been a bone of contention to me already much earlier. And then, to my surprise, I realized that it looked as if it might explain electricity. You clap your hands above your head and shout: But physics is not made this way ! (Weyl to Einstein 10.12.1918).

Weyl's reply to Einstein's criticism was, generally speaking, this: The real behavior of measuring rods and clocks (atoms and atomic systems) in arbitrary electromagnetic and gravitational fields can be deduced only from a dynamical theory of matter.

Not all leading physicists reacted negatively. Einstein transmitted a very positive first reaction by Planck, and Sommerfeld wrote enthusiastically to Weyl that there was "*...hardly doubt, that you are on the correct path and not on the wrong one.*"

In his encyclopedia article on relativity [13] Pauli gave a lucid and precise presentation of Weyl's theory, but commented on Weyl's point of view very critically. At the end he states[8]

> ...In summary one may say that Weyl's theory has not yet contributed to getting closer to the solution of the problem of matter.

Also Eddington's reaction was at first very positive but he soon changed his mind and denied the physical relevance of Weyl's geometry.

The situation was later appropriately summarized by F. London in his 1927 paper [4] as follows:

> In the face of such elementary experimental evidence, it must have been an unusually strong metaphysical conviction that prevented Weyl from abandoning the idea that Nature would have to make use of the beautiful geometrical possibility that was offered. He stuck to his conviction and evaded discussion of the above-mentioned contradictions through a rather unclear re-interpretation of the concept of "real state", which, however, robbed his theory of its immediate physical meaning and attraction.

In this remarkable paper, London suggested a reinterpretation of Weyl's principle of gauge invariance within the new quantum mechanics: The role of the metric is taken over by the wave function, and the rescaling of the metric has to be replaced by a phase change of the wave function.

In this context an astonishing early paper by Schrödinger [14] has to be mentioned, which also used Weyl's "World Geometry" and is related to Schrödinger's later invention of wave mechanics. This relation was discovered by Raman and Forman [15]. (See also the discussion by Yang [18].)

Even earlier than London, Fock [5] arrived along a completely different line at the principle of gauge invariance in the framework of wave mechanics. His approach was similar to the one by Klein [6].

The contributions by Schrödinger [14], London [4] and Fock [5] are commented in [17], where also English translations of the original papers can be found. Here, we concentrate on Weyl's seminal paper "Electron and Gravitation".

---

[8] "Zusammenfassend kann man sagen, dass es der Theorie von Weyl bisher nicht gelungen ist, das Problem der Materie der Lösung näher zu bringen".

5 Weyl's 1929 Classic: "Electron and Gravitation"

Shortly before his death late in 1955, Weyl wrote for his *Selecta* [19] a postscript to his early attempt in 1918 to construct a 'unified field theory'. There he expressed his deep attachment to the gauge idea and adds (p. 192)[9]

> Later the quantum-theory introduced the Schrödinger-Dirac potential ψ of the electron-positron field; it carried with it an experimentally-based principle of gauge-invariance which guaranteed the conservation of charge, and connected the ψ with the electromagnetic potentials A_μ in the same way that my speculative theory had connected the gravitational potentials $g_{\mu\nu}$ with the A_μ, and measured the A_μ in known atomic, rather than unknown cosmological units. I have no doubt but that the correct context for the principle of gauge-invariance is here and not, as I believed in 1918, in the intertwining of electromagnetism and gravity.

This re-interpretation was developed by Weyl in one of the great papers of the 20th century [3]. Weyl's classic does not only give a very clear formulation of the gauge principle, but contains, in addition, several other important concepts and results—in particular his two-component spinor theory.

The modern version of the gauge principle is already spelled out in the introduction:

> The Dirac field-equations for ψ together with the Maxwell equations for the four potentials f_p of the electromagnetic field have an invariance property which is formally similar to the one which I called gauge-invariance in my 1918 theory of gravitation and electromagnetism; the equations remain invariant when one makes the simultaneous substitutions
>
> $$\psi \text{ by } e^{i\lambda}\psi \quad \text{and} \quad f_p \text{ by } f_p - \frac{\partial \lambda}{\partial x^p},$$
>
> where λ is understood to be an arbitrary function of position in four-space. Here the factor $\frac{e}{ch}$, where $-e$ is the charge of the electron, c is the speed of light, and $\frac{h}{2\pi}$ is the quantum of action, has been absorbed in f_p. The connection of this "gauge invariance" to the conservation of electric charge remains untouched. But a fundamental difference, which is important to obtain agreement with observation, is that the exponent of the factor multiplying ψ is not real but pure imaginary. ψ now plays the role that Einstein's ds played before. It seems to me that this new principle of gauge-invariance, which follows not from speculation but from experiment, tells us that the electromagnetic field is a necessary accompanying phenomenon, not of gravitation, but of the material wave-field represented by ψ. Since gauge-invariance involves an arbitrary function λ it has the character of "general" relativity and can naturally only be understood in that context.

We shall soon enter into Weyl's justification which is, not surprisingly, strongly associated with General Relativity. Before this we have to describe his incorporation

[9]Später führte die Quantentheorie die Schrödinger-Diracschen Potentiale ψ des Elektron-Positron-Feldes ein; in ihr trat ein aus der Erfahrung gewonnenes und die Erhaltung der Ladung garantierendes Prinzip auf, das die ψ mit den elektromagnetischen Potentialen φ_i in ähnlicher Weise verknüpft wie meine spekulative Theorie die Gravitationspotentiale g_{ik} mit den φ_i, wobei zudem die φ_i in einer bekannten atomaren statt in einer unbekannten kosmologischen Einheit gemessen werden. Es scheint mir kein Zweifel, dass das Prinzip der Eichinvarianz hier seine richtige Stelle hat, und nicht, wie ich 1918 geglaubt hatte, im Zusammenspiel von Gravitation und Elektrizität".

of the Dirac theory into General Relativity which he achieved with the help of the tetrad formalism.

One of the reasons for adapting the Dirac theory of the spinning electron to gravitation had to do with Einstein's recent unified theory which invoked a distant parallelism with torsion. Wigner [20] and others had noticed a connection between this theory and the spin theory of the electron. Weyl did not like this and wanted to dispense with teleparallelism. In the introduction he says:

> I prefer not to believe in distant parallelism for a number of reasons. First my mathematical intuition objects to accepting such an artificial geometry; I find it difficult to understand the force that would keep the local tetrads at different points and in rotated positions in a rigid relationship. There are, I believe, two important physical reasons as well. The loosening of the rigid relationship between the tetrads at different points converts the gauge-factor $e^{i\lambda}$, which remains arbitrary with respect to ψ, from a constant to an arbitrary function of space-time. In other words, only through the loosening the rigidity does the established gauge-invariance become understandable.

This thought is carried out in detail after Weyl has set up his two-component theory in special relativity, including a discussion of P and T invariance. He emphasizes thereby that the two-component theory excludes a linear implementation of parity and remarks: "*It is only the fact that the left-right symmetry actually appears in Nature that forces us to introduce a second pair of ψ-components.*" To Weyl the mass-problem is thus not relevant for this.[10] Indeed he says: "Mass, however, is a gravitational effect; thus there is hope of finding a substitute in the theory of gravitation that would produce the required corrections."

5.1 Tetrad Formalism

In order to incorporate his two-component spinors into General Relativity, Weyl was forced to make use of local tetrads (Vierbeine). In Sect. 2 of his paper he develops the tetrad formalism in a systematic manner. This was presumably independent work, since he does not give any reference to other authors. It was, however, mainly E. Cartan who demonstrated with his work [21] the usefulness of locally defined orthonormal bases –also called moving frames– for the study of Riemannian geometry.

In the tetrad formalism the metric is described by an arbitrary basis of orthonormal vector fields $\{e_\alpha(x); \alpha = 0, 1, 2, 3\}$. If $\{e^\alpha(x)\}$ denotes the dual basis of 1-forms, the metric is given by

$$g = \eta_{\mu\nu} e^\mu(x) \otimes e^\nu(x), \quad (\eta_{\mu\nu}) = diag(1, -1, -1, -1). \tag{23}$$

[10] At the time it was thought by Weyl, and indeed by all physicists, that the 2-component theory requires a zero mass. In 1957, after the discovery of parity nonconservation, it was found that the 2-component theory could be consistent with a finite mass. See K. M. Case, [22].

Weyl emphasizes, of course, that only a class of such local tetrads is determined by the metric: the metric is not changed if the tetrad fields are subject to spacetime-dependent Lorentz transformations:

$$e^{\alpha}(x) \rightarrow \Lambda^{\alpha}_{\ \beta}(x) e^{\beta}(x). \tag{24}$$

With respect to a tetrad, the connection forms $\omega = (\omega^{\alpha}_{\ \beta})$ have values in the Lie algebra of the homogeneous Lorentz group:

$$\omega_{\alpha\beta} + \omega_{\beta\alpha} = 0. \tag{25}$$

(Indices are raised and lowered with $\eta^{\alpha\beta}$ and $\eta_{\alpha\beta}$, respectively.) They are determined (in terms of the tetrad) by the first structure equation of Cartan:

$$de^{\alpha} + \omega^{\alpha}_{\ \beta} \wedge e^{\beta} = 0. \tag{26}$$

(For a textbook derivation see, e.g., [9], especially Sects. 2.6 and 8.5.) Under local Lorentz transformations (24) the connection forms transform in the same way as the gauge potential of a non-Abelian gauge theory:

$$\omega(x) \rightarrow \Lambda(x)\omega(x)\Lambda^{-1}(x) - d\Lambda(x)\Lambda^{-1}(x). \tag{27}$$

The curvature forms $\Omega = (\Omega^{\mu}_{\ \nu})$ are obtained from ω in exactly the same way as the Yang-Mills field strength from the gauge potential:

$$\Omega = d\omega + \omega \wedge \omega \tag{28}$$

(second structure equation).

For a vector field V, with components V^{α} relative to $\{e_{\alpha}\}$, the covariant derivative DV is given by

$$DV^{\alpha} = dV^{\alpha} + \omega^{\alpha}_{\ \beta} V^{\beta}. \tag{29}$$

Weyl generalizes this in a unique manner to spinor fields ψ belonging to representations ρ of $SL(2, \mathbf{C})$:

$$D\psi = d\psi + \rho_*(\omega)\psi = d\psi + \frac{1}{4}\omega_{\alpha\beta}\sigma^{\alpha\beta}\psi. \tag{30}$$

Here, ρ_* denotes the induced representation of the Lie algebra. For a Dirac field $\sigma^{\alpha\beta}$ are the familiar matrices

$$\sigma^{\alpha\beta} = \frac{1}{2}[\gamma^{\alpha}, \gamma^{\beta}]. \tag{31}$$

(For 2-component Weyl fields one has similar expressions in terms of the Pauli matrices.)

With these tools the action principle for the coupled Einstein-Dirac system can be set up. In the massless case the Lagrangian is

$$\mathcal{L} = \frac{1}{16\pi G} R - i\bar{\psi}\gamma^{\mu} D_{\mu}\psi, \tag{32}$$

where the first term is just the Einstein-Hilbert Lagrangian (which is linear in Ω). Weyl discusses, of course, immediately the consequences of the following two symmetries:

(i) local Lorentz invariance,
(ii) general coordinate invariance.

5.2 *The New Form of the Gauge-Principle*

All this is a kind of a preparation for the final section of Weyl's paper, which has the title "electric field". Weyl says:

> We come now to the critical part of the theory. In my opinion the origin and necessity for the electromagnetic field is in the following. The components ψ_1 ψ_2 are, in fact, not uniquely determined by the tetrad but only to the extent that they can still be multiplied by an arbitrary "gauge-factor" $e^{i\lambda}$. The transformation of the ψ induced by a rotation of the tetrad is determined only up to such a factor. In special relativity one must regard this gauge-factor as a constant because here we have only a single point-independent tetrad. Not so in General Relativity; every point has its own tetrad and hence its own arbitrary gauge-factor; because by the removal of the rigid connection between tetrads at different points the gauge-factor necessarily becomes an arbitrary function of position.

In this manner Weyl arrives at the gauge-principle in its modern form and emphasizes: "*From the arbitrariness of the gauge-factor in ψ appears the necessity of introducing the electromagnetic potential.*" *The first term $d\psi$ in* (30) *has now to be replaced by the covariant gauge derivative $(d - iA)\psi$ and the nonintegrable scale factor* (19) of the old theory is now replaced by a phase factor:

$$\exp\left(-\int_{\gamma} A\right) \rightarrow \exp\left(-i\int_{\gamma} A\right),$$

which corresponds to the replacement of the original gauge group $\mathbb{R}$ by the compact group $U(1)$. Accordingly, the original Gedankenexperiment of Einstein translates now to the Aharonov-Bohm effect, as was first pointed out by Yang [16]. The close connection between gauge invariance and conservation of charge is again uncovered. The current conservation follows, as in the original theory, in two independent ways: On the one hand it is a consequence of the field equations for matter plus gauge invariance, at the same time, however, also of the field equations for the electromagnetic field plus gauge invariance. This corresponds to an identity in the coupled system of field equations which has to exist as a result of gauge invariance. All this is nowadays familiar to students of physics and does not need to be explained in more detail.

Much of Weyl's paper penetrated also into his classic book "*The Theory of Groups and Quantum Mechanics*" [23]. There he mentions also the transformation of his early gauge-theoretic ideas: "*This principle of gauge invariance is quite analogous to that previously set up by the author, on speculative grounds, in order to arrive at a unified theory of gravitation and electricity. But I now believe that this gauge invariance does not tie together electricity and gravitation, but rather electricity and matter.*" When Pauli saw the full version of Weyl's paper he became more friendly and wrote [24]:

> In contrast to the nasty things I said, the essential part of my last letter has since been overtaken, particularly by your paper in Z. f. Physik. For this reason I have afterward even regretted that I wrote to you. After studying your paper I believe that I have really understood what you wanted to do (this was not the case in respect of the little note in the Proc.Nat.Acad.). First let me emphasize that side of the matter concerning which I am in full agreement with you: your incorporation of spinor theory into gravitational theory. I am as dissatisfied as you are with distant parallelism and your proposal to let the tetrads rotate independently at different space-points is a true solution.

In brackets Pauli adds:

> Here I must admit your ability in Physics. Your earlier theory with $g'_{ik} = \lambda g_{ik}$ was pure mathematics and unphysical. Einstein was justified in criticizing and scolding. Now the hour of your revenge has arrived.

Then he remarks in connection with the mass-problem

> Your method is valid even for the massive [Dirac] case. I thereby come to the other side of the matter, namely the unsolved difficulties of the Dirac theory (two signs of m_0) and the question of the 2-component theory. In my opinion these problems will not be solved by gravitation ...the gravitational effects will always be much too small.

This remark indicates a major physical problem with classical spinor fields. Soon afterwards, beginning with Dirac's hole theory that led to the quantization of such fields with anticommutation relations, the problem was solved within special relativity, but remains in GR.

Many years later, Weyl summarized this early tortuous history of gauge theory in an instructive letter [25] to the Swiss writer and Einstein biographer C. Seelig, which we reproduce in an English translation.

> The first attempt to develop a unified field theory of gravitation and electromagnetism dates to my first attempt in 1918, in which I added the principle of gauge-invariance to that of coordinate invariance. I myself have long since abandoned this theory in favour of its correct interpretation: gauge-invariance as a principle that connects electromagnetism not with gravitation but with the wave-field of the electron. —Einstein was against it [the original theory] from the beginning, and this led to many discussions. I thought that I could answer his concrete objections. In the end he said "Well, Weyl, let us leave it at that! In such a speculative manner, without any guiding physical principle, one cannot make Physics." Today one could say that in this respect we have exchanged our points of view. Einstein believes that in this field [Gravitation and Electromagnetism] the gap between ideas and experience is so wide that only the path of mathematical speculation, whose consequences must, of course, be developed and confronted with experiment, has a chance of success. Meanwhile my own confidence in pure speculation has diminished, and I see a need for a

closer connection with quantum-physics experiments, since in my opinion it is not sufficient to unify Electromagnetism and Gravity. The wave-fields of the electron and whatever other irreducible elementary particles may appear must also be included.

Independently of Fock [26] also incorporated the Dirac equation into General Relativity by using the same method. On the other hand, Tetrode [27], Schrödinger [28] and Bargmann [29] reached this goal by starting with space-time dependent γ-matrices, satisfying$\{\gamma^\mu, \gamma^\nu\} = 2\, g^{\mu\nu}$. A somewhat later work by Infeld et al. [30] is based on spinor analysis.

6 Gauge Invariance and QED

Gauge invariance became a serious problem when Heisenberg and Pauli began to work on a relativistically invariant Quantum Electrodynamics that eventually resulted in two important papers "*On the Quantum Dynamics of Wave Fields*" [31, 32]. Straightforward application of the canonical formalism led, already for the free electromagnetic field, to nonsensical results. Jordan and Pauli on the other hand, proceeded to show how to quantize the theory of the *free field* case by dealing only with the field strengths $F_{\mu\nu}(x)$. For these they found commutation relations at different space-time points in terms of the now famous invariant Jordan-Pauli distribution that are manifestly Lorentz invariant.

The difficulties concerned with applying the canonical formalism to the electromagnetic field continued to plague Heisenberg and Pauli for quite some time. By mid-1928 both were very pessimistic, and Heisenberg began to work on ferromagnetism.[11] In fall of 1928 Heisenberg discovered a way to bypass the difficulties. He added the term $-\frac{1}{2}\varepsilon(\partial_\mu A^\mu)^2$ to the Lagrangian, in which case the component π_0 of the canonical momenta

$$\pi_\mu = \frac{\partial L}{\partial(\partial_0 A_\mu)}$$

does no more vanish identically ($\pi_0 = -\varepsilon\partial_\mu A^\mu$). The standard canonical quantization scheme can then be applied. At the end of all calculations one could then take the limit $\varepsilon \to 0$.

In their second paper, Heisenberg and Pauli stressed that the Lorentz condition cannot be imposed as an operator identity but only as a supplementary condition

[11]Pauli turned to literature. In a letter of 18 February 1929 he wrote from Zürich to Oskar Klein: "For my proper amusement I then made a short sketch of a utopian novel which was supposed to have the title 'Gulivers journey to Urania' and was intended as a political satire in the style of Swift against present-day democracy. [...] Caught in such dreams, suddenly in January, news from Heisenberg reached me that he is able, with the aid of a trick ... to get rid of the formal difficulties that stood against the execution of our quantum electrodynamics" [31].

selecting admissible states. This discussion was strongly influenced by a paper of Fermi from May 1929.

For this and the further main developments during the early period of quantum field theory, we refer to Chap. 1 of [33].

7 On Pauli's Invention of Non-Abelian Kaluza-Klein Theory in 1953

There are documents which show that Wolfgang Pauli constructed in 1953 the first consistent generalization of the five-dimensional theory of Kaluza, Klein, Fock and others to a higher dimensional internal space. Because he saw no way to give masses to the gauge bosons, he refrained from publishing his results formally. This is still a largely unknown chapter of the early history of non-Abelian gauge and Kaluza-Klein theories (Fig. 3).

Fig. 3 Wolfgang Pauli around 1956. *Source* ETH-Bibliothek Zürich, Bildarchiv. Licensed under the Creative Commons Attribution-Share Alike 3.0 Unported license

Pauli described his detailed attempt of a non-Abelian generalization of Kaluza-Klein theories extensively in some letters to A. Pais, which have been published in Vol. IV, Part II of Pauli's collected letters [34], as well in two seminars in Zürich on November 16 and 23, 1953. The latter have later been written up in Italian by Pauli's pupil Gulmanelli [35]. An English translation of these notes by P. Minkowski is now available on his home page. By specialization (independence of spinor fields on internal space) Pauli got all important formulae of Yang and Mills, as he later (Feb. 1954) pointed out in a letter to Yang [41], after a talk of Yang in Princeton. Pauli did not publish his study, because he was convinced that "*one will always obtain vector mesons with rest mass zero*" (Pauli to Pais, 6 Dec., 1953).

7.1 The Pauli Letters to Pais

At the Lorentz-Kammerlingh Onnes conference in Leiden (22–27 June 1953) A. Pais talked about an attempt of describing nuclear forces based on isospin symmetry and baryon number conservation. In this contribution he introduced fields, which do not only depend on the spacetime coordinates x, but also on the coordinates ω of an internal isospin space. The isospin group acted, however, globally, i.e., in a spacetime-independent manner.

During the discussion following the talk by Pais, Pauli said:

> ...I would like to ask in this connection whether the transformation group with constant phases can be amplified in a way analogous to the gauge group for electromagnetic potentials in such a way that the meson-nucleon interaction is connected with the amplified group...

Stimulated by this discussion, Pauli worked on the problem, and wrote on July 25, 1953 a long technical letter to Pais [36], with the motto: "Ad *usum Delfini* only". This letter begins with a personal part in which Pauli says that "*the whole note for you is of course written in order to drive you further into the real virgin-country*". The note has the interesting title:

"*Written down July 22–25 1953, in order to see how it looks. Meson-Nucleon Interaction and Differential Geometry.*"

In this manuscript, Pauli generalizes the original Kaluza-Klein theory to a six-dimensional space and arrives through dimensional reduction at the essentials of an $SU(2)$ gauge theory. The extra-dimensions form a two-sphere S^2 with space-time dependent metrics on which the $SU(2)$ operates in a space-time-dependent manner. Pauli emphasizes that this transformation group "*seems to me therefore the* natural generalization of the gauge-group *in case of a two-dimensional spherical surface.*" He then develops in 'local language' the geometry of what we now call a fibre bundle with a homogeneous space as typical fiber (in this case $SU(2)/U(1)$).

Since it is somewhat difficult to understand exactly what Pauli did, we give some details, using more familiar formulations and notations [7].

Pauli considers the six-dimensional total space $M \times S^2$, where S^2 is the two-sphere on which $SO(3)$ acts in the canonical manner. He distinguishes among the diffeomorphisms (coordinate transformations) those which leave the space-time manifold M pointwise fixed and induce space-time-dependent rotations on S^2:

$$(x, y) \to [x, R(x) \cdot y]. \tag{33}$$

Then Pauli postulates a metric on $M \times S^2$ that is supposed to satisfy three assumptions. These led him to what is now called the non-Abelian Kaluza-Klein ansatz: The metric $\hat{g}$ on the total space is constructed from a space-time metric g, the standard metric γ on S^2, and a Lie-algebra-valued 1-form,

$$A = A^a T_a \, , \; A^a = A^a_\mu dx^\mu, \tag{34}$$

on M (T_a, $a = 1, 2, 3$, are the standard generators of the Lie algebra of $SO(3)$) as follows: If $K^i_a \partial/\partial y^i$ are the three Killing fields on S^2, then

$$\hat{g} = g - \gamma_{ij}[dy^i + K^i_a(y)A^a] \otimes [dy^j + K^j_a(y)A^a]. \tag{35}$$

In particular, the non-diagonal metric components are

$$\hat{g}_{\mu i} = A^a_\mu(x)\gamma_{ij}K^j_a. \tag{36}$$

Pauli does not say that the coefficients of A^a_μ in Eq. (36) are the components of the three independent Killing fields. This is, however, his result, which he formulates in terms of homogeneous coordinates for S^2. He determines the transformation behavior of A^a_μ under the group Eq. (33) and finds in matrix notation what he calls "*the generalization of the gauge group*":

$$A_\mu \to R^{-1}A_\mu R + R^{-1}\partial_\mu R. \tag{37}$$

With the help of A_μ, he defines a covariant derivative, which is used to derive "*field strengths*" by applying a generalized curl to A_μ. This is exactly the field strength that was later introduced by Yang and Mills. To our knowledge, apart from Klein's 1938 paper, it appears here for the first time. Pauli says that ' *this is the* true *physical field, the analog of the* field strength" *and he formulates what he considers to be his* "*main result*":

The vanishing of the field strength is necessary and sufficient for the $A^a_\mu(x)$ in the whole space to be transformable to zero.

It is somewhat astonishing that Pauli did not work out the Ricci scalar for $\hat{g}$ as for the Kaluza-Klein theory. One reason may be connected with his remark on the Kaluza-Klein theory in Note 23 of his relativity article [37] concerning the five dimensional curvature scalar (p. 230):

> There is, however, no justification for the particular choice of the five-dimensional curvature scalar P as integrand of the action integral, from the standpoint of the restricted group of the cylindrical metric (gauge group). The open problem of finding such a justification seems to point to an amplification of the transformation group.

In a second letter [38], Pauli also studies the dimensionally reduced Dirac equation and arrives at a mass operator that is closely related to the Dirac operator in internal space (S^2, γ). The eigenvalues of the latter operator had been determined by him long before [39]. Pauli concludes with the statement: "*So this leads to some rather unphysical* shadow particles".

Pauli's main concern was that the gauge bosons had to be massless, as in quantum electrodynamics. He emphasized this mass problem repeatedly, most explicitly in the second letter [38] to Pais on December 6, 1953, after he had made some new calculations and had given the two seminar lectures in Zurich already mentioned. He adds to the Lagrangian what we now call the Yang-Mills term for the field strengths and says that "*one will always obtain vector mesons with rest-mass zero (and the rest-mass if at all finite, will always remain zero by all interactions with nucleons permitting the gauge group).*" To this Pauli adds: "*One could try to find other meson fields*", and he mentions, in particular, the scalar fields which appear in the dimensional reduction of the higher-dimensional metric. In view of the Higgs mechanism this is an interesting remark.

Pauli learned about the related work of Yang and Mills in late February, 1954, during a stay in Princeton, when Yang was invited by Oppenheimer to return to Princeton and give a seminar on his joint work with Mills. About this seminar Yang reports [40]: "Soon after my seminar began, when I had written down on the blackboard $(\partial_\mu - i\epsilon B_\mu)\Psi$, Pauli asked: *What is the mass of this field* B_μ?, I said we did not know. Then I resumed my presentation, but soon Pauli asked the same question again. I said something to the effect that that was a very complicated problem, we had worked on it and had come to no conclusion. I still remember his repartee: 'That is no sufficient excuse.' I was so taken aback that I decided, after a few moments' hesitation to sit down. There was general embarrassment. Finally Oppenheimer said, 'we should let Frank proceed.' Then I resumed and Pauli did not ask any more questions during the seminar." (For more on this encounter, see [40].)

In a letter to Yang [41] shortly after Yang's Princeton seminar, Pauli repeats: "*But I was and still am disgusted and discouraged of the vector field corresponding to particles with zero rest-mass (I do not take your excuses for it with'complications' seriously) and the difficulty with the group due to the distinction of the electromagnetic field remains.*" Formally, Pauli had, however, all important equations, as he shows in detail, and he concludes the letter with the sentence: "On the other hand you see, that your equations can easily be generalized to include the ω-space" (the internal space). As already mentioned, the technical details have been written up by Pauli's pupil Gulmanelli [35] and have recently been translated by P. Minkowski from Italian to English.

References

1. H. Weyl, Gravitation und Elektrizität. Sitzber. Preuss. Akad. Wiss., 465–480 (1918). See also *Gesammelte Abhandlungen.* 6 Vols. Ed. K. Chadrasekharan, Springer (an English translation is given in [17])
2. N. Straumann, Zum Ursprung der Eichtheorien bei Hermann Weyl. Physikalische Blätter **43**(11), 414–421 (1987)
3. H. Weyl, Elektron und Gravitation. Z. Phys. **56**, 330–352 (1929)
4. F. London, Quantenmechanische Deutung der Theorie von Weyl. Z. Phys. **42**, 375–389 (1927)
5. V. Fock, Über die invariante Form der Wellen- und der Bewegungsgleichungen für einen geladenen Massenpunkt. Z. Phys. **39**, 226–232 (1926)
6. O. Klein, Quantentheorie und fünfdimensionale Relativitätstheorie. Z. Phys. **37**, 895–906 (1926); for an English translation, see [17]
7. L. O'Raifeartaigh, N. Straumann Historical Origins and Some Modern Developments. Gauge Theory. Rev. Mod. Phys. **72**, 1–23 (2000)
8. *The Collected Papers of Albert Einstein*, Vol. 1–13 (Princeton University Press, 1987). See also: http://www.einstein.caltech.edu/
9. N. Straumann, *General Relativity*, Second Edition, Graduate Texts in Physics (Springer, 2013)
10. J. Audretsch, F. Gähler, N. Straumann, Wave fields in Weyl spaces and conditions for the existence of a preferred pseudo-Riemannian structure. Comm. Math. Phys. **95**, 41–51 (1984)
11. H. Weyl, Space · Time · Matter. Translated from the 4th German Edition. London: Methuen, *Raum · Zeit · Materie, 8* (Springer-Verlag, Auflage, 1922), p. 1993
12. W. Pauli, Zur Theorie der Gravitation und der Elektrizität von H. Weyl. Physikalische Zeitschrift **20**, 457–467 (1919)
13. W. Pauli, *Relativitätstheorie.* Encyklopädie der Mathematischen Wissenschaften 5.3, Leipzig: Teubner, 539–775 (1921); W. Pauli, *Theory of Relativity* (Pergamon Press, New York, 1958)
14. E. Schrödinger, Über eine bemerkenswerte Eigenschaft der Quantenbahnen eines einzelnen Elektrons. Z. Phys. **12**, 13–23 (1922)
15. V.V. Raman, P. Forman, Why was it Schrödinger who developed de Broglie's ideas? Hist. Stud. Phys. Sci. **1**, 291–314 (1969)
16. C. N. Yang, Hermann Weyl's contribution to physics, in *Hermann Weyl*, Edited by K. Chandrasekharan, Springer (1980)
17. L. O'Raifeartaigh, *The Dawning of Gauge Theory* (Princeton University Press, Princeton, 1997)
18. E. Schrödinger, *Centenary Celebration of a Polymath* (Cambridge University Press, C. Kilmister, 1987)
19. H. Weyl, *Selecta.* Birkhäuser-Verlag (1956)
20. E. Wigner, Eine Bemerkung zu Einsteins neuer Formulierung des allgemeinen Relativitätsprinzips. Z. Phys. **53**, 592–596 (1929)
21. E. Cartan, *Leçons sur la Géométrie des Espaces de Riemann*, Gauthier-Villars, Paris 1928; 2nd edn. (1946)
22. K.M. Case, Reformulation of the Majorana theory of the Neutrino. Phys. Rev. **107**, 307–316 (1957)
23. H. Weyl, *Gruppentheorie und Quantenmechanik.* Wissenschaftliche Buchgesellschaft, Darmstadt 1981 (Nachdruck der 2. Aufl., Leipzig 1931). Engl. translation: Group Theory and Quantum Mechanics (Dover, New York, 1950)
24. W. Pauli, In 22, Vol. I, p. 518
25. In Carl Seelig: *Albert Einstein* (Europa-Verlag Zürich, 1960), p. 274
26. V. Fock, Geometrisierung der Diracschen Theorie des Elektrons. Z. Phys. **57**, 261–277 (1929)
27. H. Tetrode, Allgemein-relativistische Quantentheorie des Elektrons. Z. Phys. **50**, 336–346 (1928)
28. E. Schrödinger, Diracsches Elektron im Schwerefeld I. *Sitzber. Preuss. Akad. Wiss.*, 105–128 (1932); English translation by C. Kiefer in *Gen. Rel. Grav.* **52**, article number 4 (2020)
29. V. Bargmann, Sitzber. Preuss. Akad. Wiss. **346** (1932)

30. L. Infeld, B. L. van der Waerden, Sitzber. Preuss. Akad. Wiss., 380–474 (1932)
31. W. Heisenberg, W. Pauli, Zur Quantenelektrodynamik der Wellenfelder. I. Z. Phys. **56**, 1–61 (1929)
32. W. Heisenberg, W. Pauli, Zur Quantenelektrodynamik der Wellenfelder. II. Z. Phys. **59**, 168–190 (1930)
33. S.S. Schweber, *Quantum Electrodynamics and the Men Who Made It: Dyson, Feynman, Schwinger, and Tomonaga* (Princeton University Press, Princeton, 1994)
34. W. Pauli, *Wissenschaftlicher Briefwechsel*, Vol. IV, Part II, ed. by K. v Meyenn, (Springer-Verlag, 1999)
35. P. Gulmanelli, *Su una Teoria dello Spin Isotropico* (Pubblicazioni della Sezione di Milano dell'istituto Nazionale di Fisica Nucleare, Casa Editrice Pleion, Milano, 1954)
36. Pauli to Pais, Letter [1614] in [35]
37. W. Pauli, *Theory of Relativity* (Pergamon, New York, 1958)
38. Pauli to Pais, Letter [1682] in [35]
39. W. Pauli, Über ein Kriterium für Ein- oder Zweiwertigkeit der Eigenfunktionen in der Wellenmechanik. Helv. Phys. Acta **12**, 147–168 (1939)
40. C.N. Yang, *Selected papers 1945–1980 with Commentary* (Freeman, San Francisco, 1983), p. 525
41. Pauli to Yang, Letter [1727] in [35]

Gauging the Spacetime Metric—Looking Back and Forth a Century Later

Erhard Scholz

Abstract Hermann Weyl's proposal of 1918 for generalizing Riemannian geometry by local scale gauge (later called *Weyl geometry*) was motivated by mathematical, philosophical and physical considerations. It was the starting point of his unified field theory of electromagnetism and gravity. After getting disillusioned with this research program and after the rise of a convincing alternative for the gauge idea by translating it to the phase of wave functions and spinor fields in quantum mechanics, Weyl no longer considered the original scale gauge as physically relevant. About the middle of the last century the question of conformal and/or local scale gauge transformation was reconsidered by different authors in high energy physics (Bopp, Wess, et al.) and, independently, in gravitation theory (Jordan, Fierz, Brans, Dicke). In this context Weyl geometry attracted new interest among different groups of physicists (Omote/Utiyama/Kugo, Dirac/Canuto/Maeder, Ehlers/Pirani/Schild and others), often by hypothesizing a new scalar field linked to gravity and/or high energy physics. Although not crowned by immediate success, this "retake" of Weyl geometrical methods lives on and has been extended a century after Weyl's first proposal of his basic geometrical structure. It finds new interest in present day studies of elementary particle physics, cosmology, and philosophy of physics.

1 Introduction

In 1918 *Hermann Weyl* proposed a generalization of Riemannian geometry, which he considered as better adapted to the field theoretic context of general relativity than the latter. His declared intention was to base geometry on purely "local" concepts which

E. Scholz (✉)
Faculty of Mathematics/Natural Sciences, Interdisciplinary Centre for History and Philosophy of Science, University of Wuppertal, Wuppertal, Germany
e-mail: scholz@math.uni-wuppertal.de

S. De Bianchi and C. Kiefer (eds.), *One Hundred Years of Gauge Theory*, Fundamental Theories of Physics 199, https://doi.org/10.1007/978-3-030-51197-5_2

at the outset would not allow to compare field quantities $X(p)$ and $X(p')$ at distant points p and p' of the spacetime manifold, but only for infinitesimally close ones. The possibility of comparing directly two quantities at distant points appeared to him a remnant of Euclidean geometry, which Riemannian geometry had inherited via the Gaussian theory of surfaces. In Weyl's view Riemann had generalized the latter without putting the comparability of quantities at different locations into question. He demanded in contrast that

> ...only segments at the same place can be measured against each other. The gauging of segments is to be carried out at each single place of the world (Weltstelle), this task *cannot be delegated to a central office of standards (zentrales Eichamt)*. Weyl [190, p. 56f., emph. ES][1]

He therefore considered a geometry formalized by a conformal (pseudo-Riemannian) structure as more fundamental than Riemannian geometry itself [189, p. 13]. But it has to be supplemented by a principle which allows for comparing metrical quantities at infinitesimally close points ($p \neq p'$), realized by a principle of metrical transfer. A conformal manifold could be qualified as "metrically connected" only if a comparison of metrical quantities at different points is possible:

> A metrical connection from point to point is only being introduced into it [the manifold, ES], if a *principle of transfer for the unit of length from one point P to an infinitesimally close one* is given [189, 14].[2]

Weyl formulated this principle of transfer by introducing what today would be called a connection in the local scaling group $\mathbb{R}^+$, i.e., by a real differential 1-form. The new type of "purely infinitesimal geometry" (Weyl's terminology), later called *Weyl geometry*, was built upon the two interrelated basic concepts of a *conformal structure* and a *scale connection* as the principle of transfer. Weyl called the latter a *length connection*. Both were united by the possibility of changing the metrical scale by *gauge transformations* in the literal sense (see Sect. 2.1). For a few years Weyl tried to build a unified field theory of electromagnetism and gravity upon such a geometrical structure, and to extend it to a field theory of matter.[3] But in the second half of the 1920s he accepted and even contributed actively to reformulating the gauge idea in the context of the rising new quantum mechanics.

---

[1]"...nur Strecken, die sich an der gleichen Stelle befinden, lassen sich aneinander messen. An jeder einzelnen Weltstelle muß die Streckeneichung vorgenommen werden, diese Aufgabe kann nicht einem zentralen Eichamt übertragen werden" [190, p. 56f.]. (Translation here and in the following by ES, if no reference to a published English translation is given.).

[2]"Ein metrischer Zusammenhang wird erst dann in sie hineingetragen, wenn ein *Prinzip der Übertragung der Längeneinheit von einem Punkte P zu seinem unendlich benachbarten* vorliegt" [189, p. 14]. Emphasis in the original, here and in other places, where not explicitly stated that it is by ES.

[3]Cf. [69, 179].

Several decades later this idea was generalized to non-abelian groups and became a fundamental conceptual ingredient even for the later development of high energy physics.[4] In the meantime (during the 1940s) Weyl had recanted the importance of scaling transformations (localized "similarities" as he used to call them) for the search of what he called the "physical automorphisms of the world" (see Sect. 2.2).

The basic idea underlying Weyl's "purely infinitesimal geometry" of 1918 reappeared independently at several occasions during the second half of the 20th century. In the early 1960s Carl Brans and Robert Dicke developed their program of a modified general relativistic theory of gravity with a non-minimally coupled scalar field. Dicke stated as an "evident" principle (which it was *not*, at least not for everyone):

> It is evident that the particular values of the units of mass, length, and time employed are arbitrary and that the *laws of physics must be invariant under a general coordinate dependent change of units*. Dicke [47, p. 2163, emph. ES]

This was very close to Weyl's view in 1918, but Dicke postulated local scale invariance without the complementary structure of a scale connection. As a result Brans, Dicke's PhD student, and Dicke himself developed a theory of gravity which had an implicit relationship to the special type of Weyl geometry with an integrable scale connection, in short *integrable Weyl geometry* (IWG). Other authors made this relationship explicit and generalized it to the non-integrable case.

This was part of a classical field theory program of gravity research, but the importance of conformal transformations got also new input from high energy physics. And even the original form of Weyl geometry had some kind of revival from the 1960s onward in scalar field theories of gravity or nuclear structures, initiated by authors in Japan (Omote/Utiyama/Kugo) and, independently Europe/USA (Dirac/Canuto/Maeder), and also in the foundational studies of gravity (Ehlers/Pirani/Schild). This restart in the last third of the 20th century of research building on, and extending, Weyl geometric methods in physics has lasted until today and shows that Weyl's disassociation from his scale gauge idea is not at all to be considered a final verdict on his geometrical ideas developed between 1918 and the early 1920s.

The following paper tries to give an account of the century long development from Weyl's original scale gauge geometry of 1918 (and the reasons why he thought it an important improvement on the earlier field theories building upon Riemannian geometry), through its temporary disregard induced by the migration of the gauge idea from metrical scale to quantum phase (ca. 1930–1960, Sect. 2), and the revival in the early 1970s indicated above (Sect. 3), to a report on selected research in physics, which uses Weyl geometric methods in a crucial way (Sect. 5). Basic concepts and notations of Weyl geometric gravity (in the moderately modernized form in which they are used in Sect. 5) are explained in an interlude between the historical account

[4]See N. Straumann's contribution to this volume.

and the survey of present studies (Sect. 4). Short reflections on this glance back and forth are given at the end of the paper (Sect. 6).

2 Weyl's Scale Gauge Geometry 1918–1950

2.1 Purely Infinitesimal Geometry, 1918–1923

Parallel to finalizing his book *Raum—Zeit—Materie* (RZM) [188] Weyl developed a generalization of Riemannian geometry, in which an inbuilt concept for a direct metrical comparison of quantities at distant points was no longer foreseen. It was substituted by a comparison in "purely local" regions, in the infinitesimal sense [187, 189]. Weyl introduced this generalization into the third and fourth editions of the book, and added a discussion of what might be the consequences for relativistic field theory and the description of matter. In this way his proposal entered the English translation *Space—Time—Matter* (STM) of the fourth edition of the book [194]. It became more widely known than his separate articles on the topic.

In a letter dated 1 March 1918 to *Albert Einstein*, before RZM was publicly available, Weyl announced that the publisher (Springer) would soon send the corrected page proofs of the book to Einstein.[5] In March the two men met in Berlin, and Weyl used the occasion for presenting his ideas on the generalized theory to Einstein. This started a friendly, but controversial discussion on Weyl's proposals which would extend through the whole year 1918 and continued to have some repercussions in the years to come.[6] Weyl submitted his first publication on the topic to the Berlin Academy of Science through Einstein who appended a famous critical note to it [187]. As Einstein queried, atomic clocks (spectral lines of atoms/molecules) would become dependent on their history, if one assumes that their internal time is subject to Weyl's local length transfer. Weyl did not share this assumption; he considered the length connection as a part of the general field theoretic structure, which is reflected only indirectly in the behaviour of measuring instruments. This was one difference among others which the two scientists debated in the next few years.[7]

Other physicists, among them *Arthur Eddington* and the young *Wolfgang Pauli*, reacted differently. For a period of a few years they contributed to the dissemination and an elaboration of Weyl's theory. The latter might have been of particular interest to physicists at the time, because it seemed to open a geometrical road towards a unification of the then known fundamental interactions, electromagnetism and

[5] Einstein [58, vol. 8, doc 472].

[6] Straumann [170], Lehmkuhl [100], all the letters in Einstein [58, vol. 8B].

[7] The debate between Einstein and Weyl after 1918 is being dealt with in Lehmkuhl [100].

gravity, and was related to the even more ambitious goal of a fundamental theory of matter based on a unified classical field theory along the lines of G. Mie and D. Hilbert.

A few years later this early program of a geometrical unification became considered obsolete, when the gauge idea was transformed from geometrical scale to the quantum mechanical phase degree of freedom. In this form it was the starting point for the gauge field theories of the second half of the last century, which became central for the standard model of particle physics. The early parts of this interesting story has been told from different angles[8]; it will not be repeated here. The later part—the rise of the standard model—would probably still deserve more detailed historical investigations.[9] Here we concentrate on Weyl's basic conceptual ideas for gauging the metric of spacetime, which lay at the basis of his new geometry. Not unlike the more general idea of gauge structures (with groups operating on dynamical spaces of field variables) it proved of a wider range and impact than Weyl's first approach for a physical application in unified field and matter theory.

Weyl considered the *orthogonality* relation specified by a symmetric, non-degenerate (but not necessarily definite) bilinear form in the coordinate differentials as an *objective* element for introducing a *metric* ("Maßbestimmung") in a differentiable manifold M. It is given independently of any further choice of the description; for a Lorentzian signature it corresponds to the causal structure of spacetime. On the other hand, the scale factor ("Proportionalitätsfaktor") of the bilinear form is part of the choice of a *reference frame* ("Bezugssystem") and defines a metrical *gauge* ("Eichung") in the literal sense [190, p. 58].[10] In this sense it is a subjective element of the description like the choice of a coordinate system and complementary to it [189, p. 13].

Thus far M carries a *conformal* structure, or in Weyl's words a "conformal geometry" (ibid.).[11] But this is insufficient for establishing a full-fledged metrical concept in the manifold. Without further stipulations the different points of the manifold would "be completely isolated from each other from the metrical point of view". Weyl concluded:

> A metrical connection between points is being introduced only if a *principle of transfer for the unit of length measurement from a point P to an infinitesimally close one* is given. Weyl [189, p.14][12]

---

[8]See, among others, [2, 3, 69, 143, 149, 163, 179].

[9]See [95, Chap. 22] [22–24, 66, 133].

[10] Page references here and in the following refer to Weyl's *Gesammelte Abhandlungen.*

[11]The terminology of different *structures* on a manifold appeared only in the 1940s, essentially due to protagonists of the French community around Charles Ehresmann.

[12]"Machen wir keine weiteren Voraussetzugen, so bleiben die einzelnen Punkte der Mannigfaltigkeit in metrischer Hinsicht vollständig gegeneinander isoliert. Ein metrischer Zusammenhang wird erst

He specified such a principle of transfer by adding a differential 1-form $\varphi = \varphi_\mu dx^\mu$ to the metrical coefficients g_{ik} of the chosen reference frame.[13] For him the differential form codified the change of squared length quantities $l^2 = l^2(p)$ measured at a point $p \in M$ to the respective squared lengths $l'^2 = l^2(p')$ measured at a infinitesimally close point p' expressed by the relation

$$l'^2 = (1 + \varphi(\xi))\, l^2 \ . \tag{1}$$

Here the infinitesimal displacement of the points is denoted in terms of a tangent vector $\xi \in T_pM$. But then, of course, one has to gain clarity about the transformation φ undergoes if the local scale is changed, i.e., if the metric $g_{\mu\nu}$ is conformally changed to $\tilde{g}_{\mu\nu} = \lambda(x) g_{\mu\nu}$. Weyl realized that consistency of the length transfer idea demands to transform the scale connection φ as follows:

$$\text{If} \quad g \longmapsto \tilde{g} = \lambda g \,, \quad \text{then} \quad \varphi \longmapsto \tilde{\varphi} = \varphi - \frac{1}{2} d \log \lambda \,, \tag{2}$$

i.e., $\tilde{\varphi}_\mu = \varphi_\mu - \frac{1}{2}\lambda^{-1}\partial_\mu \lambda$. Weyl called it a "modification of the gauge (Abänderung der Eichung)" [190, p. 59]; this was the first *gauge transformation* considered explicitly in the history of mathematics/physics.

He showed that a genuine geometry on a manifold M can be built upon such a generalized concept of a gauge metric (later *Weylian metric*). Slightly reformulated, Weyl characterized his metric by an equivalence class $[(g, \varphi)]$ of pairs (g, φ) consisting of of a (pseudo-) Riemannian metric g on M and a real valued differential form φ on M, equivalences being given by gauge transformations (2). Gauging the metric boils down to choosing a representative (g, φ); its first component g will be called the *Riemannian component* of the respective gauge, the second one φ its *scale connection form*.

A clue to this geometry was Weyl's finding that $[(g, \varphi)]$ possesses a *uniquely determined* (in particular gauge independent), *metric compatible, affine connection* Γ (in the sense of a symmetric linear connection) like in Riemannian geometry [187, p. 33].[14] Metric "compatibility" is here understood as the consistency of the length change of vectors under parallel transport by Γ with the length change demanded by the length transfer φ. Γ is gauge independent, although in any gauge (g, φ) it can be decomposed into two gauge dependent terms, the Levi-Civita connection $\Gamma(g)$ of g and a scale connection term $\Gamma = \Gamma(\varphi)$,

---

dann in sie hineingetragen, wenn ein *Prinzip der Übertragung der Längeneinheit von einem Punkt P zu einem unendlich benachbarten* vorliegt. Weyl [189, p. 14].

[13]In fact, Weyl wrote "$d\varphi = \varphi_i dx_i$" for the differential form, e.g., [189, p. 15]. In order to avoid confusion with the present notational conventions influenced by Cartan's symbolism of differential forms, we denote it by φ because in general it is not an exact form.

[14]In [189, p. 17] Weyl referred also to [79] in this regard.

$$\Gamma = \Gamma(g) + \Gamma(\varphi) \quad \text{where} \quad \Gamma^{\mu}_{\nu\lambda}(\varphi) = \frac{1}{2}(\delta^{\mu}_{\nu}\varphi_{\lambda} + \delta^{\mu}_{\lambda}\varphi_{\nu} - g_{\nu\lambda}\varphi^{\mu}). \qquad (3)$$

Γ induces a gauge independent *covariant derivative* and has geodesics which are no longer extremal lines of the (gauged) metric but are autoparallels of the invariant affine connection.

Weyl derived the *curvature* tensors for the connections Γ and φ. With regard to the affine connection he formed the Weyl geometric curvature tensor analogous to the Riemann tensor in the classical case. Because of this analogy it will be denoted in the following by *Riem* (with components $R^{\mu}{}_{\nu\lambda\kappa}$) although it (more precisely its (0, 4) type version $R_{\mu\nu\lambda\kappa}$) is not antisymmetric in the first pair of indices. Its contraction *Ric* is the Weyl geometric Ricci tensor, and R its scalar curvature. Like Γ, the Weyl geometric curvatures can be decomposed, in any gauge (g, φ), into contributions derived from the Riemannian part of the metric $g_{\mu\nu}$ alone, that is the Riemannian part of the curvature $Riem\,(g) = Riem\,(\Gamma(g))$, and the scale connection part of the curvature $Riem\,(\varphi)$, which is formally built like the usual Riemann tensor expression from $\Gamma(\varphi)$, but with covariant derivatives ${}_g\nabla_\kappa$ of g in place of the partial derivatives ∂_κ.[15]

The curvature formed with regard to the scale (length) connection is $f = d\varphi$, written in Cartan notation of outer derivatives (which Weyl did not use). It is called the *length (scale) curvature*. Weyl decomposed the curvature tensor $Riem$ into a contribution $\overline{R}^{\mu}{}_{\nu\lambda\kappa}$ with the same symmetry properties as the classical Riemann tensor and a contribution derived from the length curvature, which is not antisymmetric in the first two indices:

$$R^{\mu}{}_{\nu\lambda\kappa} = \overline{R}^{\mu}{}_{\nu\lambda\kappa} - \frac{1}{2}\delta^{\mu}_{\nu} f_{\lambda\kappa} \qquad (4)$$

Also this decomposition is gauge independent. Weyl called the first term the *directional curvature* ("Richtungskrümmung") [189, p. 20].

From the directional curvature he constructed a tensor C of type (1, 3) like $Riem$, which "depends only on the g_{jk}". It is conformally invariant and vanishes in the dimensions $n = 2, 3$. For $n \geq 4$, so Weyl announced, it vanishes if the manifold is conformally flat [189, p. 21]. Later it would be called *conformal curvature* or the *Weyl tensor*.[16]

For vanishing length curvature, $f = d\varphi = 0$, the scale connection can be integrated locally. Then the corresponding gauge reduces to a Riemannian metric; it will therefore be called the *Riemann gauge* of the Weylian metric. In this case the curvature tensor reduces to the directional curvature and is identical to the Riemann

---

[15]Cf. footnote 50; more details in Yuan [207] or in Eddington [55, p. 218f.], Perlick [130, p. 150ff.] etc.

[16]That $C = 0$ is also *sufficient* for conformal flatness was not clear to Weyl in 1918; he even seemed to exclude it in a side remark. It was proven by Schouten [157] and later by himself [192].

curvature of the Riemann gauge. Weyl therefore considered Riemannian geometry as a special case of his scale gauge geometry. This was his perspective in RZM from the third edition onward and also in its English version STM [194, Chap. 2].

The generalization of the metrical structure demanded an extension of tensor (of course also of vector and scalar) fields with regard to their scaling properties under a change of metrical gauge. In the third edition of RZM Weyl wrote

> In a generalised sense we shall, however, also call a linear form which depends on the co-ordinate system and the *calibration* a tensor, if it is transformed in the usual way when the co-ordinate system is changed, but becomes multiplied by the factor λ^e (where $\lambda =$ the calibration ratio) when the calibration is changed; we say that it is of *weight e*. Weyl [194, p. 127][17]

Weyl considered these fields "merely as an auxiliary conception, which is introduced to simplify calculations" (ibid.); but physicists among his readers realized that it would become of physical importance if Weyl geometry is accepted as a framework for field theory. Einstein called them "Weyl tensors" [57, p. 200], Eddington introduced the terminology of "co-tensors" or "co-invariants" [55, p. 202], which is still in use in parts of the physical literature. Here they will be called *scale covariant* quantities (tensor, vector, scalar fields).

Weyl's conjectural unification of gravity and electromagnetism built essentially on the idea of using the length connection φ as a symbolic representative of the electromagnetic potential. Then the Maxwell action density $f_{\nu\lambda} f^{\nu\lambda}\sqrt{|g|}$ for the electromagnetic field was of weight 0, if and only if the dimension is $n = 4$. This was an intriguing argument for the necessary specification of the spacetime dimension 4 in Weyl geometric field theory [187, p. 31], [186, p. 37]. In an early response to Weyl's theory it was praised by Einstein also.[18]

As an even more important feature of his theory Weyl considered the fact that his (localized) scale gauge symmetry seemed to imply the conservation of electric charge. He showed this by an argument which in hindsight might be read as an exemplar of Emmy Noether's second theorem (*ante letteram*).[19] Starting from a Lagrangian density $\mathfrak{L}$, invariant under diffeomorphism and (local) scale symmetry, Weyl derived 5 identities between the Euler-Lagrange expressions with regard to the metric and the scale connection, $\mathfrak{E}[g]_{\mu\nu} = \frac{\delta \mathfrak{L}}{\delta g^{\mu\nu}}$ and $\mathfrak{E}[\varphi]^{\mu} = \frac{\delta \mathfrak{L}}{\delta \varphi_{\mu}}$, one of which was due to the scale symmetry:

---

[17] In the 5th edition [196, p. 127].

[18] "Ihr Gedankengang ist von wunderbarer Geschlossenheit. Auch der Schluss auf die Dimensionszahl 4 hat mir sehr imponiert." [58, vol. 8B, Doc 499, 8 Apr. 1918]. Weyl could not keep up this argument after 1927; for his later deliberations see [41].

[19] At the time of writing [187] and the first edition of *Raum, Zeit, Materie* [188] Noethers seminal paper [115] was not yet written. Weyl connected up to the Hilbert's, Klein's and his own considerations in Weyl [185] on the conservation of energy-momentum in general relativity. Before late summer 1918 Noether's thoughts on this topic were not publicly accessible.

$$\mathfrak{E}[g]^{\nu}_{\nu} \equiv \partial_\nu \mathfrak{E}[\varphi]^{\nu} \quad (5)$$

The symbol $\equiv$ (used by Weyl) expresses an identity which holds independently of the dynamical equations (i.e., also "off shell" in later terminology). For a Lagrange density $\mathfrak{L}$ which contains the scale connection only in a Maxwell-like term $\mathfrak{L}[\varphi] \sim f_{\mu\nu} f^{\mu\nu}$ the Euler-Lagrange expression for φ acquires the form $\mathfrak{E}[\varphi]^{\mu} \equiv \partial_\nu(\sqrt{|g|} f^{\mu\nu}) - \mathfrak{s}^{\mu}$ with $\mathfrak{s}^{\mu} \equiv \frac{\delta\mathfrak{L} - \mathfrak{L}[\varphi]}{\delta\varphi_\mu}$ a 4-current, i.e., a covector (1-form) density $\mathfrak{s}^{\mu} = s^{\mu}\sqrt{|g|}$. In his framework Weyl interpreted it, of course, as the electric current density. If only the gravitational equations are satisfied (independent of the electromagnetic ones) the current is conserved,

$$\partial_\nu \mathfrak{s}^{\nu} = 0\,, \quad (6)$$

and with it the integral over 3-dimensional spacelike submanifolds intersecting a 4-dimensional timelike "channel" close to the boundary of which (and outside of it) the current vanishes [187, p. 37ff.]. This conservation property seemed of utmost importance to Weyl.[20]

In order to get a scale invariant gravitational action density, Weyl replaced the Hilbert term $R\sqrt{|g|}$ by quadratic expressions in Weyl geometric curvature expressions of the form $(\beta_1\, Riem^2 + \beta_2\, R^2)\sqrt{|g|}$, where he set $\beta_2 = 0$ in Weyl [187] and $\beta_1 = 0$ in the fifth edition of *Raum, Zeit, Materie*. In such an approach the Weylian metric seemed to establish a unified description of gravity ($g_{\mu\nu}$ and affine connection Γ) and electromagnetism (φ and its curvature $f = d\varphi$). But the question of the physical interpretation of the scale invariant geodesics and its relation to the free fall trajectories of neutral and of charged particles had to be answered. This was part of the discussion between Weyl and Einstein, aside from Einstein's spectral line objection to Weyl's geometrical generalization. For Weyl the latter was no compelling (knock-out) argument. He answered that one has to distinguish between the field theoretic metric and the measure indicated by clocks and rods. The latter would be realized by a specific gauge and ought to depend on the physical behaviour of atomic systems in the local (infinitesimal) neighbourhood of the atom, not on their history.

For the time being he proposed to consider the hypothesis that atoms adapt to the local field constellation of the gravitational field in such a manner that the Weyl geometric scalar curvature becomes constant, $R = const$ [196, p. 298f.].[21] At the

[20] Cf. [27, 28, 141, 142]; for the overall development and slow reception of the Noether theorems see [93].

[21] Weyl mentioned this argument already in his letter of 18 Sep. 1918 to Einstein [58, vol. 8B, Doc. 619, p. 877], but introduced it into RZM only in the 5th edition; it is thus not contained in the English translation [194]. The Weyl geometric scalar curvature R scales with the inverse metric $g^{\mu\nu}$. If it is different from zero it can thus be scaled to a constant like any other non-vanishing scalar field of the same scale weight.

time of the fifth edition of RZM (1923) he still kept to the position that the final judgement on his theory would depend on a theory of measurement. Three years later he gave up this idea and joined the interpretation of the electromagnetic potential as a phase connection in the new quantum mechanics.

In the two years before the fifth edition, on the other hand, Weyl looked for a deeper philosophical-conceptual underpinning of his geometry in a new *analysis of the problem of space* (PoS). Half a century earlier Helmholtz had analysed the principles which the motions of a rigid body have to satisfy in order for being able to establish empirical measuring rules. The evaluation of these principles led to classical Euclidean or non-Euclidean geometry. Soon Helmholtz's principles for free mobility of rigid rods were rephrased in terms of conditions for the automorphism group of space.[22] This analysis became known as the classical PoS.

The fusion of space and time into a unified spacetime in special relativity and the loss of the concept of rigid body in the general theory let Helmholtz's analysis appear obsolete in the light of relativity. At the end of 1920 Weyl set out for formulating conceptual principles for any type of geometry in a smooth manifold, which builds upon congruence and similarity operations in the infinitesimal neighbourhoods only. As a unifying ("synthetical") principle he included the postulate of a *uniquely determined affine connection* and formulated the result of his conceptual analysis in the form of axioms for those subgroups of the general linear group, which may serve as candidates for defining the local similarity and congruence relations in the manifold in the mentioned sense [191, 193]. In 1922 and early 1923 he was able to show that the only groups satisfying his axioms in a manifold of dimension n are the special pseudo-orthogonal groups $SO(p, q; \mathbb{R})$ as "congruence" groups ($p + q = n$), with similarities $SO(p, g; \mathbb{R}) \times \mathbb{R}^+$ [195]. If one accepts his characterization of the principles for infinitesimal congruence and similarity relations this can be read as a strong conceptual underpinning for the structure of Weylian manifolds.[23]

Another conceptual insight of long range was gained by Weyl in a paper which arose as a by-product to a report he wrote for F. Klein on a manuscript by J.A. Schouten [192]. The manuscript had been rejected by L.E.J Brouwer for the *Mathematische Annalen*, but was published a little later in *Mathematische Zeitschrift* [157].[24] Schouten showed that in dimension $n > 3$ the vanishing of the conformal tensor C of a Riemannian metric $g_{\mu\nu}$ implies local conformal flatness. For Weyl this was new and gave him an incentive to study the projective and conformal view-

[22] Cf. [19, 109]. At first Helmholtz believed to have derived Euclidean geometry alone, but soon learned that non-Euclidean geometry satisfies his principles, too.

[23] More details in Bernard [18], Bernard and Lobo [20], Scholz [151]. A short version of the argument can also be found in the fifth edition of RZM [196], a provisional sketch already in the 4th edition and in Weyl [194].

[24] Klein to Weyl 6.10.1920, Weyl to Klein 28.12.1920, University Library Göttingen, Codex Ms Klein 296, 297.

point which he had "touched upon only marginally" in his previous discussions of infinitesimal geometry [192, p. 201]. He started the paper with the remark:

> The construction of pure infinitesimal geometry, which I have described most consequentially in the 3rd and 4th edition of my book *Raum, Zeit, Materie*, is naturally implemented in the three levels which are characterized by the catchwords *continuous connection, affine connection, metric*. *Projective* and *conformal* geometry originate by means of abstraction from the affine respectively the metric one. Weyl [192, p. 195][25]

Under "continuous connection (stetiger Zusammenhang)" Weyl understood the possibility to consider infinitesimal neighbourhoods; in present terms it is usually expressed as a smooth structure on the underlying manifold.[26]

Weyl defined the "projective property (projektive Beschaffenheit)", in our terms the *projective structure* of a smooth manifold M, by means of an equivalence class $[\Gamma]_p$ of affine connections Γ which have "geodesics" (autoparallels) with identical traces. Γ and $\overline{\Gamma}$ are projectively equivalent, if

$$\overline{\Gamma}^{\mu}_{\nu\lambda} = \Gamma^{\mu}_{\nu\lambda} + \delta^{\mu}_{\nu}\psi_{\lambda} + \delta^{\mu}_{\lambda}\psi_{\nu} \tag{7}$$

for some differential form $\psi = \psi_\nu dx^\nu$. He characterized the *conformal structure* ("konforme Beschaffenheit") of M analogously by an equivalence class of affine connections $[\Gamma]_c$. Here he considered two affine connections Γ and $\widetilde{\Gamma}$ as belonging to the same conformal class $[\Gamma]_c$, if both are scale invariant affine connections of Weylian metrics $[(g, \varphi)]$ and $[(\tilde{g}, \tilde{\varphi})]$ with conformal Riemannian components, $g \sim \tilde{g}$. Because of (3) this boils down to the condition [192, p. 196]

$$\widetilde{\Gamma}^{\mu}_{\nu\lambda} - \Gamma^{\mu}_{\nu\lambda} = \frac{1}{2}(\delta^{\mu}_{\nu}\overline{\varphi}_{\lambda} + \delta^{\mu}_{\lambda}\overline{\varphi}_{\nu} - g_{\nu\lambda}\,\overline{\varphi}^{\mu}) \tag{8}$$

for some differential form $\overline{\varphi}$. It is important to realize that this differential form need *not be exact* (locally integrable), as one would expect if the characterization is read in the context of Riemannian geometry and their Levi-Civita connections.[27]

On this basis Weyl easily proved the following

---

[25]"Der Aufbau der reinen Infinitesimalgeometrie, wie ich ihm am folgerichtigsten in der 3. und 4. Auflage meine Buches *Raum, Zeit, Materie* geschildert habe, vollzieht sich natürlicherweise in drei Stockwerken, welche durch die Schlagworte *stetiger Zusammenhang, affiner Zusammenhang, Metrik* gekennzeichnet sind. Die *projektive* und *konforme* Geometrie enspringen durch Abstraktion aus der affinen bzw. der metrischen." Weyl [192, p. 195].

[26]For a philosophical analysis of this idea see [17]. A modern reconstruction which may even be closer to Weyl's intentions than their reformulation in terms of standard differential topology, may be possible by differential geometry with infinitesimals; see [92].

[27]Matveev and Trautman [107, 833] makes this restriction, see Sect. 5.2.

Theorem 1 [192] *A Weylian metric is uniquely determined by its projective and its conformal structures.*[28]

The theorem deals with the comparison of two different classes $[\Gamma]_p$, $[\Gamma]_c$ of affine connections with non-empty intersection (because both arise by abstraction from the same presupposed Weylian metric). If Γ and Γ' are now any two in $[\Gamma]_p \cap [\Gamma]_c$, both arising as affine connections of Weylian metrics $[(g, \varphi)]$, $[(\tilde{g}, \tilde{\varphi})]$, it has to be shown that the latter are identical. In any case the Riemannian components of such two Weylian metrics are conformally equivalent, and the Γ, Γ' are related by (8). Both Weylian metrics can be gauged to identical Riemannian components, i.e., be given in gauges of the form (g, φ), resp. (g, φ'). The differential form $\overline{\varphi}$ in (8) is then $\overline{\varphi} = \varphi - \varphi'$. As Γ and Γ' are also projectively equivalent, their difference $\Delta = \Gamma - \Gamma'$ changes a vector ξ such that $\Delta^{\mu}_{\nu\lambda}\, \xi^{\nu}\xi^{\lambda} \sim \xi^{\mu}$ (here $\sim$ stands for proportionality). Plugging in (8) for Δ shows that $g_{\nu\lambda}\xi^{\nu}\xi^{\lambda}\, \overline{\varphi}^{\mu}$ is proportional to ξ^{μ} for all vectors ξ. This implies $\overline{\varphi} = 0$.

Weyl emphasized the physical importance of this theorem: The conformal structure characterizes the *causality* relations in spacetime. The projective properties of space are an expression of, in later terminology, the *gravito-inertial* structure of spacetime. Weyl described it in the following way:

> …the tendency of persistence of the direction for a moving material particle, which impresses a certain 'natural' motion on it, once it has been set free in a specified world-direction, is the very unity of inertia and gravity, which Einstein put in the place of both, although a suggestive name for it is still lacking [192, p.196].[29]

At the time of the 5th edition of RZM Weyl could claim to have a well-rounded generalized concept of infinitesimal gauge geometry. This geometry was mathematically well developed (uniquely determined affine connection, curvature properties), had a convincing conceptual underpinning in terms of basic physical structures (causality, inertio-gravitational persistence), and could even be given a transcendental philosophical backing.

On the other hand, Weyl's early enthusiasm for the capacity of his theory for leading to a unified theory of fields and matter had become cracks because of internal technical difficulties of the program (extremely complicated differential equations). Even stronger doubts arose from the growing impression of physicists in

[28]"Satz 1. *Projektive und konforme Beschaffenheit eines metrischen Raums bestimmen dessen Metrik eindeutig*" Weyl [192, p. 196].

[29]"…die Beharrungstendenz der Weltrichtung eines sich bewegenden materiellen Teilchens, welche ihm, wenn es in bestimmter Weltrichtung losgelassen ist, eine bestimmte 'natürliche' Bewegung aufnötigt, ist jene Einheit von Trägheit und Gravitation, welche Einstein an Stelle beider setzte, für die es aber bislang an einem suggestiven Namen mangelt" [192, p.196].

the environment of Sommerfeld that quantum physics might make the attempts for a classical field theory of matter obsolete anyhow. Weyl was well aware of such doubts and tended to share them.

2.2 *Withdrawal of Scale Gauge by Weyl After 1927/29*

Already in the early 1920s E. Schrödinger had the idea that Weyl's gauge principle might be useful in a modified form for dealing with the phase of complex wave functions in quantum mechanics, rather than for scale in gravity. This idea was taken up and elaborated by F. London and independently by V. Fock after the turn towards the "new" quantum mechanics. Weyl endorsed it in the first edition of his book on *Gruppentheorie und Quantenmechanik* [198, 1st ed., p. 87]. A year later Weyl extended this to a general relativistic approach to spinor fields. This led to the now well known representation of the electromagnetic potential by a connection with values in the Lie-algebra of the phase group $U(1)$.[30] At the end of the decade Weyl was quite fond of this migration of the gauge idea from scale to phase and considered it as a definitive answer to the question how electromagnetism ought to be understood as a gauge theory.

The 1930 Rouse Ball lecture at Cambridge university gave him an opportunity for explaining his view of the program of geometrical unification to a wider scientific audience. He explained his own theory of 1918 and summarized its critical reception by physicists. He reviewed Eddington's approach to unification by affine connections and Einstein's later support for that program, always in comparison with his own "metrical" unification of 1918. He concluded that in hindsight one could consider both theory types merely as "geometrical dressings (geometrische Einkleidungen) rather than proper geometrical theories of electricity". He added with irony that the struggle between the metrical and affine unified field theories (UFT), i.e., his own 1918 theory versus Eddington/Einstein's, had lost importance. In 1930 it could no longer be the question which of the theories would "prevail in life", but only "whether the two twin brothers had to be buried in the same grave or in two different graves" [200, 343]. All in all Weyl perceived a scientific devaluation of the UFT's of the 1920s, resulting from developments in the second part of the decade:

> In my opinion the whole situation has changed during the last 4 or 5 years by the detection of the matter field. All these geometrical leaps (geometrische Luftsprünge) have been premature, we now return to the solid ground of physical facts. Weyl [200, 343]

He continued to sketch the theory of spinor fields, their phase gauge and its inclusion into the framework of general relativity along the lines of his 1929 articles.

[30] See [179, p. 274ff.], [2, 148, 171] and N. Straumann's contribution to this volume.

Weyl emphasized that, in contrast to the principles on which the classical unified field theories had been built, the new principle of phase gauge "has grown from experience and resumes a huge treasury of experimental facts from spectroscopy" (ibid. 344). He still longed for safety, just as much as at the time after the First World War, when he had designed his first gauge unification. Now he no longer expected to achieve it by geometric speculation, but tried to anchor it in more solid grounds:

> By the new gauge invariance the *electromagnetic field now becomes a necessary appendix of the matter field, as it had been attached to gravitation in the old theory.* Weyl [200, 345, emphasis in original]

In this way Weyl made it clear that he had changed his perspective. He no longer saw a chance in attempts to derive matter in highly speculative approaches from mathematical structures devised to geometrize force fields; he now set out searching for mathematical representations of matter which was based on the "huge treasury" of experimental knowledge. For him, this was reason enough to prefer the view that the electrical field"follows the ship of matter as a wake, rather than gravitation" (ibid.).

This paper indicates a re-evaluation of Weyl's view of geometry with regard to those of the early 1920s. This change of mind took place at the turn to the 1930s and is not yet present in the first German edition of his book *Philosophie der Mathematik und Naturwissenschaften* [197], but it is in the English translation for which the author formulated text amendments and changes during 1948/49 [201]. These changes were taken over into the German third edition (after Weyl's death).

In a talk with the title *Similarity and congruence* given during the time when he worked on the changes for *Philosophy of Mathematics and Natural Science* Weyl discussed the topic of *automorphism groups* as a clue for establishing objectivity of symbolic knowledge in mathematics and in physics.[31] Weyl hoped to be able to clearly distinguish between *physical* and *mathematical automorphisms*. The latter were characterized by him as the normalizer of the former—leaving open the question in which larger group the normalizer was to be taken. For classical physics this was relatively simple: the physical automorphisms of classical physics are given by the Galilei group (including Euclidean congruences), and the mathematical automorphisms are the similarity transformations extending the Galilei group. Weyl therefore used the pair similarities/congruences also in the general case as synonymous with the dichotomy mathematical/physical automorphisms.

Of course it is a difficult question to decide what the physical automorphisms are; but Weyl was sure that this is a central task of physics:

> The physicist will question Nature to reveal him her true group of automorphisms [205, p. 156].

[31] This talk is published in Weyl [205].

For relativistic physics, so Weyl argued in 1949, the physical automorphisms were given by the diffeomorphisms *Diff*(M) of the spacetime manifold M, extended by point dependent operations of $G = SO(1,3) \times U(1)$. In hindsight Weyl's physical automorphisms of relativistic physics may be considered as an informal characterization of a gauge group $\mathfrak{G}(P)$ *ante letteram* with respect to a principal bundle $\mathfrak{P}$ over M with the group G. In his view the mathematical automorphisms were then the corresponding "similarities" given by the extension from G to $\widetilde{G} = G \times \mathbb{R}^+$.[32] In the view of the mature Weyl, the contemporary knowledge of the 1930/40s in quantum physics clearly spoke for reducing the physical automorphisms to the group G.

> The atomic constants of charge and mass of the electron atomic constants and Planck's quantum of action $\hbar$, which enter the universal field laws of nature, fix an absolute standard of length, that through the wave lengths of spectral lines is made available for practical measurements [205, p. 161].[33]

In the 1940s Weyl no longer considered the local scale extension from G to $\widetilde{G}$ as part of physics, but of mathematics only. Taking up his language of 1919 in the discussion with Einstein, the laws of quantum mechanics and the universal constants $(\hbar, e, m_e)$ had now taken over the role of the *central office of standards*. For Weyl, this was a definitive good bye to the idea of localized standards of length/time due to an adaptation of atomic oscillators to local field constellations and from the view that his scale gauge geometry is an adequate conceptual framework for gravity and field theory.

He did not care about an interesting observation made by *Jan A. Schouten* and *Jan Haantjes* in the 1930s, who argued that not only the (vacuum) Maxwell equations, but also the equations of motion of test mass particles (with or without Lorentz forces) can be written in a scale covariant form, if only the mass parameter m is being transformed scale covariantly with quadratic Weyl weight $-\frac{1}{2}$ [71, 158, 159]. At the time, this was an unusual point of view. Pauli argued that only the massless Dirac equation could be considered as scale invariant [125]; a similar point of view had been expressed by Weyl in the context of the general relativistic Dirac equation [199]. Schouten's and Haantjes' common proposal to consider mass parameters as scaling quantities indicated a path toward including massive spinor fields into a basically conformal—or correspondingly a Weyl geometric—framework; but at the time it was not widely noted.

[32]For more details see [154].

[33]Similarly in PMN.

3 A New Start for Weyl Geometric Gravity in the 1970s

3.1 New Interests in Local Scale and Conformal Transformations

In the early 1950s and 1960s *Pascual Jordan*, later *Robert Dicke* and *Carl Brans* (JBD) proposed a widely discussed modification of Einstein gravity [29, 47, 86]. Essential for their approach was a (real valued) scalar field χ, coupled to the Hilbert action term. Its Lagrangian density was

$$\mathcal{L}_{JBD} = (\chi R - \frac{\omega}{\chi} \partial^{\mu} \chi \, \partial_{\mu} \chi) \sqrt{|det \, g|} \,, \tag{9}$$

where ω is a free parameter of the theory and R the scalar curvature of the Riemannian metric g. Jordan had started from a projective version of a 5-dimensional Kaluza-Klein approach. He arrived at a Lagrangian of form (9) only after several steps of simplifications and interpreted the scalar field as a varying gravitational parameter (replacing the gravitational constant).[34] For $\omega \rightarrow \infty$ the theory has Einstein gravity as limiting case. All three authors allowed for conformal transformations of the metric, $\tilde{g} = \lambda g$; but only Brans and Dicke understood them as an expression for a local scale transformation under which also the scalar field χ transforms with (quadratic length) weight -1 (matter fields and energy tensors T with weight $w(T) = -\frac{1}{2}$ etc.).

Jordan started considering conformal transformations only after Pauli had made him aware of such a possibility; he discussed them in the second edition of his book [86]. Pauli must have been aware of the closeness of this principle to Weyl's scale geometry; in his youth he had been one of the experts for it. But neither he nor Jordan, looked at the new scalar tensor theory from this point of view. The migration of the gauge idea from scale to phase geometry seems to have been considered by them, like by Weyl himself, as definitive. The US-American authors nearly a decade later were probably not even aware of the parallel.[35]

Maybe this ignorance was an advantage. Dicke did not hesitate when advocating conformal rescaling. He frankly declared it as an obvious postulate that the "*laws of physics must be invariant under a general coordinate dependent change of units*" (see quote in the introduction, p. 27). He did not mention that this would demand a basic

[34] For technical details see [70, pp. 31ff.], for the physical interpretation and the historical context [96, pp. 45ff, 65ff.] and [97]; a conceptual analysis will be given in Lehmkuhl [101].

[35] In retrospect C. Brans wrote regarding this question:"I believe (but am not sure) that I knew of other UFT's, especially Kaluza-Klein, but do not know if I was aware of Weyl's conformal work. I wish I could be more definite, but the best answer I can give to your question of whether Bob or I was aware of Weyl's conformal scalar field is 'probably not'." (e-mail of C. Brans to the author, 19 June, 2012).

restructuring, or at least reformulation of fundamental theories, although he must have been aware of it. The postulate itself agreed with Weyl's intentions of 1918. In a long appendix to his letter of 16 Nov. 1918 to Einstein, published in extended form during the next year, Weyl had stated

> Einstein's present theory of relativity (...) only deals with the arbitrariness of the coordinate system; but it is important to gain a comparable foundational stance with regard to (...) the arbitrariness of the measurement of units [190, p. 55].[36]

Knowing more about Weyl's trajectory, who had recanted this viewpoint in the 1940s, might only have been a hindrance for developing the new scalar tensor theory, or at least the inclusion of the conformal transformations viewpoint into it.

By "coordinate dependent change of units" Dicke and Brans indicated a point dependent rescaling of basic units. In the light of the relations established by the fundamental constants (velocity of light c, Planck constant $\hbar$, elementary charge e and Boltzmann constant k) all units can be expressed in terms of one independent fundamental unit, e.g., time, and the fundamental constants (which, in principle can be given any constant numerical value, which then fixes the system).[37] Thus only one essential scaling degree of units remains, and Dicke's principle of an arbitrary point dependent unit choice comes down to a "passive" formulation of Weyl's localized similarities in his scale gauge geometry, where dimensional constants are to be treated as scale covariant scalar fields with the respective Weyl weights.

A closer look shows that Dicke's boastful postulate that the "laws of physics must be invariant" under point dependent rescaling was not fully realized in JBD theory. The *modified Hilbert term* of (9) is formulated in terms of the Riemannian scalar curvature and *is not scale invariant*. The practitioners of JBD theory understand it as defined in a specific scale (the Jordan frame) and apply well known correction terms under conformal rescaling. This defect can easily be cured if one reformulates the theory in terms of a simple version of Weyl geometry (see Sect. 5.1).

---

[36]"Die bisherige Einsteinsche Relativitätstheorie bezieht sich nur auf (...) die Willkürlichkeit des Koordinatensystems; doch gilt es eine ebenso prinzipielle Stellungnahme zu (...) der Willkürlichkeit der Maßeinheiten zu gewinnen" [190, p. 55].

[37]For the recent revision of the international standard system SI see [75]. It has implemented measurement definitions with time as only fundamental unit, $u_T = 1\,s$ such that "the ground state hyperfine splitting frequency of the caesium 133 atom $\Delta\nu(^{133}Cs)_{\mathrm{hfs}}$ is exactly 9 192 631 770 hertz" [31, 24f.]. In the *New SI*, four of the SI base units, namely the kilogram, the ampere, the kelvin and the mole, will be redefined in terms of invariants of nature (www.bipm.org/en/si/new_si/). The redefinition of the meter in terms of the basic time unit by means of the fundamental constant c was implemented already in 1983. Point dependence of the time unit because of the locally varying gravitational potential on the surface of the earth is inbuilt in this system. For practical purposes it can be outlevelled by reference to the *SI second on the geoid* (standardized by the International Earth Rotation and Reference Systems Service IERS). Some, only seemingly paradoxical, consequences in the description of astronomical distances are being discussed by T. Schücker, this volume..

In spite of such a (minor) formal deficiency, the three proponents of JBD theory unknowingly brought their approach quite close to Weyl geometry by fixing a *unique affine connection* rather than changing it with the conformal transformation. They postulated the Levi-Civita connection $\Gamma := \Gamma(g)$ of the Riemannian metric g in (9), called the *Jordan frame*, as unchanging under conformal transformations. In a different scale gauge, or *frame*, $\tilde{g} = \lambda g$ they just had to express Γ in terms of the Levi-Civita connection $\Gamma(\tilde{g})$ plus correction terms in derivatives of λ (equivalent to 7). Let us summarily denote these additional terms by $\Delta(\partial\lambda)$, then

$$\Gamma = {}_{\tilde{g}}\Gamma + \Delta(\partial\lambda)\ .$$

Roger Penrose noticed that the additional terms of the (Riemannian) scalar curvature are exactly cancelled by the partial derivative terms of the kinematical term of χ if and only if $\omega = -\frac{3}{2}$. In this case, and in JBD theory only in this case, the Lagrangian (9) is conformally invariant [128].

Probably the protagonists considered the invariance of the affine connection Γ as a consequence of the principle that the "laws of nature" have to be considered as invariant under conformal rescaling. If the trajectories of test bodies are governed by the gravito-inertial "laws of physics" they should not be subject to change under a transformation of units. The same must then hold for the affine connection which can be considered a mathematical concentrate of these laws. Thus passive *conformal rescaling*, in addition to *fixing an affine connection*, have become *basic tools of JBD theory*.[38]

JBD theory originated at a time when symmetry aspects attracted more and more attention by high energy particle physics. Here symmetries were often not considered as exactly realized, but as somehow "hidden" or even "broken".[39] Also conformal transformations were now being reconsidered in field physics. *Werner Heisenberg* took up the idea of Haantjes and Schouten and proposed to consider rescaling of mass parameters as a legitimate symbolic procedure [78]; and so did other elementary particle physicists.[40] In this context it became a standard procedure to associate scale weights to physical fields. For a detailed historical report see [90].

Of course, elementary particle physicists were not interested in conformal transformations of general Lorentzian manifolds; they dealt exclusively with the group $Conf(\widetilde{\mathbb{M}})$ of conformal transformations of the (conformal) compactification $\widetilde{\mathbb{M}}$, of Minkowski space. The group can be expressed as the subgroup of projective

[38] For an outline of conformal transformation as used in JBD see [180, app. D]; surveys on the actual state of JBD theory and its applications to cosmology are given in Fujii and Maeda [65], Faraoni [61].

[39] See [22, 24].

[40] Among them Wess [181, 182] and Kastrup [87, 88].

transformations in real projective space $P^5(\mathbb{R})$ with projective coordinates $[u_0, \dots u_5]$, which leaves the hyperbolic quadric $Q(2,4)$

$$u_0^2 + u_5^2 - \sum_{j=1}^{4} u_j^2 = 0$$

invariant. The embedding of Minkowski space $\mathbb{M} \hookrightarrow Q(2,4) \subset P^5(\mathbb{R})$ is given by

$$(x_0, \dots x_3) \longmapsto [u_0, \dots, u_3, \frac{1}{2}(|x|^2 + 1), \frac{1}{2}(|x|^2 - 1)],$$

with $|x|^2 = x_o^2 - \sum_{j=1}^{3} x_j^2$. This was well known at the time.[41] But the understanding of the reciprocal transformations (inversions) with regard to Lorentzian hyperboloids rather than to euclidean spheres was hampered for some time because of confused attempts to interpret them as an expression for transformations of relatively accelerated observer systems. Such proposals had been made in the 1930/40s, among others by Haantjes [71]. They lived until the mid 1960s, although it was clear to a part of the community that this was misleading.[42] *Hans Kastrup* for example, developed an interpretation with point dependent measuring standards for describing the effects of hyperboloidal inversions and studied the representations of the full conformal group on scalar and spinor fields with the hope that the inversions would, at least, turn out to be approximative symmetries of a conformal field theory [89]. In the discussion of the point dependent standards he explicitly referred back to Weyl's scale gauge method of 1918 [89, p. 150], although the wider scope of Weyl geometry was of no importance for his investigations.

The questions regarding the inversions posed in the GR community were different. Penrose gave a detailed geometrical analysis of the light cone at infinity in Minkowski space [127]. From such a point of view it became clear that the reciprocal transformations are mappings between conformal infinity and finite light cones, which induce a conformally deformed metric (with regard to the Minkowski metric) in the neighbourhood of the latter. In general relativity such transformations were useful for studying fields in asymptotic flat spacetimes. They even became a step towards better understanding the asymptotic behaviour of general relativistic spacetimes also in more general cases.[43] In this sense, hyperboloid inversions, or their relatives in more general spacetimes, remained at best "mathematical automorphisms" in the language of Weyl, useful for technical reasons rather than for physical ones.

---

[41] Such a representation of *Conf*$(\widetilde{\mathbb{M}})$ is sketched, e.g., in Weyl [196, p. 302f.].

[42] Kastrup [90, p. 659ff.].

[43] Frauendiener [64].

In the high energy context, on the other hand, reciprocal transformations had a different appeal, in particular if applied to the energy-momentum space (in a physical sense dual to Minkowski spacetime). If everything went well, they seemed to allow conformal field theory searching for relationships between field states in extremely high energies/momenta and extremely small energy/momentum states close to a finite light cone, even though such inversion symmetries might perhaps be "hidden" or "broken" and indicate the relations only in a modified form. In this sense, the understanding of conformal symmetries in high energy physics was characterized by an active understanding in the perspective of extending the "physical" automorphisms in Weyl's description of 1948/49.

Such differences of outlook my have contributed to the unhappy constellation that the interchange between general relativity and conformal field theory (in Minkowski space) remained rather weak. As far as I can see, there were mutual methodological challenges between the fields, but no closer exchange.

3.2 Weyl Geometric Gravity with a Scalar Field and Dynamical Scale Connection: Omote, Utiyama, Dirac

In 1971 *Minoru Omote*, Tokyo, introduced a scale covariant scalar field ϕ coupling to the Hilbert term and scaling like in JBD theory into the framework of Weyl geometry [121].[44] A second paper by Omote followed after the publication of a paper by P.A.M. Dirac with a similar proposal [48] and after R. Utiyama had jumped in Omote [122] (for both authors see below). *Alexander Bregman*, at that time working at Kyoto, was inspired by Omote's proposal to separate localized rescaling from Weyl's geometrical interpretation of the infinitesimal length transport [30]. He argued that the point-dependent scale transformations could be treated "analogous to the introduction of a space-time dependence into the constant parameters of Isospin or Poincaré transformations" (ibid. p. 668). This brought the approach closer to what high energy physicists were doing at the time, although only the scale extended Poincaré group was localized, not the complete conformal group of Minkowski space. The global scale dimensions d of a physical field X could then be taken over as "Weyl weight" of X to the localized theory [30, p. 668]. With such a proposal he followed the lines of the research program for constructing general relativistic theories of gravity by "localizing" the symmetries of the Poincaré group, which had been opened by Sciama and Kibble [91, 160].[45]

---

[44]This was more than a year before the Trieste symposium at which Dirac talked about the same question (see below), while it seems that Omote's paper remained unknown to him.
[45]See [21].

A little later, more or less parallel to Dirac, *Ryoyu Utiyama*, Toyonaka/Osaka, joined the new start of a Weyl geometric approach to gravity [175]. Bregman's approach seemed to have triggered his interest which lay mainly in elementary particle physics.[46] Different from Dirac and, of course also from Weyl, he saw in the nontrivial Weylian scale connection φ a candidate for a new fundamental field of high energy physics, the Weyl gauge field or, later, the *Weyl boson*. In a series of papers [175–177] he ventured toward a bosonic interpretation of φ and presented his results at the Seventh International Conference on Gravitation and Relativity (Tel Aviv, June 1974). Utiyama emphasized that in Weyl geometry a scalar field ϕ of weight -1 could serve as a kind of *measure field* (Utiyama's terminology). With respect to it gauge invariant measurable quantities could be defined for physical observable without assuming a breaking of the scale gauge symmetry [175, 177].

Utiyama proposed to explore the ordinary Yang-Mills Lagrangian term for a Weylian scale connection

$$\mathcal{L}_\varphi = -\varepsilon \frac{1}{4} f_{\mu\nu} f^{\mu\nu} \sqrt{|\det g|} \quad \text{(here with } \varepsilon = 1) \tag{10}$$

Utiyama [177, Eq. (2.4)].[47] He studied conditions under which "Weyl's gauge field" admitted plane wave solutions, and came to the conclusion that they would be tachyonic, allowing superluminal propagation of perturbations. In Utiyama's view the "boson" had therefore to be confined to the interior of matter particles. Nevertheless he thought that this "unusual field φ_μ might play some role in establishing a model of a stable elementary particle" [175, 2089]. This view was not accepted by all his readers. *Kenji Hayashi* and *Taichiro Kugo*, two younger colleagues from Tokyo resp. Kyoto, reanalysed Utiyama's calculations and argued that, with slight adaptations of the parameters, the sign ε in (10) could be switched. Then an ordinary, at least non-tachyonic, field would result [73, 340f.].

Even then the scale connection would still have strange physical properties. After introducing Weyl geometric spinor fields and their Lagrangians in terms of scale covariant derivatives, the two physicists showed that the scale connection terms cancel in the spinor action. In their approach neither the scalar ϕ-field nor the scale connection φ coupled directly to spinor field or to the electromagnetic field. The new fields ϕ, φ seemed to characterize an extension of the gravitational sector with no direct interaction with the known elementary particles. At the very moment that a Weylian scale connection φ was interpreted as a physical field beyond electromagnetism, it started to puzzle its investigators. It seemed to pose more riddles than it

---

[46] In his first paper of 1973 Utiyama did not mention Omote. This changed in later papers, see the references of Utiyama [176].

[47] Dirac included a similar scale curvature term in his Lagrangian, but with another interpretation (see below).

was able to solve. It did not to couple to matter fields (Hayashi/Kugo), looked either tachyonic (Utiyama) or, as we shall see below (Smolin, Nieh, Hung Cheng), appeared to be of Planck mass, far beyond anything observable.

Independent of the Japanese physicists, and more or less at the same time, *Paul Adrien Maurice Dirac* brought Weyl geometry back into the rising field scalar-tensor theories, although with different interpretations from his Japanese colleagues in mind [48, 49]. His motivation had two components which may look strange from today's point of view. He started from the speculation that some long noticed interrelations of certain large numbers in physics indicated a deep structural law of the universe (the "large number hypothesis") and the idea of a varying gravity. Both ideas had their origin in the 1930s, and also P. Jordan had taken them up.[48] Dirac presented his ideas at the occasion of a symposium at Trieste 1972, honouring his 70th birthday. The talk remained unpublished but participants report that its content was close to a publication in the following year [48].[49]

Dirac's paper was important for the transmission of knowledge between generations. He introduced his readers to Weyl geometry which was no longer generally known among younger physicists, following Eddington's notation and terminology of "co-invariants" for scale covariant fields [55]. He also added the important concept of the *scale-covariant derivative* D, respectively D_μ (in later terminology and notation), for scale covariant fields to the methodological arsenal of Weyl geometry (see below, Eq. 17). It is a necessary modification of the covariant derivative of scale covariant fields for arriving again at scale covariant fields. He called it, virtually stuttering, the "co-covariant" derivative.

Similar to Jordan/Brans/Dicke—and like Omote/Bregman/Utiyama, the publications of which he apparently did not know—Dirac introduced a scalar field which he called β. Like the other authors he rescaled it with the (non-quadratic) length weight -1 and coupled it to the Hilbert term expressed by the sign-inverted Weyl geometric scalar curvature R.[50] Thus he replaced Weyl's gravity *Ansatz* in the Lagrangian, using square curvature terms, by a scale invariant Lagrangian ($\mathcal{L}_{Dir} = L_{Dir}\sqrt{|g|}$) of first order in R,

$$L_{Dir} = -\beta^2 R + kD^\lambda\beta\, D_\lambda\beta + c\beta^4 + \frac{1}{4} f_{\mu\nu} f^{\mu\nu} \,. \tag{11}$$

[48] For Dirac's and Jordan's ideas on the large number hypothesis, varying gravity and a surprising link to geophysics in the 1930s see the detailed study in Kragh [96].

[49] Charap and Tait [37, p. 249 footnote].

[50] The qualifications "sign inverted" refers to the sign convention which agrees with the definition $Riem\,(Y, Z)\,X = \nabla_Y \nabla_Z X - \nabla_Z \nabla_Y X - \nabla_{[Y,Z]} X$, i.e., $R^\mu{}_{\nu\lambda\kappa} = \partial_\lambda \Gamma^\mu_{\nu\kappa} - \partial_\kappa \Gamma^\mu_{\nu\lambda} + \Gamma^\alpha_{\nu\kappa}\Gamma^\mu_{\alpha\lambda} - \Gamma^\alpha_{\nu\lambda}\Gamma^\mu_{\alpha\kappa}$. It is preferred in the mathematical literature including [188, 5th ed., 131] and also used in the majority of recent physics books. .

The quartic potential for β was demanded by scale invariance. $f = d\varphi$ denoted the Weylian scale curvature. Different from the Japanese physicists, Dirac stuck to the outdated interpretation of the scale connection φ as the potential of the electromagnetic (Maxwell) field $F_{\mu\nu}$

$$F_{\mu\nu} = f_{\mu\nu} \,. \tag{12}$$

In the sequel I call this the *electromagnetic (em) dogma* .

For the coupling constant of the kinetic term, $k = 6$, the contributions of the scale connection to the Lagrangian essentially cancel.[51] Dirac knew that for $k \neq 6$ a large mass term for the Weyl field φ arises, which would destroy the *em*-dogma interpretation; so he chose $k = 6$. As a result Dirac could write the Lagrangian in a form using only the (sign inverted) Riemannian component ${}_gR$ of the scalar curvature

$$\mathcal{L}_{Dir\,1} = -\beta^2\, {}_gR + 6\partial^\lambda \beta\, \partial_\lambda \beta + c\beta^4 + \frac{1}{4} f_{\mu\nu} f^{\mu\nu} \,, \tag{13}$$

It was already known to be conformally invariant [128], and only the specific Weyl geometric interpretation of the *em* potential was added.

Referring back to a passage in Eddington[52] Dirac derived relations between the Euler-Lagrange expressions of his action (13), due to its invariance under diffeomorphisms and under scale transformations [48, Eqs. (7.1, 7.2)]. He called them "conservation laws". Similar to Weyl's argument in Weyl [187], they implied a vanishing divergence of the electric current if the gravitational and the scalar field equations are satisfied. Still we find no mention that such a derivation could be considered as a special case of Noether's second theorem.

If it had not been Dirac, such an approach would probably not have attracted much interest in the physics community. But he also derived dynamical equations and the Noether identities for diffeomorphisms and scale transformations. For a vanishing *em* field, $f_{\mu\nu} = 0$, he distinguished the *Riemann gauge* with $\varphi = 0$ (called by him "natural gauge") from the *Einstein gauge* (with the gravitational parameter constant, $\beta = 1$) and a hypothetical *atomic gauge* characterized as "the metric gauge that is measured by atomic apparatus" (Weyl's "natural gauge") and warned that "all three gauges are liable to be different" [48, 411].

At the end of his article Dirac discussed why one should believe in the proposed "drastic revision of our ideas of space and time". He announced another part of his research agenda, which was *independent* of the large number hypothesis:

> There is one strong reason in support of the theory. It appears as one of the fundamental principles of Nature that the equations expressing basic laws should be invariant under the widest possible group of transformations ...The passage to Weyl's geometry is a further step

[51] They reduce to boundary terms and thus are variationally negligible.

[52] Eddington [55, Sect. 61].

> in the direction of widening the group of transformations underlying physical laws. Dirac [48, 418]

So far, Dirac's explanations were close to the view of Brans and Dicke. He followed a tendency of the time for probing possible extensions of the symmetries (automorphisms) of fundamental physics and saw a new chance for Weyl geometry to play its part in such an endeavour.

Dirac's proposal for reconsidering Weyl geometry in a modified theory of gravity was taken up by field theorists, gravitational physicists and a few astronomers. An immediate and often quoted paper by *Vittorio Canuto* and coauthors gave a broader and more detailed introduction to Dirac's view of Weyl geometry in gravity and field theory [33]. The opening remark of the paper motivated the renewed interest in Weyl geometry with actual developments in high energy physics:

> In recent years, owing to the scaling behavior exhibited in high-energy particle scattering experiments there has been considerable interest in manifestly scale-invariant theories. Canuto et al. [33, 1643]

With the remark on "considerable interest in manifestly scale-invariant theories" in high energy physics the authors referred to scaling in high energy physics and, in particular, the seminal paper [32]. But the authors were careful not to claim field theoretic reality for Dirac's scalar function β [33, 1645]. They rather developed model consequences for the approach in several directions: cosmology, including the "large number hypothesis as a gauge condition" (ibid. 1651), modification of the Schwarzschild solution in the Dirac framework, consequences for planetary motion, and stellar structure. At the end the authors indicated certain heuristic links to gauge fields in high energy physics of the late 1970s.

Dirac's retake of Weyl geometric gravity and field theory had wider repercussions than might be expected if one takes his insistence on the *em*-dogma into account. The scope of the approach was much wider without it, and a lot of authors with different backgrounds and diverging research interests started again to explore the potential of Weyl's geometry for widening the geometric framework of gravity, most of them without subscribing to the *em*-dogma. Here I can only hint summarily at a sample from different groups or individual authors who took part in this exploration during the following decades until, roughly, the end of the century. Among them were P. Bouvier and A. Maeder, Geneva, from *astronomy* [25, 26, 102],[53] N. Rosen, the former collaborator of Einstein, with his PhD student M. Israelit from *gravitational theory* [80–82, 140], and M. Novello with a growing group of researchers in Brazil from *cosmology* [117, 118]. Other authors like L. Smolin, W. Drechsler, his PhD student H. Tann and H. Cheng, working in *fundamental* or *high energy physics*, also took up Weyl geomteric methods [38, 51–53, 165]. Most of these authors worked

[53] In recent publications A. Maeder has taken again up this line of research [104–106], see Sect. 5.4.

at different places and separated from each other; only in rare cases they knew and cited the respective works of their colleagues. In spite of the rising numbers of papers and authors during the last third of the 20th century one cannot speak of the birth of a literary network or even of a scientific subcommunity of Weyl geometrically oriented work in theoretical physics.

Once in a while researchers studying gravitational gauge fields in the *Cartan geometric approach* considered Weyl geometry as a special case of their wider program, with or without the additional feature of translational curvature; among them notably Charap and Tait [37], Hehl et al. [76, 77]. These papers called attention to both Noether theorems.[54] The source current of the equation for the Weyl field (scale connection) was no longer seen as an expression for the charge current of the electromagnetical field, but rather as a new dynamical entity called "dilational current". It was related to the canonical current of the localized scale symmetry and the Noether identities of the second theorem. This was similar to what Weyl and Dirac had indicated in their respective approaches, but it was now part of a much more general framework embracing also, e.g., localized affine symmetries and their corresponding disformal currents expressing some kind of "hypermomenta" [76, pp. 275f., 283f.].—These remarks are, of course, far from exhaustive. A more extended, although still incomplete survey can be found in Scholz [153].

3.3 *Foundations of General Relativity: EPS*

About the same time at which Dirac and the Japanese physicists were using Weyl geometry in the context of scalar tensor theories, new interest in Weyl's geometry of 1918 arose also in the research on the foundations of gravity. Weyl's argument of Weyl [192] that the combination of projective and conformal structures suffice for uniquely characterizing a Weyl geometric structure was taken up, abstracted and extended half a century later by *Jürgen Ehlers, Felix Pirani and Alfred Schild* (EPS in the sequel) [56]. The EPS paper was written for a *Festschrift* in the honour of J.L. Synge. Synge was known for his proposal of basing general relativity on the behaviour of standard clocks rather than length measurements (*chronometric approach*).[55] From the foundational point of view, however, clocks are no less unproblematic for defining the metric of spacetime, because they are realized by complicated material systems and would need a validated theory of time measurement for a foundational justification. The question suggested itself, whether the physical metric can be determined on

[54]In Hehl et al. [76] to the Noether theorem I (global scale symmetry in Minkowski space) and in Hehl et al. [77] and many other papers to Noether theorem II. For a condensed information on the two Noether theorems see Sect. 7.

[55]Compare similar remarks in T. Schücker's contribution to this volume.

the basis of more elementary signal structures of gravitational theory than material clocks, e.g., by like light rays, particle trajectories etc. Weyl's paper of 1921 had sketched such a type of approach.

EPS analysed Weyl's idea of 1921 in detail, using the mathematical language of differentiable manifolds and mimicking, at least to a certain degree, Hilbert's axiomatic method. They started from three sets, $\mathcal{M} = \{p, q, \ldots\}$, $\mathcal{L} = \{L, N, \ldots\}$, $\mathcal{P} = \{P, Q, \ldots\}$ representing the collections of *events, light rays* and *particle trajectories* respectively. By postulates close to physical experimental concepts of light signal exchange between particles EPS formulated different groups of axioms, which allowed them to conclude that the event set $\mathcal{M}$ could be given the structure of a (C^3-) differentiable manifold M endowed with a conformal structure $\mathfrak{c}$, the null-lines of which agree with $\mathcal{L}$, and a projective structure of (C^2-) differentiable paths in agreement with $\mathcal{P}$. The latter can equivalently be described by a projective equivalence class $\mathfrak{p} = [\Pi]$ of affine connections $\Pi = \Pi^{\lambda}_{\mu\nu}$ (ibid. p. 77). In an additional axiom C (ibid. p. 78) they secured the *compatibility* between the structures $\mathfrak{c}$ and $\mathfrak{p}$, with the upshot (C') that the null lines of the conformal cones of $\mathfrak{c}$ are projective geodesics of $\mathfrak{p}$ (ibid. 78–80).[56]

In the light of this result the criterion for *compatibility of* $\mathfrak{c}$ *and* $\mathfrak{p}$ *in the sense of EPS* can be stated as follows:

> A projective structure $\mathfrak{p}$ and a conformal structure $\mathfrak{c}$ will be called *EPS-compatible*, if the null lines of $\mathfrak{c}$ are projective geodesics (autoparallels) of $\mathfrak{p}$.

This criterion was new. For Weyl the compatibility of $\mathfrak{p}$ and $\mathfrak{c}$ was no independent problem, because he supposed both to be abstracted from the affine connection of a Weylian metric (assumed to exist in advance). From the viewpoint of the EPS context a modernized definition of *Weyl compatibility* can be stated as follows (compare Sect. 2.1):

> *Definition.* The conformal and projective structures $\mathfrak{c}$ and $\mathfrak{p}$ are said to be *Weyl compatible*, if for some $g \in \mathfrak{c}$ a differential 1-form φ can be found such that the affine connection $\Gamma(g, \varphi)$ of the Weyl metric $[(g, \varphi)]$ satisfies $\Gamma(g, \varphi) \in \mathfrak{p}$.[57]

In distinction to the Riemannian case φ need, of course, not be integrable.

Using their new concept of compatibility, EPS derived their main statement:

> A light ray structure $\mathcal{L}$ and a set of particle trajectories $\mathcal{P}$ defined on an event set $\mathcal{M}$ which satisfy the EPS axioms endow $\mathcal{M}$ with the structure of a (C^3-) differentiable manifold M and specify a (C^2-) Weylian metric $[(g, \varphi)]$ upon M. The metric is uniquely determined by the condition that its causal and geodesic structures coincide with $\mathcal{L}$ and $\mathcal{P}$ respectively.

---

[56]For a more detailed description of the paper, including references to follow up papers and some carefully critical remarks, see [174].

[57]If this holds for *some* $g \in \mathfrak{c}$, then for *any* $g \in \mathfrak{c}$ (due to Weyl's rescaling result and the scale invariance of $\Gamma(g, \varphi)$).

If one accepts the argumentation, this was an important result. The authors were cautious enough, however, to qualify their result by the remark:

> A fully rigorous formalization has not yet been achieved, but we nevertheless hope that the main line of reasoning will be intelligible and convincing to the sympathetic reader [56, p. 69f.].

In his commentary to a recent re-edition of the EPS paper Trautman hinted at the desideratum of a formal proof of the existence of the affine connection $\Gamma(g, \varphi)$ (in our notation) [174, p. 1584]. In a joint paper with Matveev the two authors even suggest to give a counter example to EPS' existence claim [107]; but this is due to the *quid pro quo* of Riemannian metrics and Weyl metrics (see Sect. 5.2).

EPS gave arguments that assuming some physically plausible axioms, spacetime may be described by a differentiable manifold M. Moreover, according to the authors light rays and particle trajectories can be used to *establish* a metric on M in the sense of *Weyl geometry*, rather than assuming the latter as a result of rod and clock measurements, chronometric prescriptions, or even only as a structural property like Weyl had done in Weyl [192].

Such a result was important also from the point of view of Einstein gravity; but there remained a gap to the latter, namely how, or whether at all, one can arrive at a (pseudo-) Riemanniang metric from the Weylian one. EPS argued that this gap might be closed by adding a single additional *Riemannian axiom*, postulating the vanishing of the scale curvature, $d\varphi = 0$, i.e., the integrability of the Weylian metric [56, p. 82]. Such a postulate would not seem implausible, as Weyl's interpretation of the scale connection φ as electromagnetic potential, the *em*-dogma, was obsolete anyhow; of course EPS did not adhere to it. But the authors did not exclude the possibility that a scale connection field φ of non-vanishing scale curvature might play the role of a "true", although still unknown, field $f = d\varphi$ which would relate "the gravitational field to another universally conserved current" ([56, p. 83]).

The paper of Ehlers, Pirani and Schild triggered a line of investigations in the foundations of general relativity, sometimes called the *causal inertial approach* (Coleman/Korté), sometimes subsumed under the more general search for a *constructive axiomatics* of GRT (Majer/Schmidt, Audretsch, Lämmerzahl, Perlick and others). The debate was opened by Audretsch [6]. It was soon continued by a collective paper written by three authors [7] and had follow up studies, among others [8]. The authors of the first mentioned paper, Audretsch, Straumann and Gähler, argued that the "gap" between Weylian and Riemannian geometry can "be closed if quantum theory as a theory of matter is made part of the total scheme" [6, 2872]. By this they referred to an investigation of scalar and spinor fields on a Weylian manifold and the assumption that the flow lines of their WKB-approximation coincide with the geodesics of the underlying metric. In this way these authors could underpin the additional

Riemannian postulate of EPS which was considered as sufficient for closing the gap between Weyl and Riemannian geometry.

But the integrability of a Weylian metric is only a necessary condition for locally establishing a Riemannian framework, not a sufficient one. Although the choice of the Riemann gauge is always possible in an integrable Weyl geometry and appears natural from a purely mathematical point of view, the question remains which of the different scale gauges expresses time and other physical quantities directly. If it differs from Riemann gauge, even the conceptually minor difference between the Riemannian structure and an integrable Weylian structure *matters for the physics of spacetime*.

The investigations of the causal inertial approach turned towards a basic conceptual analysis from the point of view of foundations of inertial geometry [40], some even looking for Desargues type characterization of free fall lines [132]. How a kind of "standard clocks" can be introduced in the Weyl geometric setting without taking refuge to atomic processes, by just using the observation of light rays and inertial trajectories, was studied by Perlick [129–131]. Another line of follow up works explored the extension of the foundational argument of the causal inertial approach to quantum physics, where particle trajectories might no longer appear acceptable as a foundational concept.

3.4 Geometrizing Quantum Mechanical Configuration Spaces

A completely different perspective on Weyl geometry was taken by *Enrico Santamato* in Naples. In the 1980s he proposed a new approach to quantum mechanics based on studying weak random processes of ensembles of point particles moving in a Weylian modified configuration space [144–146]. He compared his approach to that of *Madelung-Bohm* and to the stochastic program of *Feynès-Nelson*.[58] While the latter dealt with stochastic (Brownian) processes, Santamato's approach was closer to the view of Madelung and Bohm because it assumed only random initial conditions, with classical trajectories given in Hamilton-Jacobi form (this explains the attribute "weak" above). One can interpret Bohm's particle trajectories as deviating from those expected in Newtonian mechanics by some "quantum potential". Santamato found this an intriguing idea, but he deplored its "mysterious nature" which "prevents carrying out a natural and acceptable theory along this line". He hoped to find a rational explanation for the effects of the "quantum force" by means of a geometry

[58] For E. Nelson's program to re-derive the quantum dynamics from classical stochastic processes and classical probability see [10].

with a modified affine connection of the system's configuration space. Then the deviation from classical mechanics would appear as the outcome of "fundamental properties of space" [144, p. 216], understood in the sense of *configuration* space.

In his first paper Santamato started from a configuration space with coordinates $(q^1, \ldots, q^n)$ endowed with a Euclidean metric. More generally, his approach allowed for a general positive definite metric g_{ij}, and later even a metric of indefinite signature, for dealing with general coordinates of n-particle systems, and perhaps, in a further extension, with spin. The Lagrangian of the system, and the corresponding Hamilton-Jacobi equation, contained the metric, either explicitly or implicitly. This Euclidean, or more generally Riemannian, basic structure was complemented by a Weylian scale connection. Santamato's central idea was that the modification of the Hamilton-Jacobi equation induced by a properly determined *scale connection* could be used to express the *quantum modification* of the classical Hamiltonian, much like was done in the Madelung-Bohm approach. Then the quantum aspects of the systems would be geometrized in terms of Weyl geometry, surely a striking and even beautiful idea, if it works.

Santamato thus headed towards a new program of *geometrical quantization sui generis*. This had nothing to do with the better known geometric quantization program initiated more than a decade earlier by J.-N. Souriau, B. Kostant and others, which was already well under way in the 1980s [94, 164, 167]. In the latter, geometrical methods underlying the canonical quantization were studied. Starting from a *symplectic phase space* manifold of a classical system, the observables were "prequantized" in a Hermitian line bundle, and finally the Hilbert space representation of quantum mechanics was constructed on this basis.[59] Santamato's geometrization was built upon a different structure, Weyl geometry rather than symplectic geometry, and had rather different goals.

Like other proposals in the dBMB (de Broglie-Madelung-Bohm) family, Santamato's program encountered little positive response. In the following decades he shifted his research to more empirically based studies in nonlinear optics of liquid crystals and quantum optics. Perhaps a critical paper by *Carlos Castro Perelman*, a younger colleague who knew the program nearly from its beginnings, contributed to what would be an extended period of interruption. Castro discussed "a series of technical points" which seemed important for Santamato's program from the physical point of view [35, p. 872]. This criticism may have contributed to a first interruption of work on the research program. It was taken up again and extended by Santamato, Martini and other researchers in the last decade.

---

[59] See, e.g., [206] or [72, Chaps. 22/23]. A classical monograph on the symplectic approach to *classical* mechanics is [1]; but this does not discuss spinning particles. In the 1980s the symplectic approach was already used as a starting platform for (pre-)quantization to which proper quantization procedures could then hook up, see e.g., [166]. Souriau was an early advocate of this program. In his book he discussed relativistic particles with spin [168, Sect. 14].

After the turn of the century, Santamato came back to foundational questions, working closely together with his colleague *Franceso De Martini* from the University of Rome. Both had cooperated in their work on quantum optics already for many years. In the 2010s they turned to geometrical quantization in a series of joint publications that continued the program Santamato had started three decades earlier. They showed how to deal with spinor fields in this framework, in particular with the Dirac equation [147], and they discussed the famous *Einstein-Podolsky-Rosen* (EPR) non-locality question [42, 43]. Moreover, they analyzed the helicity of elementary particles and showed that the *spin-statistics relationship* of relativistic quantum mechanics can be derived in their framework without invoking arguments from quantum field theory [44, 45]. In this new series of papers, the authors took Minkowski space as the starting point for their construction of the configuration spaces, which could be extended by internal degrees of freedom. They also enlarged the perspective by making a transition from point dynamical Lagrangians to a dynamically equivalent description in terms of scale invariant field theoretic Lagrangians in two scalar fields. For more details one may consult [153] from which this short survey has been adapted.

With these papers we have already entered present work on Weyl geometric methods. In the next chapters we turn towards contemporary researches in greater breadth.

4 Interlude

Let us first lay open the notations and concepts used in the rest of our presentation. They are oriented at the historical literature but deviate from it where it seems advisable.

4.1 Basic Concepts, Notations, and Integrable Weyl Geometry

In the following we use scale weights such that quantities of dimension length L scale with weight 1, $w(L) = 1$ (different from Weyl's practice and, correspondingly, the preceding part of this article in which the convention of "quadratic" scale weights with $w(L) = \frac{1}{2}$ is used). Like usual we assume a non-scaling vacuum velocity of light c and a non-scaling Planck constant $\hbar$; time T, mass M, and energy E scale then respectively by $w(T) = 1,\ w(M) = w(E) = -1$.[60] These weights will be called

[60] If F. Hehl's suspicion that the velocity of light may vary with the gravitational potential turns out to be right (if one considers more precise solutions of the Maxwell equation in GR than assumed in

Weyl weights; they are the negatives of the mass/energy weights used by elementary particle physicists.

The simplest way to define a *Weylian metric* on a differentiable manifold M is by specifying an equivalence class $\mathfrak{g} = [(g, \varphi)]$, consisting of a pseudo-Riemannian metric g (locally $g = g_{\mu\nu}dx^\mu dx^\nu$), the *Riemannian component* of the Weylian metric, and a real-valued differential 1-form φ (locally $\varphi = \varphi_\nu dx^\nu$) representing the *scale connection*. Equivalence $(g, \varphi) \sim (\tilde{g}, \widetilde{\varphi})$ is defined by the gauge transformation

$$\tilde{g} = \Omega^2 g, \qquad \widetilde{\varphi} = \varphi - d\log\Omega = \varphi - \frac{d\Omega}{\Omega}, \tag{14}$$

with a strictly positive re-scaling function Ω on M.[61] Choosing a representative $(g, \varphi) \in \mathfrak{g}$ means to *gauge* the metric.

If the scale connection is closed in some gauge (g, φ), $d\varphi = 0$, it is so in any gauge and is locally (i.e., in simply connected regions) exact with potential, say $-\omega$, such that $\varphi = -d\omega$. Weyl's length transfer can then be integrated

$$\Omega = e^{\int_\gamma \varphi(\dot{\gamma})} = e^{-\omega}$$

independently of the path γ. One therefore speaks of an *integrable Weyl geometry* (IWG). Locally a gauge of the form $(\tilde{g}, 0)$ with

$$\tilde{g} = e^{2\int_\gamma \varphi(\dot{\gamma})} = e^{-2\omega} g \quad \text{and} \quad \tilde{\varphi} = \varphi + d\omega = 0 \tag{15}$$

can be chosen, the *Riemann gauge* of an IWG. Because of the remaining freedom of choosing different scale gauges *this does not mean*, however, a structural *reduction to Riemannian geometry*. That may be important for physical applications if there are reasons to assume or to explore the possibility that measuring instruments ("clocks") don't adapt to the Riemann gauge. The Jordan frame of Brans-Dicke theory, e.g., corresponds to Riemann gauge if it is being formulated in terms of IWG. The still ongoing debate on the question which frame, respectively gauge, expresses measured quantities best in BD theory shows that an a priori preference for the Riemann gauge may be a shortcut shadowing a physically important question.

A more fundamental difference to Riemannian geometry arises if $f = d\varphi \neq 0$. f is the curvature of the scale connection, shorter the *scale curvature* (not to be confused with the scalar curvature R of the affine connection). Physically it expresses the field strength of the scale connection, often called the *Weyl field* of the structure.

the optical limit), a specification like, e.g., a constraint to conformally flat Riemannian metric, has to be included in the definition of c.

[61] Note the difference to Eq. (2), due to the different conventions for scaling weights.

The uniquely defined *affine connection* compatible with $[(g, \varphi)]$ will often be written as Γ, if context makes clear what is meant. In more complicated contexts we write more explicitly $\Gamma = \Gamma\,(g, \varphi)$, although the second expression might suggest a scale gauge dependence of Γ, which is not the case. Γ decomposes into gauge dependent contributions like in (3), $\Gamma\,(g, \varphi) = \Gamma = \Gamma\,(g) + \Gamma\,(\varphi)$ with $\Gamma(g)$ the Levi-Civita connection of g and

$$\Gamma^{\mu}_{\nu\lambda}(\varphi) = \delta^{\mu}_{\nu}\varphi_{\lambda} + \delta^{\mu}_{\lambda}\varphi_{\nu} - g_{\nu\lambda}\,\varphi^{\mu}\,, \tag{16}$$

here without the factor $\frac{1}{2}$ in comparison to (3) because of the modified weight convention. Γ itself is gauge invariant.

The covariant derivative ∇ and the curvature tensors *Riem*, *Ric* will usually denote the ones derived from the Weyl geometric affine connection Γ. They are scale invariant. In any gauge they decompose into a gauge dependent Riemannian component, *Riem* (g), *Ric* (g), and another one expressing the contribution of the scale connection *Riem* (φ), *Ric* (φ), etc.[62] The scalar curvature scales with weight $w(R) = -2$.

The scaling behavior of (scalar, vector, tensor, spinor) fields X on a Weylian manifold $(M, [(g, \varphi)]$ has to be specified by a corresponding scale weight $w(X)$ according to invariance principles of the Lagrangian or, from a more empirical point of view, according to its physical dimension. From a mathematical point of view $w(X)$ specifies the representation type of the scale group for the adjoint bundle (to the scale bundle) in which X lives.

The full covariant derivative of fields X in the sense of the Weyl geometric structure in a gauge (g, φ) has to take into account the scale connection in addition to the Levi-Civita connection of g. It will be called *scale covariant derivative* D (Dirac's "co-covariant derivative, p. xxx) and is given by

$$DX = \nabla(g)\,X + w(X) \otimes X\,. \tag{17}$$

For example, one gets for a a vector field ξ of weight w in coordinates $D_{\mu}\xi^{\nu} = (\nabla_{\mu} + w\varphi_{\mu})\,\xi^{\nu}$. Weyl understood the compatibility of a Weylian metric represented by (g, φ) with an affine connection Γ in the sense that parallel transport by Γ has to "respect the length transfer". This condition can be stated more formally as vanishing of the scale covariant derivative of g:

---

[62]Compare footnote 50. For the scale connection component *Riem* (φ) of the Weyl geometric Riemann curvature one finds $R(\varphi)^{\mu}{}_{\nu\lambda\kappa} = \nabla(g)_{\lambda}\,\Gamma^{\mu}_{\nu\kappa} - \nabla(g)_{\kappa}\,\Gamma^{\mu}_{\nu\lambda} + \Gamma^{\alpha}_{\nu\kappa}\Gamma^{\mu}_{\alpha\lambda} - \Gamma^{\alpha}_{\nu\lambda}\Gamma^{\mu}_{\alpha\kappa}$ [207, eq. (9)]. There is also a gauge invariant decomposition of *Riem* into what Weyl called directional curvature and length curvature, which must not be confused with the gauge dependent decomposition just mentioned. The *length curvature* F as part of the Riemann tensor is closely related to the scale connection 2-form $f = d\varphi$; its components are $F^{\mu}_{\nu\kappa\lambda} = -\delta^{\mu}_{\nu}\,f_{\kappa\lambda}$. As F and f can easily be translated into one another; they are often identified in loose speech.

$$Dg = 0 \iff \nabla(\Gamma)\, g = -2\varphi \otimes g \tag{18}$$

The term Q with $Q_{\lambda\mu\nu} := -2\varphi_\lambda g_{\mu\nu}$ on the r.h.s. of the equation is a necessary conceptual feature for ensuring the *metric compatibility of the affine connection* in the sense of Weyl geometry. From the viewpoint of Riemannian geometry, in contrast, it appears as a deviation from the norm, and has been termed "non-metricity" by Schouten, in this special form containing g as "semi-metricity". Using the scale covariant derivative the Weyl geometric affine connection can be written in a form close to the one for the Levi-Civita connection [84, Eq. (46)]:

$$\Gamma(g,\varphi)^\lambda_{\mu\nu} = \frac{1}{2} g^{\lambda\alpha}(D_\mu g_{\nu\alpha} + D_\nu g_{\alpha\mu} - D_\alpha g_{\mu\nu}) \tag{19}$$

The condition (18) is sometimes used as an entry point for defining Weyl structures (pseudo-Riemannian, Hermitian, quaternionic etc.).

Definition: A (pseudo-Riemannian) *Weyl structure* is given by a triple $(M, \mathfrak{c}, \nabla)$ of a differentiable manifold M, a pseudo-Riemannian conformal structure $\mathfrak{c} = [g]$ and a covariant derivative ∇, such that for all $g \in \mathfrak{c}$ there is a differential 1-form φ for which $\nabla g + 2\varphi \otimes g = 0$.

Gauge transformations of type (14) for changes of the representative of $\mathfrak{c}$ are a consequence of this definition [124].

Weyl introduced *scale invariant geodesics* by the usual condition $\nabla_u\, u = 0$ for the covariant derivative $\nabla = \nabla(g, \varphi)$ of the tangent field $u = \dot\gamma(\tau)$ of a curve $\gamma(\tau)$. In coordinates and with $\Gamma = \Gamma(g, \varphi)$ this is

$$\ddot\gamma^\lambda + \Gamma^\lambda_{\mu\nu}\dot\gamma^\mu\dot\gamma^\nu = 0\,. \tag{20}$$

Different to Riemannian geometry, scale invariant geodesics are (usually) not parametrized by curve length. This can be changed by introducing gauge dependent re-parametrizations of γ, that is a class of curves with the same trace as Weyl's scale invariant geodesics such that the tangent fields scale with weight $w(\dot\gamma) = -1$. This leads to introducing *scale covariant geodesics* of weight $w(\dot\gamma) = -1$, characterized by the condition $D_{\dot\gamma}\dot\gamma = \nabla_{\dot\gamma}\,\dot\gamma - \varphi(\dot\gamma)\,\dot\gamma = 0$, or in coordinates

$$\ddot\gamma^\lambda + \Gamma^\lambda_{\mu\nu}\dot\gamma^\mu\dot\gamma^\nu - \varphi_\nu\dot\gamma^\nu\dot\gamma^\lambda = 0\,. \tag{21}$$

The difference between the solutions of (20) and (21) (in coordinates $\varphi_\nu\dot\gamma^\nu\dot\gamma^\lambda$) is proportional to $\dot\gamma$. It thus does not change the direction of the curve, but only its parametrization. This results in $\nabla_{\dot\gamma}(g(\dot\gamma, \dot\gamma)) = 0$ for a scale covariant geodesic, and a constant length function $l = g(\dot\gamma, \dot\gamma)$ like in Riemannian geometry.[63]

[63] For the Riemannian case see, e.g., [98, p. 92].

The length function l is not only constant along a scale covariant geodesic but also scale invariant, because of $w(l) = w(g(\dot{\gamma}, \dot{\gamma})) = 2 - 1 - 1 = 0$. For $l = 1$ the integrated length of a geodesic (with regard to the Riemannian g component of a gauge (g, φ)) can be read off from the parametrization. It is, of course, scale dependent.

Similarly the (squared) *Riemannian length* of any differentiable curve $l_R^2(\gamma) = \int g(\dot{\gamma(\tau)}, \dot{\gamma(\tau)}) d\tau$ is a scale dependent quantity. If one corrects the integrand by the Weylian length transfer function, one finds that the integral

$$l_W^2(\gamma) = \int e^{2\int \varphi(\dot{\gamma})} g(\dot{\gamma(\tau)}, \dot{\gamma(\tau)}) d\tau \tag{22}$$

leads to a scale invariant curve length l_W (up to a constant factor depending on the value of the rescaling function Ω at the initial point of the curve $e^{\Omega(\tau_0)}$). This holds for any φ. It can be considered as the *Weylian length* of the curve. For an integrable Weyl geometry this boils down to measuring curve lengths in Riemann gauge (15).

4.2 General Features of Weyl Geometric Gravity

The following remarks refer mainly to the gravitational sector. Theories building on such an approach will add matter fields and their couplings to the gravitational sector (some special case follow in the next sections). A scalar field coupling to the Hilbert term will be treated here as part of the gravitational sector, although in certain respects, in particular its energy-momentum, it also carries features of matter fields. A matter-like contribution to the energy-momentum results from its kinetic term which is not considered in this section; but see Sect. 5.4.

Weyl geometric gravity is a scale in/co-variant theory. Its Lagrangians L are *scale covariant* functions of weight $w(L) = -n$ with a *scale invariant* Lagrange density $\mathfrak{L} = L\sqrt{|\det g|}$. In the following we assume $n = 4$. Typical gravitational Lagrangians are then of the form

$$L_{grav} = \alpha_1 \phi^2 R + \beta_1 R^2 + \beta_2 Ric^2 + \beta_3 Riem^2 + \dots , \tag{23}$$

where ϕ is a scale covariant real scalar field of weight $w(\phi) = -1$, $Riem$, Ric, and R denote the Weyl geometric Riemann, Ricci and scalar curvatures, $f = d\varphi$ is the field strength of the scale connection (Weyl field), $Riem^2 = R^{\mu}{}_{\nu\kappa\lambda} R_{\mu}{}^{\nu\kappa\lambda}$, $Ric^2 = R_{\mu\nu} R^{\mu\nu}$, and the dots indicate other quadratic terms in Weyl geometric cuvature of weight -4 including, if one wants, the (conformal) Weyl tensor of the Riemannian component in any gauge. Weyl never considered the α_1 and β_2 terms (i.e., for him always $\alpha_1 = \beta_2 = 0$); moreover he considered $\beta_1 = 0$ in Weyl [187], respectively $\beta_3 = 0$ in

Weyl [196]; Dirac assumed all $\beta_i = 0$ $(i = 1, 2, 3)$ etc. There are indications that the quadratic curvature terms of the classical Lagrangian and the Weyl field term may be of importance in strong gravity regimes and also for the quantization of gravity (Sects. 5.3 and 5.4).

With $\beta_1 = \beta_3 = 1$, $\beta_2 = -4$ the explicitly given quadratic curvature terms of (23) in $\mathfrak{L}$ reduce to the scalar density κ of the Gauss-Bonnet theorem as generalized by Chern:

$$\kappa = (R^2 - 4Ric^2 + Riem^2)\sqrt{|\det g|} \tag{24}$$

For orientable compact differentiable manifolds in dimension $n = 4$ the theorem tells us that

$$32\pi\ \chi(M) = \int_M \kappa(g)\, dx\,,$$

with $\chi(M)$ the Euler-characteristic of M and g any Riemannian metric g on M. Interestingly, recent authors realized that Weyl geometric field theories for which the Riemannian component of the quadratic curvature Lagrange densities reduce to the topologically preferred Gauss-Bonnet form κdx behave particularly well with regard to unitarity [172] and stability [84].

Consider a Weyl geometric field theory with the Lagrangian density

$$\mathfrak{L} = \mathfrak{L}_{grav} + \mathfrak{L}_{\phi kin} + \mathfrak{L}_{V(\phi)} + \mathfrak{L}_f + \mathfrak{L}_{mat}\ \ [+\mathfrak{L}_{constr}]\,, \tag{25}$$

where $\mathfrak{L}_{grav}$ is a gravitational Lagrangian of the form (23), $\mathfrak{L}_{\phi kin}$ the kinetic term of a gravitational (i.e., non-minimally coupled) scalar field ϕ, $\mathfrak{L}_f$ a Yang-Mills-Maxwell term for the scale connection (Weyl field) (10), the matter term $\mathfrak{L}_{mat}$ is scale invariantly written and $\mathfrak{L}_{constr}$ denotes—where appropriate—a Lagrangian constraint, e.g., like in (33) below. An infinitesimal scale transformation with $\Omega = e^{d\rho} = 1 + d\rho + O(d\rho^2)$ leads to a corresponding variation of the fields. The Noether identity (46) for the scale symmetry can be read off from the functional expressions $f_X[\ ]$ and $f_X^{\mu}[\ ]$ in the linear combinations of (45); they depend on the respective field X and can easily be calculated.[64] For a matter field ψ of scale weight $w(\psi) = k$ (in the most important cases $k = \frac{3}{2}$ or 0) the Noether identity with regard to local scale transformations becomes

$$\partial_\nu \mathfrak{E}_\varphi^\nu + 2g^{\mu\nu}\mathfrak{E}_g - \phi\,\mathfrak{E}_\phi + k\psi\mathfrak{E}_\psi \equiv 0\,, \tag{26}$$

[64] Varying a scale covariant field X of weight w by an infinitesimal scale transformation gives functional expressions $f_X[\ ] = w\,[\ ]$, $f_X^{\mu} = 0$; the gauge transformation for the scale connection implies $f_\varphi[\ ] = 0$, $f_\varphi^{\mu}[\ \ ] = -1$.

with the respective Euler-Lagrange expressions of the fields, $\mathfrak{E}^{\nu}_{\varphi} = \frac{\delta \mathfrak{L}}{\delta \varphi_{\nu}}$, $\mathfrak{E}_{g} = \frac{\delta \mathfrak{L}}{\delta g^{\mu\nu}}$ etc. In the special case of the Lagrangian (13) this boils down to the "conservation laws" discussed by Dirac (see Sect. 3.2).

Even leaving the influences on perturbative quantization out of sight, the Noether identity of the scale symmetry has at least two remarkable consequences. Like in other cases it implies, at first, a dependence between the dynamical equations of the fields. In our case, (26) shows that an admissible field configuration satisfying the gravitational equation (g), the Weyl field equation (φ), and the matter equation (ψ) "automatically" also satisfies the scalar field equation (ϕ). In this sense the scalar field equation is no additional restriction for fields satisfying the other dynamical conditions of the theory.

Secondly, a consideration similar to Weyl's in 1918 (see Sect. 2.1) leads to the conservation of the dynamical scale current even with the fields "off shell" with regard to the Weyl field equation. If the fields are "on shell" with respect to the dynamical equations of g, ϕ, ψ, there remains only

$$\partial_{\nu} \mathfrak{E}^{\nu}_{\varphi} = 0 \,. \tag{27}$$

Because of

$$\mathfrak{E}^{\mu}_{\varphi} = (\sqrt{|g|}) f^{\mu\nu}) - \mathfrak{s}^{\mu} \,,$$

where $\mathfrak{s}^{\mu} = \frac{\delta(\mathfrak{L} - \mathfrak{L}_f)}{\delta \varphi_{\mu}}$ is the total *dynamical scale current*, the latter is conserved by the same reasons as in Weyl's theory:

$$\partial_{\mu} \mathfrak{s}^{\mu} = 0 \,. \tag{28}$$

In our framework the scale connection, and with it the Weyl field, *does not couple* effectively to the matter fields or to the potential term; it only does so to the scalar field, the gravitational sector and, eventually, to the action of the Lagrangian constraint,

$$\mathfrak{s}^{\mu} = \frac{\delta \mathfrak{L}_{grav}}{\delta \varphi_{\mu}} + \frac{\delta \mathfrak{L}_{\phi kin}}{\delta \varphi_{\mu}} \; [+ \frac{\delta \mathfrak{L}_{constr}}{\delta \varphi_{\mu}}] \,. \tag{29}$$

For a kinetic term like in (38) below the contribution $\mathfrak{s}^{\mu}_{\phi} = s^{\mu}_{\phi} \sqrt{|g|}$ of the scalar field to the dynamical scale current is easy to determine[65]:

$$\mathfrak{s}^{\mu}_{\phi} = \frac{\alpha}{2} D^{\mu} \phi^2 \sqrt{|g|} \,.$$

[65]This corresponds to the result in Hehl et al. [77, p. 283f.], while the contribution of the gravitational sector to the scale ("dilational") current is not mentioned there. It has been overlooked also in Ohanian [120, Eq. (13)f.].

It even turns out to be scale invariant. The gravitional contribution to the current is more involved; in the case of the Dirac-Weyl theory with $\beta_1 = \beta_2 = \beta_3 = 0$ in (23) it is

$$\mathfrak{s}^{\mu}_{grav} = (3D^{\mu}\phi^2 - \frac{3}{2}\phi^2\partial^{\mu}\ln|g|)\sqrt{|g|}\,.$$

The contribution of a Lagrangian constraint of type (33) is

$$\mathfrak{s}^{\mu}_{constr} = \frac{\delta\mathfrak{L}_{\phi constr}}{\delta\varphi_{\mu}} = -\,\partial_{\nu}(\sqrt{|g|}\,\lambda^{\mu\nu})\,. \tag{30}$$

As the Riemannian component g of the Weylian metric is the only field which enters the Lagrangian with second partial derivatives, the canonical scale current of Weyl geometric gravity according to (47) is

$$\begin{aligned} J^{\mu}_{scale} &= \frac{\partial\mathfrak{L}_{\phi kin}}{\partial(\partial_{\mu}\phi)}\delta\phi + \frac{\partial\mathfrak{L}_{mat}}{\partial(\partial_{\mu}\psi)}\delta\psi \\ &+\frac{\partial\mathfrak{L}_{grav}}{\partial(\partial_{\mu}g^{\nu\lambda})}\delta g^{\nu\lambda} + \frac{\partial\mathfrak{L}_{grav}}{\partial(\partial_{\kappa}\partial_{\mu}g^{\nu\lambda})}\delta(\partial^{\kappa}g^{\nu\lambda}) - \partial^{\kappa}\frac{\partial\mathfrak{L}_{grav}}{\partial(\partial_{\kappa}\partial_{\mu}g^{\nu\lambda})}\delta g^{\nu\lambda}\,. \end{aligned} \tag{31}$$

The contributions of the Riemannian metric to the scale current are highly involved. Although they don't look particularly attractive it may be worthwhile to take them seriously, because the conservation condition of the canonical scale current could turn out useful for a perturbative quantization of the theory, analogous to the role of the Ward identity for QED (cf. Sects. 5.3 and 7). At the moment, however, this seems still quite unclear.

4.3 Weyl Geometric Gravity in IWG

Scale covariant theories with $\alpha_1 \neq 0$ and a nowhere vanishing gravitational scalar field have a scale gauge in which the scalar field is constant, $\phi \doteq \phi_0$. Here the following convention is used:

$\doteq$ denotes equalities which only hold in specific scale gauges.

The gauge with $\phi \doteq \phi_0$ will be called the *scalar field gauge* with regard to ϕ (in short ϕ-gauge). If the coefficient of the modified Hilbert term assumes the value $\alpha_1\phi^2 \doteq (16\pi G)^{-1} \sim \frac{1}{2}E_p^2$ (E_P the reduced Planck energy), it specifies the *Einstein gauge*. Theories with a distinguished scalar field ϕ allow to define:

With regard to ϕ *scale invariant observable quantities* of a scale covariant field X with weight $w = w(X)$ are given by forming the proportion with ϕ^{-w}, i.e., by $\check{X} = \phi^{w}X$.

This boils down to determining the field values in ϕ-gauge (up to the constant factor ϕ_o^w). Probably that was the reason for Utiyama to consider ϕ as a "measuring field" (see p. 45). These theoretical considerations may get underpinned physically if reasons turn up which allow to conclude that the result of measuring processes (e.g., atomic clocks) are displayed in the scalar field gauge of ϕ. The scaling of the Higgs field Φ indicates in this direction (if it underlies a common biquadratic potential with the gravitational field ϕ; Sect. 5.3).

If one spells out the modified Hilbert term of (23) in terms of the Riemannian and the scale connection contributions to R,

$$R = R(g) + R(\varphi) \quad \text{with} \quad R(\varphi) = -(n-1)(n-2)\varphi_\nu \varphi^\nu - 2(n-1)\nabla(g)_\nu \, \varphi^\nu \tag{32}$$

a mass term $\frac{1}{2} m_\varphi^2 \varphi_\nu \varphi^\nu$ for the Weyl field turns up, which in dimension $n = 4$ would indicate the impressive value $m_\varphi^2 = 6\, E_P^2$. Even if one allows for quantum corrections, or includes other terms in an approach of this type, one arrives at a mass far beyond experimental access (e.g., at the LHC). Bosonic couplings of the Weyl field thus occur at extremely small distances (close to the Planck length); their effects integrate out at longer distances, comparable to the invisibility of the weak interaction at microscopic scales.

For investigations in *low energy regions*, i.e., at laboratory or astronomical scales, Weyl geometric gravity can *effectively be modelled* in terms of scale connections with vanishing curvature, $d\varphi = 0$, i.e., in the framework of *integrable Weyl geometry*. The reduction to the effective theory can be expressed by Lagrange multiplier functions $\lambda_{\mu\nu}$ and the constraint

$$L_{constr} = -\frac{1}{2} \lambda_{\mu\nu} f^{\mu\nu} \, . \tag{33}$$

This condition enforces a vanishing scale curvature (Weyl field), $f_{\mu\nu} = 0$. Obviously the dynamical scale current (29) vanishes if the multiplier fields $\lambda_{\mu\nu}$ satisfy

$$\partial_\nu (\sqrt{|g|} \lambda^{\mu\nu}) = \frac{\delta \mathfrak{L}_{grav}}{\delta \varphi_\mu} + \frac{\delta \mathfrak{L}_{\phi kin}}{\delta \varphi_\mu} \, . \tag{34}$$

This condition fixes the Lagrangian multipliers, ensures the vanishing of the scale current, $\mathfrak{s}^\mu = 0$, and a trivialization of the Weyl field, $d\varphi = f = 0$.[66]

In the following we simplify this approach to Weyl geometric gravity in the framework of IWG even further by setting the external constraint of a vanishing scale curvature, $d\varphi = 0$. The scalar field ϕ contains an additional degree of freedom in comparison with Einstein gravity. In the Riemann gauge it can be expressed as

---

[66] In this way Friedrich Hehl's objection that the scale current "drifts around uncontrolled by any field equation" in the IWG approach to gravity [74, p.166] can easily be dispelled.

$\phi \underset{Rg}{\doteq} \phi_0 e^{-\sigma}$, where $\underset{Rg}{\doteq}$ denotes equality in the Riemann gauge; similarly $\underset{Eg}{\doteq}$ for the Einstein gauge.

The simplified version of Weyl geometric gravity is a *scalar-tensor theory* in which the new scalar degree of freedom is expressed partially by the scalar field and partially by the (integrable) scale connection. This can be seen as follows: The data in an arbitrary scale gauge arise from Riemann gauge by (length-) rescaling with a real valued function $\Omega = e^{\omega}$:

$$g \doteq e^{2\omega} g, \qquad \varphi_\nu \doteq -\partial_\nu \omega \ \ (\varphi \doteq -d\omega), \qquad \phi \doteq \phi_0\, e^{-(\sigma+\omega)} \tag{35}$$

The information of the new dynamical degree of freedom is now no longer encoded by the scalar field alone, but is being distributed among the scalar field and the scale connection, i.e., encoded by the pair (ϕ, φ). In the Einstein gauge $(\hat{g}, \hat{\varphi}, \hat{\phi})$, specified by $\hat{\phi} \doteq \phi_0$, the respective values are

$$\hat{g} \underset{Eg}{\doteq} e^{-2\sigma} g \qquad \hat{\varphi}_\nu \underset{Eg}{\doteq} \partial_\nu \sigma \qquad \hat{\phi} \underset{Eg}{\doteq} \phi_0 \ \ \text{(constant)}\,. \tag{36}$$

The dynamical information of the new degree of freedom is here completely contained in the integrable scale connection with integral σ, while the scalar field has become a constant. This has interesting dynamical consequences for inertial paths if they follow Weyl geometric geodesics (Sect. 5.4).

At least test bodies do so. This is the case, because the doubly covariant energy momentum tensor $T_{\mu\nu}$ scales with weight -2 (also in the non-integrable case).[67] The "conservation" condition of GR translates, at least in IWG, to a vanishing scale covariant divergence

$$D_\mu T^{\mu}_{\ \nu} = 0\,. \tag{37}$$

This allows to generalize the classical Geroch-Jang theorem establishing the *geodesic principle* in Einstein gravity *for test bodies* [67] without great effort to the *framework of integrable Weyl geometry* [155, Appendix 5.3]. The case of extended bodies is more complicated and needs further clarification.[68]

Let us be content with these short remarks. In the next section some topics in the foundations of gravity and Weyl geometric researches in other subfields of physics will be reviewed.

---

[67] For the Hilbert energy tensor $T^{(m)}_{\mu\nu} = -\frac{2}{\sqrt{|g|}} \frac{\delta \mathfrak{L}_m}{\delta g^{\mu\nu}} = -\left(2\frac{\partial L_m}{\partial g^{\mu\nu}} - L_m\, g_{\mu\nu}\right)$ this results from the scaling weights $w(L_m) = -4$ and of g (similarly for the canonical energy tensor). It agrees with the "phenomenological" weight read off from the physical dimension of the energy density $T^{0}_{\ 0}$; in dimension 4 $w(T^0_0) = w([EL^{-3}]) = -4$.

[68] The detailed study in Ohanian [119] discusses, among others, the problem for the geodesic principle of extended bodies in BD theory. This cannot be translated 1 : 1 to the IWG case.

5 Interest Today

After the reinterpretation of the original gauge idea as a phase factor in quantum mechanics in the second half of the 1920s, Weyl's scale geometry (purely infinitesimal geometry) of 1918 lay dormant for nearly half a century, enhanced by Weyl's own clear disassociation from this idea in the 1940s. Surprisingly, it lived up again rejuvenated in the early 1970s, enriched by new concepts and triggered by new interests in conformal transformations in gravity theory, cosmology, quantum physics, and field theory. The following section discusses selected topics in which Weyl geometric methods enter crucially into ongoing present research. Of course this selection is quite subjective; it is necessarily biased by my own interests and delimited by the restrictions in perspective and knowledge of the author.

5.1 *Philosophical Reflections on Gravity*

Extending the Riemannian framework for gravity theories may help to clarify conceptual questions. This is, of course, a truism and holds not only for the Weyl geometric generalization but also, e.g., for Cartan geometry. Here we deal with the first option. A classical subject for this type of questions is the discussion of the problem which is the "physical frame" in Brans-Dicke theory.

Brans-Dicke theory presupposes the framework of Riemannian geometry and uses conformal transformations not only for the metric but also for the scalar field, while keeping the affine connection arising as a solution of Eq. (9) fixed.[69] In a technical sense it works in the framework of an integrable Weyl structure $(M, \mathfrak{c}, \nabla)$, where the Jordan frame corresponds to the Riemann gauge of the Weylian metric. For most practitioners this goes unnoticed. Exceptions are [136] and the authors of the Brazilian group of Weyl geometric gravity, [4, 5, 138].[70] This theory is nearly a Weyl geometric scalar tensor theory arising from (23) with $\alpha_2 = \beta_i = 0,\ i = 1, 2, 3$ by adding a quadratic scale covariant kinetic term of the scalar field

$$L_{\partial\phi 2} = -\frac{\alpha}{2} D_\nu \phi \, D^\nu \phi \,, \tag{38}$$

[69] For modern presentations of BD theory see, e.g., [65]; the transformation of the Levi-Civita geodesics of the Jordan frame to other frames are discussed in (ibid, Sect. 3.4).

[70] Note, however, that Romero et al. have a peculiar view of BD theory, different from the usual one and influenced by their paradigm called WIST (*Weyl integrable spacetime*). This is a theory of gravity placed in an integrable Weyl geometric framework, but with a broken scale symmetry and Levi-Civita geodesics of the distinguished frame. In their view also in BD the free fall trajectories are determined by the Levi-Civita geodesics of a distinguished frame (which can be the Einstein frame) [139, Sect. 6]. This modified version of BD theory is equivalent to WIST [139, Sect. 8]. .

even though the $L_{\partial\phi 2}$ is usually not written explicitly in a scale covariant form but given in the Jordan frame in the form $L_{\partial\phi 2} \doteq -\frac{\alpha}{2}\partial_\nu\,\phi\partial^\nu\phi$. Additional terms arise in other frames from the respective conformal transformations. I consider such an approach as *nearly* Weyl geometric gravity, because often a matter term L_m is added, which breaks the scale invariance of the Lagrange density. This raises the question in which scale gauge (frame) the matter Lagrangian couples minimally to the Riemannian component of the metric. This is usually called the *physical metric* (or frame) [65, p. 227f.] and one has to decide in which relation the physical metric/frame/gauge stands to the Riemann/Jordan frame and the Einstein frame. In a Weyl geometric scalar tensor theory as conceived here, one better writes the matter Lagrangian in a scale covariant form ($w(L_m) = -4$). This leads to scale *invariant dynamical equations* (containing scale *co*variant terms).

The physical breaking of the scale symmetry is a separate question and can be discussed at a later stage of the theory development. It is related to fixing the basic parameters of the measuring processes.[71] Then one arrives at the scale gauge in which observational values are expressed directly. Let us assume, for the moment, that this is, e.g., the Einstein gauge.[72] This is then also the frame (or gauge) in which the Lagrangian matter terms are expressed most directly in the form known from non-scale invariant formulations. Insofar this is the "physical gauge". The Riemann gauge has, on the other hand, the pleasant feature that the Weyl geodesics are identical to the Levi-Civita ones of the gauge. From a more conservative viewpoint one might say that the free fall trajectories of BD gravity are those of Riemann gauge. But free fall and matter Lagrangian are both important physical features of a theory of gravity. Which one should we then consider as "physical"? The *two frames express different aspects of physical reality* in a mathematically most direct way. In this sense both can be considered as "physical"; it does not matter which way, we only need to be clear what is meant.

C. Romero, M. Pucheu and other authors of the Brazilian group mentioned above (Sect. 3.2) use the framework of Weyl geometric gravity to discuss different inter-theory relations [137, 139]. In the first paper they discuss how Einstein gravity can be expressed in this wider framework by a scale-covariantization of dimensional quantities. In particular the gravitational constant is promoted to a scalar field (proportional to ϕ^2 in our notation), which also couples to the matter term (which then becomes $\phi^4 L_m$). Their version of Weyl geometric scalar tensor theory (WIST) is shown to be equivalent to their view of BD theory.[73] Finally they discuss the relationship between original Nordström's scalar theory of gravity to the Einstein-Fokker version from the

---

[71] Cf. [136] and [152].

[72] For an argument in favour of the Einstein gauge (under certain assumptions) see Sect. 5.3, observation (∗).

[73] See footnote 70.

point of view of conformal transformations of a flat Lorentzian manifolds [137]. This has been taken up in a recent broader philosophical discussion of the usefulness of a Weyl geometric point of view for this type of theory relation by Duerr [54].

5.2 *Recent Work on the EPS Argument*

The central statement of Ehlers/Pirani/Schild allows, in principle, to read off a Weylian metric on spacetime from sufficiently detailed knowledge of the free fall trajectories of test particles and of the gravitational bending of light. If one does not prematurely keep the standard assumption of some unknown dark matter particles for granted, one has to take the possibility into account that the anomalous observations of dynamics on the galaxy or cluster level and in cosmology may be an indicator of a more general gravitational structure than Einstein gravity and Riemannian geometry. Many geometrically viable generalizations of Einstein gravity lead to an EPS compatible pair of conformal and projective structures. An EPS type of argument has recently been used as a motivation for an Einstein-Palatini approach to cosmology [62, 63]. In our perspective it would, of course, seem more natural to pass to the Weyl geometric framework directly (cf. Sect. 5.4). This presupposes that the EPS compatibility of any pair of a conformal and a projective structure does, in fact, imply the existence of a Weyl geometric structure.

Exactly this point has been questioned by A. Trautman in an editorial comment for a republication of the EPS paper as a "Golden Oldie" [174]. In the comments Trautman made clear that the arguments given in the original paper are rather vague and far from a mathematical proof; he thus raised doubts with regard to the status of the existence statement of a Weylian metric given an EPS compatible pair of projective and conformal structures $\mathfrak{p}$ and $\mathfrak{c}$. This statement thus ought to be considered a conjecture rather than a theorem, as which it had been treated in large parts of the literature up to then.

In a first investigation of the case [107] announce "a theorem giving the necessary and sufficient conditions for compatibility of conformal and projective structures" [107, p. 822]. The authors use, however, a compatibility criterion of the structures $\mathfrak{p}$ and $\mathfrak{c}$ relying on the framework of Riemannian geometry, thus different from Weyl's:

> *Definition.* The conformal and projective structures $\mathfrak{c}$ and $\mathfrak{p}$ are said to be compatible if there is $g \in \mathfrak{c}$ such that $\Gamma(g) \in \mathfrak{p}$. [107, p. 823, notation slightly adapted]

This definition restricts the underlying metrical structure to the Riemannian case. It thus expresses the *Riemann compatibility* of the two structures.[74]

[74] This is also noted in [110, Remark 2.2].

In such a case one only needs to consider a non-integrable Weyl structure to find a "counterexample" to the EPS result (it satisfies the EPS compatibility condition without being Riemann compatible), which the authors did [107, p. 822]. They comment that their result "completes a line of research initiated by Weyl and continued by physicists" (ibid. p. 824). This is wrong; their compatibility condition is tailored on the framework of Riemannian geometry. The "line of research initiated by Weyl" is not touched by the result, and even less "completed".

A veritable achievement of the paper lies, on the other hand, in formulating an analytical criterion for the compatibility of two structures $\mathfrak{p}$ and $\mathfrak{c}$, even though it is established for Riemann compatibility only [107, Eqs. (7) and (8)]. It can easily be generalized (i.e., weakened) to the case of Weyl compatibility in the following way.

For given $\mathfrak{p}$ and $\mathfrak{c}$ first choose an affine connection $\overline{\Gamma} \in \mathfrak{p}$ and a Riemannian metric $g \in \mathfrak{c}$, the latter with Levi-Civita connection $\Gamma(g)$. Any Weylian metric subordinate to the conformal structure $\mathfrak{c}$ can be expressed in a scale gauge with Riemannian component g, i.e., in a gauge of the form (g, φ) with some real differential form φ. Denote its (Weylian) affine connection by $\Gamma(g, \varphi)$.

Using the Thomas symbol $\Pi(\Gamma)$ of an affine connection Γ

$$\Pi^i_{jk}(\Gamma) = \Gamma^i_{jk} - \frac{1}{n+1}\delta^i_j\Gamma^l_{lk} - \frac{1}{n+1}\delta^i_k\Gamma^l_{lj} ,$$

a necessary and sufficient condition for the projective equivalence of two affine connections Γ and Γ' is the equality of their Thomas symbols, respectively the vanishing of the Thomas symbol for the difference:

$$\Pi(\Gamma) - \Pi(\Gamma') = \Pi(\Gamma - \Gamma') = 0 .$$

For checking the Weyl compatibility of $\mathfrak{p}$ and $\mathfrak{c}$ one has to see whether $\Gamma(g, \varphi)$ and $\overline{\Gamma}$ are projectively equivalent, i.e., whether or not the Thomas symbol of their difference vanishes.

Calculating first the Thomas symbol T for the difference of the Levi-Civita connection of g and of $\overline{\Gamma}$,

$$T(g, \overline{\Gamma}) := \Pi(\Gamma(g) - \overline{\Gamma}) , \tag{39}$$

the searched for difference $\Pi(\Gamma(g, \varphi) - \overline{\Gamma})$ can be expressed by $T(g, \overline{\Gamma})$ and φ; it will be denoted by $\tilde{T}(g, \varphi, \overline{\Gamma}) = \Pi(\Gamma(g, \varphi) - \overline{\Gamma})$. In components it is:

$$\tilde{T}(g, \varphi, \overline{\Gamma})^i_{jk} = T(g, \overline{\Gamma})^i_{jk} + \frac{1}{n+1}(\delta^i_j\varphi_k + \delta^i_k\varphi_j) - g_{jk}\varphi^i . \tag{40}$$

Vanishing of $\tilde{T}$ is necessary and sufficient for the projective equivalence of $\Gamma(g, \varphi)$ and $\overline{\Gamma}$, and thus the Weyl compatibility of the given projective and conformal structures. Let us call $\tilde{T}$ the *generalized Matveev symbol* associated with a projective structure $\mathfrak{p}$ and a Weylian metric $[(g, \varphi)]$ subordinate to the conformal structure $\mathfrak{c}$. We thus get the result:

> A projective structure $\mathfrak{p}$ and a conformal structure $\mathfrak{c} = [g]$ are Weyl compatible, if and only if there is a real differential form φ (not necessarily closed) such that for some (and thus for any) $\overline{\Gamma} \in \mathfrak{p}$ the generalized Matveev symbol (40) vanishes, $\widetilde{T}(g, \varphi, \overline{\Gamma}) = 0$.

This argumentation follows step by step the one in Matveev and Trautman [107] although, as already said, this paper discusses the restricted case of Riemann compatibility and $d\varphi = 0$ only. It leads to a formal criterion for testing Weyl compatibility of a projective and a conformal structure.

The question whether EPS-compatibility is strong enough to imply Weyl compatibility is not touched by this observation. In a recent investigation V. Matveev has taken up the problem in the Weyl geometric framework again and has shown that the original EPS conjecture is, in fact, true [108]. We therefore can trust the EPS argument and rely on it as a proven theorem in foundational studies and alternative gravity theories.

5.3 *How Does the Standard Model Relate to Gravity?*

Two mutually related aspects of the standard model of elementary particle physics (SM) cry out for considering it from a Weyl geometric perspective if one wants to know more about a possible link of the SM to gravity: the nearly scale covariance of the SM Lagrangian (of weight -4) in a Minkowskian background with only the Higgs mass term breaking the scale symmetry, and the role of the Higgs scalar field Φ (weight $w(\Phi) = -1$) itself. Since the first decade of the SM this has motivated authors to adapt the SM Lagrangian to a Weyl geometric framework and to enhance it by a scale invariant gravity sector. In the first three decades that often happened without most authors knowing about similar attempts undertaken by other colleagues. Several of them brought forward the idea that the (norm of the) initially still hypothetical Higgs field might play the role of a gravitional scalar field which couples to the generalized Hilbert term of (23) [38, 52, 53, 165].[75]

With the role of the Higgs field being established and elevated to the status of an object of empirical research, all the more after the empirical confirmation of its physical reality in 2013, the attention has shifted to models with (at least) two scalar fields, a gravitational scalar field ϕ coupling to the Hilbert term and the Higgs field Φ as part of the matter sector, both scaling with the Weyl weight -1 respectively the mass/energy weight +1 used by the elementary particle physicists. This allows

[75]Cf. [153, Sect. 11.5].

to consider models in which the Higgs field couples indirectly to the gravitational sector through a common biquadratic potential term with the gravitational scalar field (41) and/or directly (non-minimally) by an additional term $\alpha_1'|\Phi|^2 R$ added to the gravitational Lagrangian (23). These approaches are not mutually exclusive and may turn out crucial for establishing a link between the SM and gravity.

Among the more recent studies some explore the conceptual and formal framework for importing the SM into a Weyl geometric scalar tensor theory of gravity with a modified Hilbert term in the classical Lagrangian (without quadratic gravitational terms and without considering quantization) [36, 112–114, 134, 135, 150, 152]. Interesting features arise already on this level if one assumes a common biquadratic potential for the gravitational scalar field and the Higgs field:

$$\begin{aligned} V(\Phi,\phi) &= \frac{\lambda_1}{4}\left(|\Phi|^2 - \frac{\mu}{\lambda_1}\phi^2\right)^2 + \frac{\lambda}{4}\phi^4, \qquad \lambda,\, \lambda_1 > 0\,, \qquad (41)\\ &= \frac{\lambda_1}{4}|\Phi|^4 - \frac{\mu}{2}|\Phi|^2\phi^2 + \frac{\lambda'}{4}\phi^4 \qquad \lambda' = \lambda + \frac{\mu^2}{\lambda_1}\,,\\ \mathcal{L}_V &= -V(\Phi,\phi)\sqrt{|g|} \end{aligned}$$

Such a potential is considered in Shaposhnikov and Zenhäusern [161, 162] for studying global scale symmetry in the SM on Minkowski space and has been imported to conformal theories of gravity in Bars et al. [12].[76] If the dynamical terms do not disturb the ground state of the potential minimum too strongly, there is a common minimum of the potential with $(\partial_{|\Phi|}V, \partial_\phi V) = 0$ also in the locally scale symmetric framework. This "binds" the fields together in the sense of keeping them proportional, $\phi \sim |\Phi|$, whichever scale gauge is considered. In particular the gauge in which the gravitational scalar field is constant, the Einstein gauge, is the same in which the norm of the Higgs field, i.e., its expectation value, is scaled to a constant. I propose to call it the *Higgs gauge*.

Then *the Higgs gauge and the Einstein gauge are identical* in the rest state of the scalar fields, and the Lagrangian mass expressions of the form $\mu\,|\Phi|^2$ acquire the constant value $m^2 = \mu\,|\Phi_o|^2 = \mu_e v^2$ as usual in the electro-weak theory ($v \approx 246\,\mathrm{GeV}$ the reference energy of the electroweak interaction). In the literature this is sometimes considered as a"generation" of mass (as though the mass terms did not exist already before) and the choice of Higgs gauge as a "breaking" of the scale symmetry (as though the scale symmetry had to be given up in the light of the physically preferred choice of the scale gauge).

The latter perspective is unhappy because it obscures an important point for rounding off the Weyl geometric framework. As the mass term of the electron $\mu_e|\Phi|^2$ scales with the norm of Higgs field $|\Phi|$ like the masses of all elementary particles,

[76] For an introduction to a Weyl geometric framework see [36, 152].

it becomes constant with the latter, say $m_e^2 = \mu_e |\Phi_o|^2 = \mu_e v^2$. In accordance with this the Rydberg constant R_y, important for the determination of atomic frequencies, scales (classically) with m_e; this allows to infer that time measurements which rely on atomic frequencies can be read off in the Higgs gauge = Einstein gauge.[77] This leads to the

Observation (∗): If Weyl geometric gravity with a gravitational scalar field ϕ is linked to the SM via a biquadratic potential of ϕ and the Higgs field Φ, we expect that the Einstein gauge of the gravitational scalar field is identical to the Higgs gauge and corresponds to measurement units according to those of the new SI conventions.[78] This can be interpreted as an adaptation of atomic clocks to the local field constellation of the pair (Φ, ϕ), the Higgs field and the gravitational scalar field.

From a general point of view this is close to Weyl's hypothesis of an adaptation of clocks to the local constellation of the gravitational field. But because of the central role of the Higgs field in the adaptation to (Φ, ϕ) the latter gains a more solid physical foundation than Weyl's rather speculative proposal relating to the scalar curvature of spacetime (Sect. 2.1). Note also that a biquadratic potential (41) gives the Higgs field a distinguished relation among the matter fields to the gravitational sector. The other way round, the gravitational scalar field ϕ enters the so-called *Higgs portal* in a particularly simple form.

Present day physicists want to understand the establishment of the Higgs gauge and/or Einstein gauge as a breaking of symmetry. This is what, e.g., the authors of Nishino and Rajpoot [112] do. They consider the scalar field ϕ acquiring a constant values as a physical process of symmetry breaking, related to a Goldstone boson which gives mass to the Weyl field and to the Higgs field. This is assumed to happen at an extremely high energy level far beyond the electroweak scale. Other authors go a step further and discuss a possible dynamical underpinning by a mechanism of a spontaneous breaking of scale symmetry in close analogy to the Coleman/Weinberg mechanism proposed in the early 1970s for the breaking of the $U(1)$ symmetry in electroweak theory [46, 68, 120, 172].

These studies work with a quantization of at least the scalar field and the Weyl field (the scale connection). Dengiz/Tekin for example start from a scale invariant Lagrangian including quadratic curvature terms in the gravitational sector. Applying a perturbative quantization scheme for the gravitational scalar field they find a

---

[77] $R_y = \frac{e^4 m_e}{2\hbar} = \frac{\alpha_f^2}{2} m_e c$, where e denotes the charge of the electron, α_f the fine structure constant. The energy levels E_n in the Balmer series of the hydrogen atom are given by $E_n = -R_y n^{-2}$; the other atomic frequencies are influenced by the Rydberg constant similarly, although in a more complicated way. With field quantization α_f becomes dependent on the energy scale under consideration; the Higgs gauge as discussed here is therefore clearly a feature of the classical theory emerging from the quantum level at long ranges.

[78] For the new SI conventions see footnote 37; the connecting link to Weyl geometric gravity is discussed in Scholz [152, Sect. 5].

Coleman/Weinberg effect, after a "tedious renormalization and regularization procedure" at the two loop-level [46, p. 4]. Although there remains the caveat that at the minimum of the quantum corrected potential the perturbative calculation loses validity, they conclude that "the Weyl symmetry of the classical Lagrangian will not survive quantization" (ibid. p. 5). Ohanian and Ghilencea argue inversely. Ohanian assumes a scale invariant Lagrangian close to the Planck scale without the quadratic curvature terms, while Ghilencea uses a combined linear (modified Hilbert term) and quadratic gravitational Lagrangian with a prominent contribution of Weyl's conformal curvature, which does not satisfy the Gauss-Bonnet constraint (24). Both authors find in their respective approaches that a Coleman-Weinberg-like consideration for the scalar field leads to breaking of scale symmetry for the low energy classical Lagrangian. This reduces the geometry to a Riemannian framework and the gravity to Einstein's theory.[79]

In a follow up paper to Dengiz and Tekin [46] Tahanyi and the authors of the first paper come to a different conclusion. They undertake a more fundamental study of quantum properties of the Weyl geometrically extended gravitational sector including quadratic expressions in the curvature and derive the full gravitational particle spectrum of the theory for an (Anti-) de-Sitter and a Minkowski background. They find the Gauss-Bonnet relation (24) as a necessary constraint for unitarity of the quantized field theory with gravitational Lagrangian (23) and arrive at the conclusion:

> …, the only unitary theory, beyond three dimensions, among the Weyl-invariant quadratic theories is the Weyl-invariant Einstein-Gauss-Bonnet model [our Lagrangian (23) constrained by (24), ES] which propagates a massless spin-2 particle as well as massive spin-1 and massless spin-0 particles [172, p. 8]

This is a highly interesting result. The mentioned particles are, of course, the graviton, the boson of the Weyl field (scale connection), and the boson of the gravitational scalar field. In this model the Weyl boson is extremely heavy as usual, and *the Weyl field has only short range effects*; on larger scales it can be represented classically by an integrable scale connection. *The scalar field*, on the other hand (and different from Ohanian's model), *can support long range gravitational interactions* in addition to the ones mediated by the graviton. For the classical theory this boils down to a Lagrangian constraint (33) with Lagrangian multiplier functions $\lambda_{\mu\nu}$ enforcing $\alpha_2 = 0$ in (23). In the light of Tanhanyi et al.'s derivation the classical Einstein gravity limit of Ohanian and Ghilencea seems to result from the special choices of their gravitational Lagrangians, both violating the Gauss-Bonnet constraint.

The question still remains, whether for a Weyl symmetric bare Lagrangian—a kind of classical template for a quantum field theory under construction—the local scale symmetry is broken under quantization. Often the so-called *trace anomaly* is considered as an indicator that this is in fact the case. This means that the energy-momentum

[79]Ohanian [120, p. 10f.], Ghilencea [68, p. 7].

tensor of (pseudo-) classical quantum matter fields, represented by scalar fields or Dirac spinors, has a vanishing trace, whereas the expectation value of the quantized trace no longer vanishes. The trace anomaly has puzzled theoretical physicists for a long time, and is often interpreted as implying the breaking of scale invariance at the quantum level.

The authors of [39] come to a different conclusion. The trace anomaly is, of course, present also in their approach, but it *no longer signifies breaking of the local scale invariance*. The reason lies in a cancellation of the trace terms of the quantized fields by corresponding counter-terms that arise from a *complex* gravitational scalar field which they call a "dilaton". In order to achieve this the authors use a coherent scale-covariantization of all terms by substituting dimensional constants with the appropriate power of the scalar field and a dimensionless coefficient. In a series of examples they show that "with a suitable quantization procedure, the equivalence between conformal frames can also be maintained in the quantum theory" [39, p. 21]. They even can show that the renormalization flow preserves the Weyl symmetry in their case studies. This stands in contrast to Dengiz/Tekin's observation mentioned above. It would be interesting to see, whether a coherent scale covariantization as used by Codello/Percacci et al. can be implemented in their scheme too, leading to a maintenance of the scale symmetry also in the context of a perturbative quantization of gravity.

5.4 Open Questions in Cosmology and Dark Matter

Weyl geometric gravity can also be used to explore the open problems in cosmology and dark matter phenomena from its sight. Already Weyl started to consider cosmological questions in his extended geometrical framework. Since the retake of it by Dirac and the Japanese authors the conjecture that the Weyl field, or now also the gravitational scalar field, might help understanding dark matter effects has been expressed by different authors time and again. This is so still today. There even is an author of the 1970/80s, A. Maeder, who has taken up the line of investigation following Dirac and Canuto after an interruption of several decades, with the goal of establishing links to the present research in cosmology and astrophysics [103–105] (cf. Sect. 3.2). This is an interesting attempt, but as I can see the approach of these papers suffers from an incoherent usage of scale transformation conventions and an unclear dynamical basis. It needs more time and astronomical expertise to judge whether the striking results claimed by the author with regard to accounting for recent empirical data (CMB, galactic rotation, cluster dynamics) are crucially affected by the theoretical deficiencies, or whether they can be upheld in an improved framework. If so, Maeder's approach would indicate a blueprint for a unifying account of cosmology and dark matter phenomena, which is rather, perhaps even fundamentally, different from the present mainstream picture of cosmology and dark matter.

In the sequel I discuss two less incisive examples of recent research approaches to the field from the point of view of Weyl geometry. A review of the cosmological investigations of the Brazilian group initiated by M. Novello and its external Greek contributor J. Miritzis (see Sect. 3.2) has been given at another place.[80] Here I add a short discussion of studies by a group of authors around J. Jiménez and T. Koivisto (later also including L. Heisenberg) [83–85] and supplement it by a remark on a recent proposal for modelling galactic and cluster dynamics in a Weyl geometric scalar tensor theory with an unconventional kinetic term [155]. A comparison of the Weyl geometric approach(es) with conformal cosmological models of a more general type, in particular C. Wetterich's "universe without expansion" [183, 184], would be an interesting task, but cannot be carried out here.

Let us start the report on the first example with an interesting remark made by the two authors of the first paper about the reason for studying a Weyl geometric framework:

> …(T)he nearly scale invariant spectrum of cosmological perturbations and the fact that the Higgs mass is the only term breaking scale invariance in the Standard Model of elementary particles are very suggestive indications that this could be an actual symmetry of a more fundamental theory that has been spontaneously broken [84, p. 2].

The paper is intended to lay the ground for further investigations of gravity and cosmology in the framework of Weyl geometry.

In the second paper, [85], the framework is extended to what the authors call a *generalized Weyl geometry*. Here a linear connection Γ with non-vanishing torsion is accepted, accompanied by a concomitant generalization of the Weyl geometric compatibility condition (18) between the Riemannian component of the Weylian metric and the covariant derivative $\nabla(\Gamma)$:

$$\nabla g = -2b\,\varphi \otimes g\,, \tag{42}$$

where b is any real valued parameter. The permissible linear connections in any scale gauge (g, φ) are then given by the 2-parameter family:

$$\Gamma^{\mu}_{\nu\lambda} = \Gamma(g)^{\mu}_{\nu\lambda} + b'\delta^{\mu}_{\nu}\varphi_{\lambda} + b\,\delta^{\mu}_{\lambda}\varphi_{\mu} - b'\varphi^{\mu} g_{\mu\nu} \tag{43}$$

For $b \neq b'$ the connection has torsion, with $\Gamma^{\lambda}_{\mu\nu} - \Gamma^{\lambda}_{\nu\mu}$ (the *distortion*) being a linear expression in the Weyl field φ. If both parameters are equal to 1 the usual (torsion free) Weyl geometric structure is included in the generalized framework.[81]

---

[80] Scholz [153, Sect. 11.6.4].

[81] The authors start from a slightly more general linear connection with three parameters b_1, b_2, b_3 but soon specialize to the generalized metric compatibility of the general form

The authors study quadratic gravity theories in this generalized framework. With an interesting argument they constrain the quadratic terms for the case $n = 4$ by the (external) condition that their Riemannian component reduces to the Chern-Gauss-Bonnet form (24):

> The reason for this restriction is the well-known fact that the Gauss-Bonnet term (...) leads to second order equations of motion for the metric tensor and, thus, avoids the potential presence of Ostrogradski's instabilities [84, Sect. III].

This adds a classical field theoretic argument to the quantum field consideration reproduced in Sect. 5.3 for constraining quadratic gravity theories by the Gauss-Bonnet relation.

Like the Brazilian group, Jiménez/Koivisto don't insist on a scale invariance of their complete Lagrange density. Vaguely appealing to some "spontaneously" breaking of the scale symmetry at high energies, they add a (Weyl geometric) Einstein term which does the job, or at least documents a broken scale symmetry:

$$\mathfrak{L}_E = -\frac{1}{2} M_p^{n-2} R \sqrt{|\det g|} \qquad (n = dim\, M)$$

In later papers they use more general considerations also violating the scale invariance of the total Lagrangian. This seems to be a precondition for constructing cosmological models which allow to establish links with the present mainstream discourse. Jiménez and Koivisto [85] starts with a general functional term of the form $f(R)$ (with Weyl geometric scalar curvature R), which is transformed via a Lagrangian constraint trick into the form of a scalar tensor theory of Brans-Dicke type, with the Lagrangian constraint function playing the role of the scalar field [85, Eq. (21)]. In very simple cases ("taking into account only the leading order quadratic correction to the Einstein-Hilbert action", $L = R + \frac{R^2}{6M^2}$) the authors arrive at a model which in the Riemannian case implies a special type of inflationary scenario ("Starobinsky inflation"). In their Weyl geometric generalization they find a 1-parameter generalization of this model due to the Weylian scale connection, similar to a vector field introduced also by other approaches in the literature.

This is only the starting point for more extensive studies of cosmological models in Jiménez et al. [83] undertaken with extended (wo-)manpower by including L. Heisenberg into the team. The three researchers study a class of quadratic gravity theories in generalized Weyl geometry satisfying the Gauss-Bonnet relation and with a scale symmetry breaking Einstein-Hilbert term. By assuming an external constraint of a purely temporal and homogeneous scale connection they reduce their geometrical "vector tensor" theory to a scalar tensor theory with a scalar field (the

(42), which implies $b_3 = 2b_1 - b_2$. The corresponding parameters used here are then given by $b = (b_1 - b_2)$, $b' = \frac{1}{2}(b_2 + b_3)$.

temporal component of the scale connection) satisfying a non-dynamical algebraic constraint (ibid. p. 8). They study perturbations around the de Sitter universe and look for bouncing solutions, which arise for non-realistic values of the paramaters only. Finally properties arising from adding a matter component to the Lagrangian are studied.

Resuming their work the authors find that

> …the class of vector-tensor theories studied in this work and which naturally arise in the framework of geometries with a linear vector distortion can give rise to a rich an[d] interesting phenomenology for cosmology. Jiménez et al. [83, p. 23]

This adds rich material to the stock of models already studied without generalizing the Weyl geometric structure by the Brazilian group and Miritzis; but it keeps still far away from the quest for an improved, or even alternative, understanding of empirical data. From a cultural point of view one might get the impression that now also part of the work in Weyl geometric gravity has joined the "postmodern" wave of studying the widest possible range of theoretical models with only vague allusions to the empirical data collected in astronomy and observational cosmology.

The approach in Scholz [155] has a different motivation. It deals with the question whether a Weyl geometric linear gravity theory ($\beta_1 = \beta_2 = \beta_3 = 0$ in 23) with scalar field can reproduce the deviation of galactic dynamics from the Newtonian expectation, i.e., imply MOND-like dynamics for low velocity trajectories, *and lead to a gravitational light deflection* which one would expect for a dark matter explanation of the anomalous galactic dynamics.[82] According to (33) the author assumes a framework of integrable Weyl geometry (external constraint $f_{\mu\nu} = 0$ in 23) and studies the consequences of an unconventional kinetic term $L_{\partial\phi}$ of the scalar field. So far this approach can be understood as a generalized BD-type scalar field theory in a Weyl geometric setting; it furthermore has to assumes a screening of the scalar field which restricts its sphere of action to regimes of extremely weak gravity like in the superfluid approach to anomalous galactic dynamics [14–16]. The kinetic term of the scalar field combines a cubic expression $L_{\partial\phi 3}$ in the partial derivatives of the scalar field $|\partial\phi|^3$ and a second order derivative (in $\partial^2\phi$) term $L_{\partial 2\phi}$, both of weight -4 and a scale invariant Lagrange density,

$$L_{\partial\phi} = L_{\partial\phi 3} + L_{\partial 2\phi}\,. \tag{44}$$

$L_{\partial\phi 3}$ is similar to the kinetic term in the first relativistic generalization of the MOND theory introduced by J. Bekenstein and M Milgrom, called RAQUAL ("relativistic aquadratic Lagrangian") [13]. $L_{\partial 2\phi}$ is adapted from the cosmological studies of Novello and the Brazilian group [118], but has been modified such that

[82] MOND stands for the *modified Newtonian dynamics* devised by Milgrom [111]. It is characterized by a new constant a_0 with the dimension of an acceleration. For a recent survey see [60].

$w(L_{\partial 2\phi}) = -4$. Due to its special form it keeps the dynamical equation for ϕ of second order.[83]

The Lagrangian density of baryonic matter is assumed to be brought into a scale invariant form such that its (twice covariant) Hilbert energy-momentum tensor scales with weight -2. Stretching the result of the Geroch-Jang theorem slightly, from test bodies to extended bulk of matter, the free fall trajectories of matter can reasonably be assumed to follow Weyl geometric geodesics (see end of Sect. 4.2). The Einstein gauge is considered as distinguished for indicating measured values of observable quantities (cf. (∗) in Sect. 5.4). The Levi-Civita part $\Gamma(g)$ of the affine connection in Einstein gauge can be computed from the Riemannian component of the Weyl geometric Einstein equation, and the scale connection contribution $\Gamma(\varphi)$ from the scalar field equation. Step by step the following properties of the model are established in Scholz [155]:

1 With the scale connection in Einstein gauge written in the form $\varphi \underset{Eg}{\doteq} d\sigma$ like in (36), the weak field approximation of the scalar field equation leads to a non-linear Poisson equation for σ, known from Milgrom's theory for the deep MOND regime,
$$\nabla \cdot (|\nabla\sigma|\nabla\sigma) = 4\pi G\, a_o\, \rho_{bar}$$
with the usual MOND constant a_0; Scholz calls it the *Milgrom equation*. It is sourced by baryonic matter only.[84]

2 σ induces an acceleration for free fall trajectories, due to the scale connection, $a_\varphi = -\nabla\sigma$. It is of deep MOND type.

3 The weak field approximation for determining the Riemannian component g of the Weylian metric in the Einstein gauge boils down to a Newton approximation exactly like in Einstein gravity. It is sourced by the baryonic matter only, not by the scale field (see 4), and induces an acceleration a_N like in Newton gravity with potential Φ_N.

4 The scalar field has an energy momentum tensor with non-negligible energy density $\rho_\phi = (4\pi G)^{-1}\, \nabla^2\sigma$. This looks like dark matter, but is has negative pressures p_1, p_2, p_3 with peculiar properties:

[83] For the variation with regard to ϕ the second order terms of $L_{\partial 2\phi}$ form a divergence expression; not so, however, while varying g. Thus the dynamical equation of ϕ remains of second order, while the $L_{\partial 2\phi}$ term contributes essentially to the source term of the Einstein equation (the Hilbert energy-momentum tensor).

[84] In complete generality, the scalar field equation in Einstein gauge is a scale covariant generalization of the Milgrom equation.

(i) The contribution of the scalar field to the Newton-Poisson source (of the Einstein weak field equation) vanishes, $\rho_\phi + \sum p_j = 0$.

(ii) On the other hand it contributes to the gravitational light deflection. In the central symmetric case it does so by an additive contribution of 2σ to the deflection potential. For baryonic matter in Einstein gravity the deflection potential is $2\Phi_N$.[85]

5 (i) 2 and 3 together imply a MOND-like phenomenology for test particles, with a deep MOND acceleration added to the Newtonian one, both with the source ρ_{bar}, in the regime of activity of the scalar field (screened for larger accelerations).

(ii) The light deflection of the model is identical to the one expected for a source given by $\rho_{bar} + \rho_\phi$ in Einstein gravity.

(iii) The "dynamical phantom mass" ρ_{phant}, i.e., the mass density expected in Newton gravity for generating the additional acceleration a_φ predicted by the model, coincides with the "optical phantom mass" ρ'_{phant} which one would expect in a dark matter approach for the same light deflection as in the model. Both agree with the mass/energy density of the scalar field, $\rho_\phi = \rho_{phant} = \rho'_{phant}$.

Resumée: The *gravitational effects* for test bodies and for light rays of the model's scalar field ϕ, observed in the Einstein gauge, *coincide with those of a halo of (pressure-less) dark matter* generating a Newtonian potential σ, although the energy-momentum of the scalar field carries negative pressures and is, in this sense, similar to "dark energy". The potential σ of the additional acceleration due to the scale connection is given by a Milgrom equation (with the baryonic matter as its source).

This is an exceptional model among the many models which propose relativistic generalizations of Milgrom's modified Newtonian dynamics. Most of them work with several scalar fields, often also with an additional dynamical vector field, and/or strange (non-conformal) deformations of the metric. The original RAQUAL theory, with only one additional scalar field, is inconsistent with the observed amount of gravitational light deflection in galaxies and clusters. To my knowledge there is only the approach of "mimetic gravity" [178] besides the present Weyl geometric scalar tensor model, which is able to derive a *MOND-like dynamics* for low velocity trajectories and a corresponding *enhancement of the gravitational light deflection* on the basis of adding just a *single scalar field* (at large scales, thus not counting the

[85]The deflection angle is computed from the first two components of the Ricci tensor like in [34, p. 288f.]. It is the gradient of $\frac{1}{2}(-h_{00} + h_{jj})$ (for any $1 \le j \le 3$), where h is the deviation of a weak field metric from the Minkowski metric, $g = \eta + h$. $\frac{1}{2}(-h_{00} + h_{jj})$ thus plays the role of a *deflection potential*.

Higgs field here).[86] But alas, the kinetic term of the Weyl geometric model looks so strange that it may remain an exercise in theory construction only. In any case, it deserves to be mentioned here as an outcome of recent and even present investigations in Weyl geometric gravity.

6 An Extremely Short Glance Back and Forth

We have seen an eventful trajectory of Weyl's proposal for generalizing Riemannian geometry during the last century: from a promising first attempt (1918 to the middle of the 1920s) at formulating a type of geometry which was intended to bring the framework of gravity closer to the principles of other field theories, at that time essentially electromagnetism, through a phase of disillusion and non-observance (1930–1970), to a new rise in the 1970s. In the last third of the 20th century dispersed contributions to various fields of physics brought Weyl geometry back as a geometrical framework for actual research, partially fed by the introduction of a hypothetical scalar field which had at least one foot in the gravitational sector. But these new researches were far from influential on the main lines of development in theoretical physics.

Although first hypothetical links between the new attempts at Weyl geometric generalizations of gravity theory and the rising standard model of elementary particle physics were soon formulated, they remained rather marginal. Also the vistas opened up by Weyl geometric scalar tensor theories for astronomy at large scales and cosmology remained without a visible impact on the respective communities. One reason for the overall reservation may have been the fact that, at that time, scalar fields played only a hypothetical, perhaps even only formal role within physical theories. This changed, of course, with the empirical confirmation of the Higgs boson.

It may be worthwhile to investigate whether an additional scalar field with links to the Higgs portal and gravity can help to answer some of the open questions of the present foundational research in physics. Some of the recent work surveyed in this paper make already some steps in this direction. As it is always difficult to discern reliable patterns of actual developments, I leave it as a question for the future to judge whether this is a fruitful enterprise, or just another contribution to screening possible theory alternatives without clear evidence of empirical support.

To end with a personal remark as a historian of mathematics who has spent some time on studying the development of this subfield of physical geometry: it is pleasant

86 Of course models have still to be checked by astronomers. Heuristic considerations indicate that the Weyl geometric model may even "predict" the dynamics of galaxy clusters without assuming additional dark matter [155, Sect. 3.2].

to see that, contrary to Weyl's disillusionment in this regard after 1920, his beautiful and conceptually convincing idea for generalizing Riemannian geometry is still alive and helps triggering new lines of research also a century after it was first devised.

7 Appendix: An Outline of the Noether Theorems

Consider a Lagrangian field theory governed by an action $S = \int \mathfrak{L}\, dx$, with $\mathfrak{L}$ depending on fields ϕ_a and their partial derivatives up to order l, which is invariant under a global operation of a k-dimensional continuous (Lie) group G. Then the *first Noether theorem* establishes conserved currents, i.e., 1-form densities with components $J^{\mu}_{g_i}$, to any generator g_i of the group ($1 \leq \mu \leq n = \dim M,\ 1 \leq i \leq k$). They are called the *Noether currents* or *canonical currents* of the symmetry. These currents are conserved, $\partial_\nu J^{\nu}_{g_i} = 0$, if the equations of motion of the field theory are satisfied (the fields are "on shell"). In the context of a Euclidean, Minkowski or (pseudo-) Riemannian metrical environment the 1-form densities can be Hodge-dualized to $(n-1)$-forms $j_{g_i} = *J_{g_i}$. The conservation is then expressed by the closure condition of the form $dj_{g_i} = 0$.

In the case of localized, i.e., point dependent operations of a continuous symmetry (group G, *dim* $G = k$) for a Lagrangian field theory like above, in particular for gauge symmetries, the *second Noether theorem* establishes the existence of k differential identities between the Euler-Lagrange expressions of the theory, which hold irrespective of whether the dynamical equations are satisfied or not (they are valid "off shell").

Let the Euler-Lagrange expressions of the field ϕ_a be $\mathfrak{E}_a = \frac{\delta \mathfrak{L}}{\delta \phi_a}$ Assume moreover that a collection of infinitesimal symmetries parametrized by one or several arbitrary functions $\lambda(x)$, respectively $\lambda_i(x)$, shifts the field ϕ_a by an amount $\delta_\lambda \phi_a$ which depends linearly on λ and its first derivative only,

$$\delta_\lambda \phi_a = f[\phi_a]\lambda + f^{\nu}[\phi_a]\partial_\nu \lambda \, , \tag{45}$$

with functional expressions $f[\ ]$, $f^{\nu}[\ ]$ $(1 \leq \nu \leq n)$ depending on λ and on ϕ_a. Then the variation with respect to λ and integration by parts leads to the following form for the *Noether identities*, one for each λ_i [9, Eq. (13)]:

$$\sum_a \left(f[\phi_a]\mathfrak{E}_a - \partial_\nu f^{\nu}[\phi_a \mathfrak{E}_a] \right) \equiv 0 \, . \tag{46}$$

In certain cases, e.g., for a Weyl geometric field theory with a Maxwell term or, more generally, a gauge theory with Yang-Mills term, the Noether identities imply a vanishing divergence for the source term of the dynamical equation of the corresponding gauge field (cf. Sect. 4.1). The latter is called the *dynamical current* $\mathfrak{s}^{\mu}$ of the gauge field. The conservation $\partial_\nu \mathfrak{s}^\nu = 0$ even holds if the equation of motion for the gauge field under consideration is not satisfied (the fields may be"off shell" regarding the corresponding gauge field equation), but the other equations of motion of the theory are satisfied.

In addition, currents analogous to those given by the first theorem can be formed for the localized symmetry also, but only after a choice of the gauge—mathematically spoken after trivializing the gauge bundle. One can then focus on 1-parameter families of locally constant operations in the gauge (the trivialization) for any generator of the group. Following the proof argumentation of the first Noether theorem one derives conserved currents which are likewise called *canonical* or *Noether currents*. For a Lagrangian density $\mathfrak{L} = \mathfrak{L}(\phi_a, \partial\phi_a, \partial\partial\phi_a)$ depending on a certain number of fields ϕ_a and their partial derivatives up to order 2 the canonical current can be obtained from partial derivatives of the Lagrangian multiplied by variations of the fields $\delta\phi_a = \delta_{g_i}\phi_a$ and their derivatives $\partial_\nu\delta\phi_a$ under the 1-parameter symmetry generated by g_i in the specified gauge:

$$J^{\mu}_{g_i} = \sum_a \left(\frac{\partial \mathfrak{L}}{\partial(\partial_\mu \phi_a)} \delta\phi_a + \frac{\partial \mathfrak{L}}{\partial(\partial_\mu \partial_\nu \phi_a)} \partial_\nu \delta\phi_a - \partial_\nu \frac{\partial \mathfrak{L}}{\partial(\partial_\mu \partial_\nu \phi_a)} \delta\phi_a \right) . \tag{47}$$

The partial derivative can be replaced by any other appropriate derivation operator on the fields (e.g., covariant derivatives) [99, p. 727f.]

One has to keep in mind however that here, in distinction to the first theorem, the currents depend on the chosen gauge and are conserved under the same condition as above ("on shell" for the other dynamical equations, "off shell" for the one of the gauge field related to the group G, respectively its Lie algebra). Because of these peculiar properties Noether took up a terminology introduced by Hilbert and called such conservation laws *improper*.

The historically first of such "improper" laws arose for the energy-momentum current in general relativity [93, 173]. More of them appeared in the gauge theories of the second half of the 20th century and became important for the renormalization of the quantized field theories, at first for QED ("Ward identity"). Later these identities were generalized to other quantum gauge field theories ("Ward-Takahashi identity"). The first Noether theorem is explained in most textbooks of field theory and of relativity theory; the second theorem is often skipped or alluded to only vaguely, a helpful exception is [9] cited above.

Acknowledgements I thank the organizers of the conference *100 Years of Gauge Theory*, Silvia De Bianchi and Claus Kiefer, for their invitation and for providing the occasion for very stimulating interactions on the topic. Friedrich Hehl encouraged me strongly to include a discussion of the Noether theorems and the conserved scale currents in this survey.

References

1. R. Abraham, J.E. Marsden, in *Foundations of Classical Mechanics* (Addison Wesley, Redwood City etc., 1978). (5-th revised ed. 1985)
2. A. Afriat, Weyl's gauge argument. Found. Phys. **43**, 699–705 (2013)
3. A. Afriat, Logic of gauge (Bernard/Lobo, 2019, p. 265–203)
4. T.S. Almeida, J.B. Formiga, M.L. Pucheu, C. Romero, From Brans-Dicke gravity to a geometrical scalar-tensor theory. Phys. Rev. D **89**(064047), 10 (2014). arXiv:1311.5459
5. T.S. Almeida, M.L. Pucheu, C. Romero, A geometrical approach to Brans-Dicke theory, in *Accelerated Cosmic Expansion*, ed. by C. Moreno Gonzales, J.E. Madriz Aguliar, L.M. Reyes Barrera. Astrophysics and Space Science Proceedings, vol. 38 (2014), pp. 33–41
6. J. Audretsch, Riemannian structure of space-time as a consequence of quantum mechanics. Phys. Rev. D **27**, 2872–2884 (1983)
7. J. Audretsch, F. Gähler, S. Norbert, Wave fields in Weyl spaces and conditions for the existence of a preferred pseudo-riemannian structure. Commun. Math. Phys. **95**, 41–51 (1984)
8. J. Audretsch, C. Lämmerzahl, Constructive axiomatic approach to spacetime torsion. Class. Quantum Gravity **5**, 1285–1295 (1988)
9. S. Avery, B. Schwab, Noether's second theorem and Ward identities for gauge symmetries. J. High Energy Phys. **2**, 031 (2016). arXiv:1510.07038
10. G. Bacciagaluppi, A conceptual introduction to Nelson's mechanics, in *Endophysics, Time, Quantum and the Subjective*, ed. by R. Buccheri, M. Saniga, A. Elitzur (Singapore: World Scientific, 2005), pp. 367–388. (Revised postprint in https://philsci-archive.pitt.edu/8853/1/Nelson-revised.pdf)
11. A. Baeumler, M. Schroeter, in *Handbuch der Philosophie. Bd. II. Natur, Geist, Gott* (Oldenbourg, München, 1927)
12. I. Bars, P. Steinhardt, N. Turok, Local conformal symmetry in physics and cosmology. Phys. Rev. D **89**, 043515 (2014). arXiv:1307.1848
13. J. Bekenstein, M. Milgrom, Does the missing mass problem signal the breakdown of Newtonian gravity? Astrophys. J. **286**, 7–14 (1984)
14. L. Berezhiani, J. Khoury, Theory of dark matter superfluidity. Phys. Rev. D **92**(103510) (2015). arXiv:1507.01019
15. L. Berezhiani, J. Khoury, Dark matter superfluidity and galactic dynamics. Phys. Lett. B **753**, 639–643 (2016). arXiv:1506.07877
16. L. Berezhiani, J. Khoury, Emergent long-range interactions in Bose-Einstein condensates. Phys. Rev. D **99**, 076003 (2019). arXiv:1812.09332
17. J. Bernard, in *L'idéalisme dans l'infinitésimal. Weyl et l'espace à l'époque de la relativité* (Presse Universitaires de Paris Ouest, Paris, 2013)
18. J. Bernard, La redécouverte des tapuscrits des conférences d'Hermann Weyl à Barcelone. Revue d'histoire des mathématiques **21**(1), 151–171 (2015)
19. J. Bernard, Riemann's and Helmholtz-Lie's problems of space from Weyl's relativistic perspective. Stud. Hist. Philos. Mod. Phys. **61**, 41–56 (2018)
20. J. Bernard, C. Lobo (eds.), *Weyl and the Problem of Space. From Science to Philosophy* (Springer, Berlin, 2019)
21. M. Blagojević, F. Hehl, in *Gauge Theories of Gravitation. A Reader with Commentaries* (Imperial College Press, London, 2013)
22. A. Borrelli, The making of an intrinsic property: "symmetry heuristics" in early particle physics'. Stud. Hist. Philos. Mod. Phys. A **50**, 59–70 (2015)
23. A. Borrelli, The story of the Higgs boson: the origin of mass in early particle physics. Eur. Phys. J. H **40**(1), 1–52 (2015)

24. A. Borrelli, The uses of isospin in early nuclear and particle physics. Stud. Hist. Philos. Mod. Phys. **60**, 81–94 (2017)
25. P. Bouvier, A. Maeder, Scale invariance, metrical connection and the motions of astronomical bodies. Astron. Astrophys. **73**, 82–89 (1979)
26. P. Bouvier, A. Maeder, Consistency of Weyl's geometry as a framework for gravitation. Astrophys. Space Sci. **54**, 497–508 (1977)
27. K. Brading, Which symmetry? Noether, Weyl, and conservation of electric charge. Stud. Hist. Philos. Mod. Phys. **33**, 3–22 (2002)
28. K. Brading, A note on general relativity, energy conservation and Noether's theorems, in *The Universe of General Relativity*, ed. by A.J. Kox, J. Eisenstaedt (Einstein Studies Basel/Boston, Birkhäuser, 2005), pp. 125–135
29. C. Brans, R.H. Dicke, Mach's principle and a relativistic theory of gravitation. Phys. Rev. **124**, 925–935 (1961)
30. A. Bregman, Weyl transformations and Poincaré gauge invariance. Prog. Theor. Phys. **49**, 667–6992 (1973)
31. Bureau International des poids et mesures, in *Resolutions adopted by the General Conference on Weights and Measures (24th meeting)* (Paris, 17–21 October 2011). www.bipm.org/en/si/new_si/
32. C. Callan, S. Coleman, J. Roman, A new improved energy-momentum tensor. Ann. Phys. **59**, 42–73 (1970)
33. V. Canuto, P.J. Adams, S.-H. Hsieh, E. Tsiang, Scale covariant theory of gravitation and astrophysical application. Phys. Rev. D **16**, 1643–1663 (1977)
34. S. Carroll, *Spacetime and Geometry* (Addison Wesley, San Francisco, 2004)
35. C. Castro, On Weyl geometry, random processes, and geometric quantum mechanics. Found. Phys. **22**, 569–615 (1992)
36. M. de Cesare, J.W. Moffat, M. Sakellariadou, Non-Riemannian geometry and the origin of physical scales in a theory with local conformal symmetry. Eur. J. Phys. C 77(605), 12 (2017). arXiv:1612.08066
37. J.M. Charap, W. Tait, A gauge theory of the Weyl group. Proc. R. Soc. Lond. A **340**, 249–262 (1974)
38. H. Cheng, Possible existence of Weyl's vector meson. Phys. Rev. Lett. **61**, 2182–2184 (1988)
39. A. Codello, G. D'Orodico, C. Pagani, R. Percacci, The renormalization group and Weyl invariance. Class. Quantum Gravity **30**(115015), 22 (2013). arXiv:1210.3284
40. R. Coleman, H. Korté, Constraints on the nature of inertial motion arising from the universality of free fall and the conformal causal structure of spacetime. J. Math. Phys. **25**, 3513–3526 (1984)
41. S. De Bianchi, From the problem of space to the epistemology of science: Hermann Weyl's reflections on the dimensionality of the world (Bernard/Lobo, 2019), p. 189–209
42. F. De Martini, E. Santamato, Interpretation of quantum-nonlocality by conformal geometrodynamics. Int. J. Theor. Phys. **53**, 3308–3322 (2014). arXiv:1203:0033
43. F. De Martini, E. Santamato, Nonlocality, no-signalling, and Bell's theorem investigated by Weyl conformal differential geometry. Phys. Scr. **2014**, T163 (2014). arXiv:1406.2970
44. F. De Martini, E. Santamato, Proof of the spin-statistics theorem. Found. Phys. **45**(7), 858–873 (2015)
45. F. De Martini, E. Santamato, Proof of the spin-statistics theorem in the relativistic regime by Weyl's conformal quantum mechanics. Int. J. Quantum Inf. **14**(04), 1640011 (2016)
46. S. Dengiz, B. Tekin, Higgs mechanism for new massive gravity and Weyl-invariant extensions of higher-derivative theories. Phys. Rev. D **84**(024033), 7 (2011)
47. R.H. Dicke, Mach's principle and invariance under transformations of units. Phys. Rev. **125**, 2163–2167 (1962)

48. P.A.M. Dirac, Long range forces and broken symmetries. Proc. R. Soc. Lond. A **333**, 403–418 (1973)
49. P.A.M. Dirac, Cosmological models and the large number hypothesis. Proc. R. Soc. Lond. A **338**, 439–446 (1974)
50. W. Drechlser, Mass generation by Weyl symmetry breaking. Found. Phys. **29**, 1327–1369 (1999)
51. W. Drechsler, Geometric formulation of gauge theories. Zeitschrift f. Naturforschung **46a**, 645–654 (1991)
52. W. Drechsler, Mass generation by Weyl symmetry breaking. Found. Phys. **29**, 1327–1369 (1999)
53. W. Drechsler, H. Tann, Broken Weyl invariance and the origin of mass. Found. Phys. **29**(7), 1023–1064 (1999). arXiv:gr-qc/98020
54. P. Duerr, Unweyling three mysteries of Nordström gravity. Stud. Hist. Philos. Sci. B **69**, 32–49 (2020)
55. A.S. Eddington, in *The Mathematical Theory of Relativity* (University Press, Cambridge, 1923). (2nd edition 1924)
56. J. Ehlers, F. Pirani, A. Schild, The geometry of free fall and light propagation, in *General Relativity, Papers in Honour of J.L. Synge*, ed. by L. O'Raifertaigh (Clarendon Press, Oxford, 1972) pp. 63–84. (Reprint *General Relativity and Gravity***44** (2012), 1587–1609)
57. A. Einstein, Über eine naheliegende Ergänzung des Fundamentes der allgemeinen Relativitätstheorie. *Sitzungsberichte der Preussischen Akademie der Wissenschaften zu Berlin, physikalisch-math. Klasse* (1921), pp. 261–264. (Einstein, 1987ff,*7 Doc. 54*, p. 411–416), in (Einstein, 2006, p. 198–201)
58. A. Einstein, in *The Collected Papers of Albert Einstein* (University Press, Princeton, 1987)
59. A. Einstein, in *Akademie-Vorträge. Herausgegeben von Dieter Simon* (Wiley-VCH, Weinheim, 2006)
60. B. Famaey, S. McGaugh, Modified Newtonian dynamics (MOND): observational phenomenology and relativistic extensions. Living Rev. Relat. **15**(10), 1–159 (2012)
61. V. Faraoni, *Cosmology in Scalar-Tensor Gravity* (Kluwer, Dordrecht, 2004)
62. L. Fatibene, M. Francaviglia, Extended theories of gravitation and the curvature of the universe–Do we really need dark matter? in *Open Questions in Cosmology*, ed. by G.J. Olmo. InTechOpen London chapter 5 (2012). https://www.intechopen.com/books/open-questions-in-cosmology/extended-theories-of-gravitation-and-the-curvature-of-the-universe-do-we-really-need-dark-matter-
63. L. Fatibene, M. Francaviglia, Weyl geometries and timelike geodesics. Int. J. Geom. Methods Mod. Phys. **9**(05) (2012). arXiv:1106.1961
64. J. Frauendiener, Conformal infinity. Living Rev. Relativ. **3**(4) (2000)
65. Y. Fujii, K.-C. Maeda, *The Scalar-Tensor Theory of Gravitation* (University Press, Cambridge, 2003)
66. P. Galison, *Image and Logic. A Material Culture of Microphyics* (University Press, Chicago, 1997)
67. R. Geroch, P.S. Jang, Motion of a body in general relativity. J. Math. Phys. **16**, 65–67 (1975)
68. D.M. Ghilencea, Spontaneous breaking of Weyl quadratic gravity to Einstein action and Higgs potential. J. High Energy Phys. **2019**, Article 49 (2019). arXiv:1812.08613
69. G. Hubert, On the history of unified field theories. Living Rev. Relativ. **2004**, 2 (2004)
70. H. Goenner, Some remarks on the genesis of scalar-tensor theories. General Relativ. Gravity **44**(8), 2077–2097 (2012). arXiv: 1204.3455
71. J. Haantjes, The conformal Dirac equation. Verhandelingen Koninklijke Nederlandse Akademie van Wetenschappen Amsterdam, Proceedings section of sciences **44**, 324–332 (1941)

72. B.C. Hall, *Quantum Theory for Mathematicians* (Springer, Berlin, 2001)
73. K. Hayashi, T. Kugo, Remarks on Weyl's gauge field. Prog. Theor. Phys. **61**, 334–346 (1979)
74. F. Hehl, Gauge theory of gravity and spacetime, in *Towards a Theory of Spacetime Theories*, ed. by D. Lehmkuhl et al. (Springer, 2017), pp. 145–170
75. F.W. Hehl, C. Lämmerzahl, Physical dimensions/units and universal constants: their invariance in special and general relativity. Annalen der Physik **531**(5), 1800407, 10 (2019)
76. F.W. Hehl, J.D. McCrea, W. Kopczy?ski, The Weyl group and ist currents. Phys. Lett. A **128**, 313–318 (1988)
77. F.W. Hehl, E. Mielke, R. Tresguerres, Weyl spacetimes, the dilation current, and creation of gravitating mass by symmetry breaking, in *Exact Sciences and their Philosophical Foundations; Exakte Wissenschaften und ihre philosophische Grundlegung*, ed. by W. Deppert; K. Hübner e.a. (Peter Lang, Frankfurt/Main, 1988), pp. 241–310
78. W. Heisenberg, Quantum theory of fields and elementary particles. Rev, Mod. Phys. **29**, 269–278 (1957)
79. G. Hessenberg, Vektorielle Begründung der Differentialgeometrie. Mathematische Annalen **79**, 187–217 (1917)
80. M. Israelit, *The Weyl-Dirac Theory and Our Universe* (Nova Science, New York, 1999)
81. M. Israelit, N. Rosen, Weyl-Dirac geometry and dark matter. Found. Phys. **22**, 555–568 (1992)
82. M. Israelit, N. Rosen, Cosmic dark matter and Dirac gauge function. Found. Phys. **25**, 763–777 (1995)
83. J.B. Jiménez, L. Heisenberg, T.S. Koivisto, Cosmology for quadratic gravity in generalized Weyl geometry. J. Cosmol. Astropart. Phys. **04**(046), 29 (2016)
84. J.B. Jiménez, T.S. Koivisto, Extended Gauss-Bonnet gravities in Weyl geometry. Class. Quantum Gravity **31**(13), 135002 (2014). arXiv:1402.1846
85. J.B. Jiménez, T.S. Koivisto, Spacetimes with vector distortion: inflation from generalised Weyl geometry. Phys. Lett. B **756**, 400–404 (2016). arXiv:1509.02476
86. P. Jordan, *Schwerkraft und Weltall* (Vieweg, Braunschweig, 1952). (2nd revised edtion 1955)
87. H. Kastrup, Some experimental consequences of conformal invariance at very high energies. Phys. Lett. **3**, 78–80 (1962)
88. H. Kastrup, Zur physikalischen Deutung und darstellungstheoretischen Analyse der konformen Transformationen von Raum und Zeit. Annalen der Physik **9**, 388–428 (1962)
89. H. Kastrup, Gauge properties of Minkowski space. Phys. Rev. **150**(4), 1183–1193 (1966)
90. H. Kastrup, On the advancement of conformal transformations and their associated symmetries in geometry and theoretical physics. Annalen der Physik **17**, 631–690 (2008)
91. T. Kibble, Lorentz invariance and the gravitational field. J. Math. Phys. **2**, 212–221 (1961). (Blagojević/Hehl, 2013, chap. 4)
92. A. Kock, in *Synthetic Differential Geometry*, 2nd edn. (London Mathematical Society, London, 2006). (Revised version of 1st edition 1981)
93. Y. Kosmann-Schwarzbach, in *The Noether Theorems. Invariance and Conservation Laws in the Twentieth Century* (Springer, Berlin, 2011)
94. B. Kostant, in *Quantization and Unitary Representations. 1. Prequantisation*. Lecture Notes in Mathematics, vol. 170 (Springer, Berlin, 1970)
95. H. Kragh, *Quantum Generations : A History of Physics in the Twentieth Century* (University Press, Princeton, 1999)
96. H. Kragh, in *Varying Gravity. Dirac's Legacy in Cosmology and Geophyics*. (Springer-Birkhäuser, Science Networks Heidelberg, 2016)
97. H. Kragh, Varying constants of nature: fragments of a history. Phys. Perspect. **21**, 257–273 (2019)

98. W. Kühnel, in *Differentialgeometrie. Kurven–Flächen–Mannigfaltigkeiten*. (Vieweg, Braunschweig/Wiesbaden, 1999). (English as *Differential Geometry–Curves–Surfaces–Manifolds* American Mathematical Society, 2000)
99. J. Lee, R.M. Wald, Local symmetries and constraints. J. Math. Phys. **31**, 725–743 (1990)
100. D. Lehmkuhl, The Einstein-Weyl correspondence on the geodesic principle. Manuscript in preparation (2020)
101. D. Lehmkuhl, The first rival: Jordan's scalar-tensor theory and its role in reconsidering the foundations of General Relativity. Manuscript in preparation (2020)
102. A. Maeder, The problem of varying G and the scale covariant gravitation and cosmology, in *Physical Cosmology. Proceedings of the Les Houches Summer School* (Les Houches, France, July 2–27, 1980). (North-Holland, R. Balian et al. Amsterdam, 1979), pp. 533–543
103. A. Maeder, An alternative to the ΛCDM model: the case of scale invariance. Astrophys. J. (834), 16 (2017)
104. A. Maeder, Dynamical effects of the scale invariance of the empty space: the fall of dark matter? Astrophys. J. (849), 13 (2017). arXiv:1710.11425
105. A. Maeder, Scale-invariant cosmology and CMB temperatures as a function of redshifts. Astrophys. J. (847), 065. 7 (2017)
106. A. Maeder, The acceleration relation in galaxies and scale invariant dynamics: another challenge for dark matter (2018). arXiv:1804.04484
107. V.S. Matveev, A. Trautman, A criterion for compatibility of conformal and projective structures. Commun. Math. Phys. **329**, 821–825 (2014)
108. V. Matveev, E. Scholz, Light cone and Weyl compatibility of conformal and projective structures (2020). arXiv:2001.01494. (Submitted to *General Relativity and Gravitation*)
109. J. Merker, in *Le problème de l'espace. Sophus Lie, Friedrich Engel et le problème de Riemann-Helmholtz* (Hermann, Paris, 2010)
110. T. Mettler, Extremal conformal structures on projective surfaces. To appear in Annali della Scuola Normale Superiore Pisa-Classe di Scienze (2015). arXiv:1510.01043
111. M. Milgrom, A modification of Newtonian dynamcis as a possible alternative to the hidden matter hypothesis. Astrophys. J. **270**, 365–370 (1983)
112. H. Nishino, S. Rajpoot, Broken scale invariance in the standard model (2004). arXiv:hep-th/0403039
113. H. Nishino, S. Rajpoot, Broken scale invariance in the standard model, in *AIP Conference Proceedings*, vol. 881 (2007), pp. 82–93. arXiv:0805.0613. (with different title)
114. H. Nishino, S. Rajpoot, Implication of compensator field and local scale invariance in the standard model. Phys. Rev. D **79**, 125025 (2009). arXiv:0906.4778
115. E. Noether, Invariante Variationsprobleme, in *Göttinger Nachrichten* (1918), pp. 235–257. (Noether, 1982, vol 1, 770ff.)
116. E. Noether, *Gesammelte Abhandlungen* (Springer, Berlin, 1982)
117. M. Novello, L.A.R. Oliveira, A marionette universe. Int. J. Mod. Phys. A **1**(4), 943–953 (1986)
118. M. Novello, L.A.R. Oliveira, J.M. Salim, E. Elbaz, Geometrized instantons and the creation of the universe. Int. J. Mod. Phys. D **1**, 641–677 (1993)
119. H. Ohanian, The energy-momentum tensor in General Relativity and in alternative theories of gravitation, and the gravitational vs. Inertial mass (2010). arXiv:1010.5557 [gr-qc]
120. H. Ohanian, Weyl gauge-vector and complex dilaton scalar for conformal symmetry and its breaking. General Relativ. Gravity **48**(25) (2016). arXiv:1502.00020
121. M. Omote, Scale transformations of the second kind and the Weyl space-time. Lettere al Nuovo Cimento **2**(2), 58–60 (1971)
122. M. Omote, Remarks on the local-scale-invariant gravitational theory. Lettere al Nuovo Cimento **10**(2), 33–37 (1974)

123. L. O'Raifeartaigh, *The Dawning of Gauge Theory* (University Press, Princeton, 1997)
124. L. Ornea, Weyl structures in quaternionic geomety. A state of the art, in *Selected Topics in Geometry and Mathematical Physics*, ed. by E. Barletta, vol. 1. (Univ. degli Studi della Basilicata, Potenza, 2001), pp. 43–80. arXiv:math/0105041
125. W. Pauli, Über die Invarianz der Dirac'schen Wellengleichungen gegenüber Ähnlichkeitstransformationen des Linienelementes im Fall verschwindender Ruhmasse. Helvetia Phys. Acta **13**, 204–208 (1940). (Pauli, 1964, II, 918–922)
126. W. Pauli, in *Collected Scientific Papers* ed. by R. Kronig, V.F. Weisskopf (Wiley, New York, 1964)
127. R. Penrose, The light cone at infinity, in *Relativistic Theories of Gravitation*, ed. by L. Infeld (Pergamon, 1964), pp. 369–373
128. R. Penrose, Zero rest-mass fields including gravitation: asymptotic behaviour. Proc. R. Soc. Lond. A **284**, 159–203 (1965)
129. V. Perlick, Characterization of standard clocks by means of light rays and freely falling particles. Gen. Relativ. Gravit. **19**, 1059–1073 (1987)
130. V. Perlick, *Zur Kinematik Weylscher Raum-Zeit-Modelle. Dissertationsschrift* (Fachbereich Physik, TU Berlin, Berlin, 1989)
131. V. Perlick, Observer fields in Weylian spacetime models. Class. Quantum Gravity **8**, 1369–1385 (1991)
132. H. Pfister, Newton's first law revisited. Found. Phys. Lett. **17**, 49–64 (2004)
133. A. Pickering, *Constructing Quarks* (University Press, Edinburgh, 1988)
134. I. Quiros, Scale invariance and broken electroweak symmetry may coexist together (2013). arXiv:1312.1018
135. I. Quiros, Scale invariant theory of gravity and the standard model of particles (2014). arXiv:1401.2643
136. I. Quiros, R. Garcìa-Salcedo, A.J. Madriz-Aguilar, T. Matos, The conformal transformations' controversy: what are we missing. Gen. Relativ. Gravit. **45**, 489–518 (2013). arXiv:1108.5857
137. C. Romero, J.B. Fonsec-Neto, M.L. Pucheu, Conformally flat spacetimes and Weyl frames. Found. Phys. **42**, 224–240 (2012). arXiv:1101.5333
138. C. Romero, J.B. Fonseca-Neto, M.L. Pucheu, General relativity and Weyl frames. Int. J. Mod. Phys. A **26**(22), 3721–3729 (2011). arXiv:1106.5543
139. C. Romero, J.B. Fonseca-Neto, M.L. Pucheu, General relativity and Weyl geometry. Class. Quantum Gravity **29**(15), 155015, 18 (2012). arXiv:1201.1469
140. N. Rosen, Weyl's geometry and physics. Found. Phys. **12**, 213–248 (1982)
141. D. Rowe, The Göttingen response to general relativity and Emmy Noether's theorems, in *The Symbolic Universe. Geometry, and Physics 189–1930* ed. by J. Gray (University Press, Oxford. 1999), pp. 189–233
142. D. Rowe, Relativity in the making, 1916–1918: Rudolf J. Humm in Göttingen and Berlin. Unpublished manuscript (2019)
143. T. Ryckman, in *The Reign of Relativity. Philosophy in Physics 1915–1925* (University Press, Oxford, 2005)
144. E. Santamato, Geometric derivation of the Schrödinger equation from classical mechanics in curved Weyl spaces. Phys. Rev. D **29**, 216–222 (1984)
145. E. Santamato, Statistical interpretation of the Klein-Gordon equation in terms of the spacetime Weyl curvature. J. Math. Phys. **25**(8), 2477–2480 (1984)
146. E. Santamato, Gauge-invariant statistical mechanics and average action principle for the Klein-Gordon particle in geometric quantum mechanics. Phys. Rev. D **32**(10), 2615–2621 (1985)
147. E. Santamato, F. De Martini, Derivation of the Dirac equation by conformal differential geometry. Found. Phys. **43**(5), 631–641 (2013). arXiv:1107.3168

148. E. Scholz, Local spinor structures in V. Fock's and H. Weyl's work on the Dirac equation (1929), in *Géométrie au vingtième siècle, 1930–2000* ed. by P. Nabonnand, J.-J. Szczeciniarz, D. Flament, J. Kouneiher (Hermann, Paris, 2005), pp. 284–301
149. E. Scholz, The changing concept of matter in H. Weyl's thought, 1918–1930, in *The interaction between Mathematics, Physics and Philosophy from 1850 to 1940*, ed. by J. Lützen, V. F. Hendricks, K.F. Jørgensen (Springer, Dordrecht, 2006). arXiv:math.HO/0409576
150. E. Scholz, Weyl geometric gravity and electroweak symmetry breaking. Annalen der Physik **523**, 507–530 (2011). arXiv:1102.3478
151. E. Scholz, The problem of space in the light of relativity: the views of H. Weyl and E. Cartan, in *Eléments d'une biographie de l'Espace géométrique*, ed. by L. Bioesmat-Martagon. (Edition Universitaire de Lorraine, Nancy, 2016), pp. 255–312. arXiv:1310.7334
152. E. Scholz, Paving the way for transitions–a case for Weyl geometry, in *Towards a Theory of Spacetime Theories*, vol. 13, ed. by D. Lehmkuhl e.a. Einstein Studies, Basel, (Birkhäuser-Springer, Berlin, 2017) pp. 171–224. arXiv:1206.1559
153. E. Scholz, The unexpected resurgence of Weyl gometry in late 20-th century physics, in *Beyond Einstein. Perspectives on Geometry, Gravitation and Cosmology*, vol. 13, ed. S. Walter, D. Rowe, T. Sauer, Einstein Studies, Basel (Birkhäuser-Springer, Berlin, 2018), pp. 261–360. arXiv:1703.03187
154. E. Scholz, Weyl's search for a difference between 'physical' and 'mathematical' automorphisms. Stud. Hist. Philos. Mode. Phys. **61**, 57–67 (2018). arXiv:1510.00156
155. E. Scholz, A scalar field inducing a non-metrical contribution to gravitational acceleration and a compatible add-on to light deflection (2019). arXiv:1906.04989. (Submitted to *General Relativity and Gravitation*)
156. E. Scholz (ed.), *Hermann Weyl's Raum-Zeit-Materie and a General Introduction to His Scientific Work* (Birkhäuser, Basel, 2001)
157. J.A. Schouten, Über die konforme Abbildung n-dimensionaler Mannigfaltigkeiten mit quadratischer Maßbestimmmung auf eine Mannigfaltigkeit mit euklidischer Maßbestimmung. Mathematische Zeitschrift **11**, 58–88 (1921)
158. J.A. Schouten, J. Haantjes, Über die konforminvariante Gestalt der Maxwellschen Gleichungen und der elektromagnetischen Energie-Impulsgleichungen. Physica **1**, 869–872(1934). https://neo-classical-physics.info/uploads/3/4/3/6/34363841/schouten_haantjes_-_conf._inv._of_maxwell.pdf. (English under)
159. J.A. Schouten, J. Haantjes, Über die konforminvariante Gestalt der relativistischen Bewegungsgleichungen, in *Verhandelingen Koninklijke Nederlandse Akademie van Wetenschappen Amsterdam, Proceedings section of sciences*, vol. 39 (1936), pp. 1059–1065. https://neo-classical-physics.info/uploads/3/4/3/6/34363841/schouten_-_conf._inv._of_rel._eom.pdf. (English under)
160. D.W. Sciama, On the analogy between charge and spin in general relativity, in *Recent Developments in General Relativity Festschrift for L. Infeld* (Oxford and Warsaw, Pergamon and PWN, 1962), pp. 415–439. (Blagojevic/Hehl, 2013, chap. 4)
161. M. Shaposhnikov, D. Zenhäusern, Quantum scale invariance, cosmological constant and hierarchy problem. Phys. Lett. B **671**, 162–166 (2009). arXiv:0809.3406
162. M. Shaposhnikov, D. Zenhäusern, Scale invariance, unimodular gravity and dark energy. Phys. Lett. B **671**, 187–192 (2009). arXiv:0809.3395
163. S. Sigurdsson, Hermann Weyl, mathematics and physics, 1900–1927, Ph.D thesis Harvard University, Department of the History of 'Science Cambridge, 1991
164. D. Simms, On the Schrödinger equation given by geometric quantization, in *Differential Geometrical Methods in Mathematical Physics II*, vol. 676, ed. by K. Bleuler; H.R. Petry; A. Reetz. Lecture Notes in Mathematics (Springer, Berlin, 1978), pp. 351–356

165. L. Smolin, Towards a theory of spacetime structure at very short distances. Nucl. Phys. B **160**, 253–268 (1979)
166. J. Śniarycki, *Geometric Quantization and Quantum Mechanics* (Springer, Berlin, 1980)
167. J.-M. Souriau, Quantification géométrique. Commun. Math. Phys. **1**, 374–398 (1966)
168. J.-M. Souriau, *Structure des systèmes dynamiques* (Duno, Paris, 1970). (English as, Souriau, 1997)
169. J.-M. Souriau, in *Structure of Dynamical Systems. A Symplectic View of Physics*. (Springer, Berlin, 1997). (Translated from (Souriau, 1970) by C.-H. Cushman-de Vries)
170. N. Straumann, Zum Ursprung der Eichtheorien. Physikalische Blätter **43**, 414–421 (1987)
171. N. Straumann, Ursprünge der Eichtheorien. Scholz **2001**, 138–160 (2001)
172. M.R. Tanhayi, S. Dengiz, B. Tekin, Weyl-invariant higher curvature gravity theories in n dimensions. Phys. Rev. D **85**(064016), 9 (2012)
173. A. Trautman, Conservation laws in general relativity, in *Gravitation: An Introduction to Current Research*, ed. by L. Witten, chapter 5 (Wiley and Sons, New York, 1962) pp. 169–198
174. A. Trautman, Editorial note to J. Ehlers, F.A.E. Pirani and A. Schild, The geometry of free fall and light propagation. Gen. Relativ. Gravity **441**, 1581–1586 (2012)
175. R. Utiyama, On Weyl's gauge field. Prog. Theor. Phys. **50**, 2028–2090 (1973)
176. R. Utiyama, On Weyl's gauge field. Gen. Relativ. Gravit. **6**, 41–47 (1975)
177. R. Utiyama, On Weyl's gauge field II. Prog. Theor. Phys. **53**, 565–574 (1975)
178. S. Vagnozzi, Recovering a MOND-like acceleration law in mimetic gravity. Class. Quantum Gravity **34**, 185006 (2017). arXiv:1708.00603
179. V. Vizgin, *Unified Field Theories in the First Third of the 20th Century. Translated from the Russian by J. B. Barbour* (Birkhäuser, Basel, 1994)
180. R. Wald, *General Relativity* (University Press, Chicago, 1984)
181. J. Wess, On scale transformations. Nuovo Cimento **14**, 527–531 (1959)
182. J. Wess, The conformal invariance in quantum field theory. Nuovo Cimento **18**, 1086–1107 (1960)
183. C. Wetterich, Cosmologies with variable Newton's constant. Nucl. Phys. B **302**, 645–667 (1988)
184. C. Wetterich, Universe without expansion. Phys. Dark Universe **2**(4), 184–187 (2013). arXiv:1303.6878
185. H. Weyl, Zur Gravitationstheorie. Annalen der Physik **54**, 117–145 (1917). (Weyl, 1968, vol. 1, p. 670–698)
186. H. Weyl, Gravitation and electricity, in *The Dawning of Gauge Theory*, ed. by L. O'Raifeartaigh (University Press, Princeton, 1918/1997) pp. 23–37. (English translation of Weyl (1918a))
187. H. Weyl, Gravitation und Elektrizität." *Sitzungsberichte der Königlich Preußischen Akademie der Wissenschaften zu Berlin* (1918), pp. 465–480. (Einstein, 1987ff., 7 Doc. 54, p. 411–416), in (Einstein, 2006, p. 198–201)
188. H. Weyl, *Raum,-Zeit-Materie. Vorlesungen über allgemeine Relativitätstheorie* (Springer, Berlin, 1918). (Weitere Auflagen: ^{2}1919, ^{3}1919, ^{4}1921, ^{5}1923, ^{6}1970, ^{7}1988, ^{8}1993)
189. H. Weyl, Reine Infinitesimalgeometrie. Mathematische Zeitschrift **2**, 384–411 (1918). (Weyl, 1968, II, 1–28)
190. H. Weyl, Eine neue Erweiterung der Relativitätstheorie. Annalen der Physik **59**, 101–133 (1919). (Weyl, 1968, II, 55-87)
191. H. Weyl, Das Raumproblem. Jahresbericht DMV **30**, 92 (1921) (Abstract of talk, extended version in (Weyl, 1968, II, 212-228), erroneously identical with Weyl (1922a))
192. H. Weyl, Zur Infinitesimalgeometrie: Einordnung der projektiven und der konformen Auffassung. *Nachrichten Göttinger Gesellschaft der Wissenschaften* (1921), pp. 99–112. (Weyl, 1968, II, 195–207)

193. H. Weyl, Das Raumproblem. Jahresbericht DMV **31**, 205–221 (1922). (Weyl, 1968, II, 328–344)
194. H. Weyl, *Space–Time–Matter. Translated from the 4th German edition by H. Brose* (Methuen, London, 1922). (Reprint New York: Dover 1952)
195. H. Weyl, *Mathematische Analyse des Raumproblems. Vorlesungen gehalten in Barcelona und Madrid* (Springer, Berlin, 1923) (Nachdruck Darmstadt: Wissenschaftliche Buchgesellschaft 1963)
196. H. Weyl, *Raum-Zeit -Materie*, 5th edn. (Springer, Berlin, 1923)
197. H. Weyl, *Philosophie der Mathematik und Naturwissenschaft* (Oldenbourg, München, 1927) ((Baeumler, 1927, Bd. II A); separat. Weitere Auflagen ^{2}1949, ^{3}1966. English with comments and appendices Weyl (1949a), French Weyl (2017))
198. H. Weyl, *Gruppentheorie und Quantenmechanik* (Hirzel, Leipzig 1928) (^{2}1931, English translation R.P. Robertson, New York: Dutten 1931)
199. H. Weyl, Elektron und Gravitation. Zeitschrift für Physik **56**, 330–352 (1929). ((Weyl, 1968, III, 245–267) [85]. English in (O'Raifeartaigh, 1997, 121–144))
200. H. Weyl, Geometrie und Physik. Die Naturwissenschaften **19**, 49–58 (1931) (Rouse Ball Lecture Cambridge, May 1930. (Weyl, 1968, III, 336–345))
201. H. Weyl, *Philosophy of Mathematics and Natural Science* (University Press, Princeton, 1949) (^{2}1950, ^{3}2009)
202. H. Weyl, *Symmetrie. Ins Deutsche übersetzt von Lulu Bechtolsheim* (Birkhäuser/Springer, Basel/Berlin, 1955) (^{2}1981, 3. Auflage 2017: Ergänzt durch einen Text aus dem Nachlass 'Symmetry and congruence', und mit Kommentaren von D. Giulini, E. Scholz und K. Volkert)
203. H. Weyl, *Gesammelte Abhandlungen*, vol. 4, ed. by K. Chandrasekharan (Springer, Berlin, 1968)
204. H. Weyl, *Philosophie des mathématiques et des sciences de la nature. Traduit de l'anglais par Carlos Lobo* (MétisPresses, Genève, 2017)
205. H. Weyl, Similarity and congruence: a chapter in the epistemology of science. ETH Bibliothek, Hs 91a:31 (2017/1949). (Published in (Weyl, 1955, 3rd edition, 153–166))
206. N.M.J. Woodhouse, *Geometric Quantization* (Clarendon, Oxford, 1991)
207. F.-F. Yuan, Y.-C. Huang, A modified variational principle for gravity in Weyl geometry. Class. Quantum Gravity **30**(19), 195008 (2013). arXiv:1301.1316

On Empirical Equivalence and Duality

Sebastian De Haro

A science can never determine its subject-matter except up to an isomorphic representation.

—Hermann Weyl (1934)

Abstract I argue that, on a judicious reading of two existing criteria—one syntactic and the other semantic—dual theories can be taken to be empirically equivalent. The judicious reading is straightforward, but leads to the surprising conclusion that very different-looking theories can have equivalent empirical content. And thus it shows how a widespread scientific practice, of interpreting duals as empirically equivalent, can be understood by a thus-far unnoticed feature of existing accounts of empirical equivalence.

1 Introduction

The phenomenon of duality has been a central topic in theoretical physics for several decades. A duality is an isomorphism between (possibly very different-looking) theories, and so dualities are powerful tools for theory construction.[1] They are also

[1]Isomorphisms, in a sense different to the one used here, are also discussed in the literature on scientific representation. On the semantic view, models represent their target systems in virtue of their being isomorphic to them (other morphisms are sometimes also used). See for example [11, 13]. For a discussion in the context of ontic structural realism, see French and Ladyman [12, p. 108].

S. De Haro (✉)
Trinity College, Cambridge CB2 1TQ, UK
e-mail: sd696@cam.ac.uk

Department of History and Philosophy of Science,University of Cambridge, Cambridge, UK

Vossius Center for History of Humanities and Sciences, University of Amsterdam, Amsterdam, Netherlands

S. De Bianchi and C. Kiefer (eds.), *One Hundred Years of Gauge Theory*, Fundamental Theories of Physics 199, https://doi.org/10.1007/978-3-030-51197-5_3

very useful in describing empirical phenomena that would otherwise be intractable.

The main question that I will address in this paper is whether dual theories are, or can be taken to be, empirically equivalent—as physicists often claim that they are—on the existing criteria of empirical equivalence. Given that a duality in physics is, in essence, an isomorphism between theories, this question reminds us of Weyl's question of whether physics can determine its own subject-matter—to which he answered that this is only possible up to isomorphism.[2]

My motivation is two-fold. First—as I will argue in a moment—empirical equivalence is an aspect of dualities that has remained relatively under-developed in the recent philosophical discussions, compared to the rich extant analyses of theoretical equivalence (sometimes also called 'physical equivalence'); and yet it is often presupposed by those analyses. Second, physicists commonly say that dual theories are empirically equivalent, and standardly use this in their theoretical constructions: so that, without an account that explains how duals can be empirically equivalent, this important scientific practice would be unintelligible.[3]

About the first motivation: while the recent philosophical discussion of dualities has focussed on the analysis of theoretical equivalence, it has for the most part *assumed* that dual theories are empirically equivalent, without normally being explicit about a detailed criterion of empirical equivalence, and without arguing that, in general, two duals are empirically equivalent under that criterion.[4] For example, Fraser [17, p. 1] writes that 'dual theories are regarded as not merely empirically equivalent, but physically equivalent', and discusses an example of a formal equivalence between Euclidean field theory and relativistic quantum field theory, in which 'the type of equivalence that obtains... goes well beyond empirical equivalence, but falls short of physical equivalence' (p. 2). Also, Read [28, p. 213] and Le Bihan and Read [18, p. 2], cite van Fraassen's [32] notion of empirical equivalence, and take empirical equivalence as a necessary condition for two theories to be dual. And, while I agree with these papers, I also suggest that one needs to explain *how* dual theories can be empirically equivalent, according to the standard criteria of empirical equivalence—or to explain how these criteria need to be modified, in order to get empirical equivalence of dual theories. In other words, empirical equivalence is not automatic, and needs to be argued for.[5]

[2]Weyl [37, pp. 95–96]; see also p. 129.

[3]Empirical equivalence is also essential in discussions of empirical under-determination in connection with scientific realism. I take this up in [7]: and so, I will not pursue it further here.

[4]Arguments for empirical equivalence have of course been given in specific examples. For electric-magnetic duality, see Dieks et al. [10, pp. 209–210] and Weatherall [36, Sect. 3]; for T duality, see Huggett [16] and Butterfield [2, Sect. 6.3]; for gauge-gravity dualities, see De Haro [5, pp. 116–117] and Dawid [3, pp. 24–26].

[5]For a discussion of empirical equivalence in the context of category theory, see Weatherall [36]. Although Weatherall appears to stress the lack of empirical equivalence more than I do, we both argue that "duals are not automatically empirical equivalent", and we both conclude that it is nevertheless possible to view dual theories as empirically equivalent. The present paper, specifically, works out how two duals can be empirically equivalent on the syntactic and semantic conceptions. Another author worth mentioning is Dawid (2017: p. 28), who argues that empirical equivalence takes on

About the second motivation: it would be desirable to have an account of empirical equivalence that can be applied more generally to modern theories of physics. To this end, there are two broad traditional accounts of the notion of empirical equivalence in the philosophical literature that are useful: of which one is syntactic (see for example Quine [25, 26]) and the other semantic (see for example van Fraassen [32, 33]). Thus it would be of considerable interest to see whether or not they give the same verdicts for dualities. For example, some—though not all—recent accounts of theoretical equivalence, notably Weatherall [35, 36], give verdicts of empirical equivalence in specific examples, but without making explicit the connection with these accounts of empirical equivalence. And thus it seems important, for these accounts of theoretical equivalence and for modern theories of physics generally, to see whether the syntactic and semantic accounts of empirical equivalence involve any significant differences.

Thus my project starts with a comparison of the criteria of empirical equivalence that are given by the syntactic and semantic conceptions of theories. I will do this in Lutz's [20] spirit of reconciliation between the two: indeed my argument will be that, although the two criteria of empirical equivalence give *prima facie* different verdicts in various cases, on a deeper—and perhaps surprising—analysis they can be reconciled. The analysis in question is suggested by dualities.

At first sight, the syntactic criterion of empirical equivalence appears to be stronger than the semantic criterion because, while the former requires *identity* of observational sentences, the latter requires only *isomorphism* of models. Indeed, this distinction between identity and isomorphism is not innocuous: and it partly *motivates* van Fraassen's [32] adoption of the more liberal semantic, as against the stricter syntactic, criterion. And since dualities are isomorphisms rather than identities, dual theories are empirically *in*equivalent by a straightforward application of the syntactic criterion: while empirically equivalent by the semantic criterion. In Sect. 4, I will motivate the *judicious reading* of van Fraassen's [32, 33] criterion which leads to this verdict, as a surprising but straightforward new application of his proposal.

But, as I announced, the two views can be reconciled: and this entails a new application of the *syntactic* criterion of empirical equivalence, to a *different pair of theories*: namely, a theory T and its "reinterpreted dual", T', such that these two theories—one of which is reinterpreted using the duality—are empirically equivalent.

Thus the semantic and syntactic criteria can be made to give the same verdicts: through a judicious reading of the semantic criterion, and applying the syntactic criterion to a different pair of theories.

Underlying this possibility of using the duality to "change the interpretation" is the idea that formal theories admit various interpretations, and that the discovery of a duality can be a good reason to reinterpret a given theory—especially if such a reinterpretation is motivated by scientific practice. This, then, not only agrees with, but also *explains*, a fruitful and widespread scientific practice. Indeed, I believe that a conception of the criteria of empirical equivalence that did not allow for dual

a different role in string theory than it traditionally does in the scientific process: namely, as an indicator of important constraints on theory construction, that he argues is not fully visible in the classical limit. For a discussion of Dawid's views in connection to mine, see De Haro [7, Sect. 4.2].

theories to be empirically equivalent would render the importance of dualities, in current theoretical physics, philosophically unintelligible.

In Sect. 2, I review Quine's syntactic and van Fraassen's semantic conceptions of empirical equivalence. Then, in Sect. 3, I briefly introduce duality, and illustrate it in some examples. Section 4 brings the two topics together, and finds that dual theories can—surprisingly—be taken to be empirically equivalent. Section 5 summarises the paper's main thesis.

2 Empirical Equivalence

In this Section, I review the two criteria of empirical equivalence that I will compare in Sect. 4: Quine's [25, 26] syntactic, and van Fraassen's [32, 33] semantic criterion.

Quine [25, p. 319] says that two theories are *empirically equivalent* if they imply the same observational sentences—also called 'observational conditionals'—for all possible observations: present, past, future[6] or 'pegged to inaccessible place-times'.[7] He puts it thus:

> The empirical content of a theory formulation is summed up in the observation conditionals that the formulation implies (Quine [26, p. 323]).

Quine does not tell us, in either of his two papers on empirical under-determination [25, 26], what he means by 'observation'. But there are of course a few general things one can say. First, his views on observation were formed against the background logical positivist view that observation ultimately reduces to human sense data. However, his views would not have entailed stronger logical positivist doctrines such as Carnap's reduction of theoretical terms to observational terms. He summarises his empiricist commitments thus: 'Two cardinal tenets of empiricism remained unassailable... and so remain to this day. One is that whatever evidence there *is* for science *is* sensory evidence. The other... is that all inculcation of meanings of words must rest ultimately on sensory evidence' (Quine [27, p. 249]). His mention of 'inaccessible place-times' (Quine [26, p. 234]) suggest that, by observation, he had something broader in mind than mere perception by the human senses—but what, he does not say.

Another influential account of the meaning of 'empirical' is by van Fraassen [32, p. 64]:

> To present a theory is... to present certain parts of those models (the *empirical substructures*) as candidates for the direct representation of observable phenomena. The structures which can be described in experimental and measurement reports we call *appearances*: the theory is empirically adequate if it has some model such that all appearances are isomorphic to empirical substructures of that model.

[6]Quine [25, p. 179].

[7]Quine [26, p. 234]. One area where inaccessible place-times appear is of course cosmology. See Glymour [14], Malament [21], and Manchak [22], who discuss the under-determination of topology by local geometric structure.

Van Fraassen famously restricts the scope of 'observable phenomena' to observation by the unaided human senses. Accordingly, his mention of 'experimental and measurement reports' is restricted in the kinds of experiments and measurements that it affords. Thus I will set van Fraassen's notion of observability aside but keep his notion of empirical adequacy as a useful *semantic* alternative to Quine's *syntactic* construal of the empirical.[8]

Van Fraassen's [32, p. 67] notion of *empirical equivalence* also involves consideration of the theory's *models:* namely, the structures that satisfy the theorems of the theory; alternatively, the structures that comprise the theory, regarded as a collection of models. It is the *empirical substructures* of those models that 'are candidates for the direct representation of observable phenomena' (1980: p. 64). Thus two theories, T and T', are *empirically equivalent* if for every model M of T there is a model M' of T' such that all empirical substructures of M are isomorphic to empirical substructures of M', and vice-versa.

To summarise: two theories are **empirically equivalent** if they imply *the same observational sentences* (on Quine's syntactic conception) or if *the empirical substructures of their models are isomorphic to each other* (on van Fraassen's semantic conception).

3 Duality

This section introduces the notion of duality, and gives two examples that I will use in my analysis of empirical equivalence in Sect. 4.[9]

A duality is an isomorphism between two theory formulations. I will denote the duality map by $d : T \to T'$, where T and T' are two bare theories, i.e. before we give them a physical interpretation. The duality maps (isomorphically) the states of T to the states of T', and likewise for their quantities: while preserving the *values* of the quantities, the dynamics, and the symmetries that are stipulated for T and T'.[10]

An interpretation can be modelled using the idea of an *interpretation map*: namely, a partial function mapping bits of a bare theory (paradigmatically: the states and the quantities) into the theory's domain of phenomena. I will denote such a map by: $i : T \to D$.[11].

Consider, for example, orthodox quantum mechanics, as often presented in textbooks. Its interpretation map, i, maps the bare theory to its domain of phenomena as follows:

[8]For an account of observation that, in my opinion, resonates better with modern science, see [19].

[9]A detailed Schema for dualities is presented in [4, 8]. Further work on dualities is in [10, 16, 29, 30].

[10]For the relation between symmetry and duality, see De Haro [6].

[11]For a detailed exposition of interpretations in terms of maps, the conditions they satisfy, and how this can be used for referential semantics semantics, see De Haro [4, 5, 8].

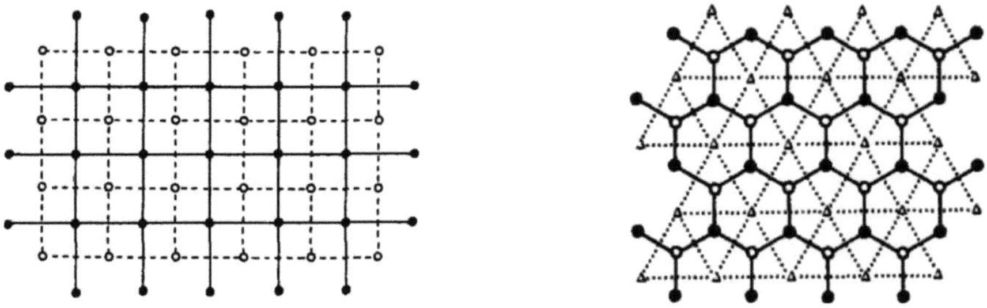

Fig. 1 A lattice (solid lines) and its dual lattice (broken lines). Left: square lattice. Right: honeycomb lattice. Baxter [1], with permission

$$
\begin{aligned}
i(x) &= \text{`the position, with value } x, \text{ of the particle upon measurement'} \\
i\left(|\psi(x)|^2\right) &= \text{`the probability density of finding the particle at position } x, \\
&\quad \text{upon measurement'}
\end{aligned}
$$

etc., where x is an eigenvalue of the position operator. Thus the eigenvalues of self-adjoint operators are interpreted as possible outcomes of individual measurements. The absolute value squared of wave-functions are interpreted as Born probabilities, etc.

Notice that not everything that is contained in the domain D (such as: outcomes of measurements, Born probabilities, physical states of a system, etc.) counts as 'empirical'. The values of quantities for a given state are typically observable, but the states themselves are not. This motivates the identification, for quantum mechanics, of van Fraassen's *empirical substructures* with the set of transition amplitudes of self-adjoint operators, and the expressions in which they appear.

I will illustrate *duality* in the example of Kramers-Wannier duality in statistical mechanics. Consider the Ising model on a square lattice (see Fig. 1).[12] Each lattice site, i.e. each vertex of the lattice, is occupied by a spin σ_i (where i labels the lattice sites) with two possible values: $+1$ or -1. Two nearest-neighbour spins σ_i and σ_j contribute a potential energy $-J\,\sigma_i\,\sigma_j$, where J is some fixed energy, and the total energy is the sum over all such pairs.

The *partition function* is defined as the sum of the exponentials of the energy (Boltzmann factors), summed over all the states. Thus for a square lattice of N sites, we sum the exponentials of the energies of the pairs:

$$
Z_N := \sum_\sigma e^{-E(\sigma)/k_B T} = \sum_\sigma \exp\left(K \sum_{(i,j)} \sigma_i\,\sigma_j \right) =: e^{-N\,F_N},
$$

[12] I follow the treatment in Baxter [1, Sect. 6.2] and Savit [31, pp. 456–457].

where we sum over pairs (i, j), i.e. all *edges* (nearest-neighbour pairs of spins). The sum over σ is the sum over all spin values, ± 1. Furthermore, I have defined $K := J/k_B T$, where k_B is Boltzmann's constant and T is the temperature. On the right is the definition of the free energy, F_N.

Kramers-Wannier duality involves a reinterpretation of the spin variables, as belonging to the 'dual lattice', which one gets by placing points at the centres of the faces of the original lattice, and connecting the points in adjacent faces (i.e. faces that share an edge): see Fig. 1. The result, for the square lattice, is that the original Ising model (with weight K) and the Ising model on the dual lattice (with weight K^*) are related to each other in the limit that the lattice is infinite, i.e. $N \to \infty$. First, one defines the free energy in the limit of infinite N:

$$\mathcal{F}(K) := \lim_{N\to\infty} F_N = -\lim_{N\to\infty} \frac{1}{N} \ln Z_N, \tag{1}$$

and then one derives the following transformation property of the free energy:

$$\begin{aligned} \mathcal{F}(K^*) &= \mathcal{F}(K) + \ln \sinh 2K \\ \tanh K^* &:= e^{-2K}. \end{aligned} \tag{2}$$

Thus the free energies of the two models are equal, up to $\ln \sinh 2K$. Notice that the $K \to \infty$ limit of one model maps, through Eq. (2), to the $K^* \to 0$ limit of the other, and vice-versa. Alternatively, since K and K^* depend on the temperatures T and T^*, respectively, *the Ising model at high temperature is dual to the Ising model at low temperature, and vice-versa.*

This is the basic idea of Kramers-Wannier duality: there is a well-defined one-to-one map between the two models (the high-temperature and the low-temperature Ising models, with their specific weights) such that the free energies of the two models and all other quantities map onto each other, through Eq. (2). Furthermore, this map generalises to other quantities of interest, such as the correlation functions, $\langle \sigma_i \, \sigma_l \rangle$, between spins σ_i and σ_l that may be arbitrarily far away from each other (see Savit [31, p. 457]). If this map generalises to all the quantities of interest in the theory, which depend on arbitrary spin states, then it is an isomorphism between the two theories—it is a *duality*.

My second example of a duality is gauge-gravity duality, which was used in the RHIC experiments in Brookhaven, NY. The duality successfully relates the four-dimensional quantum field theory (QCD, quantum chromodynamics) that describes the quark-gluon plasma, produced in high-energy collisions between lead atoms, to the properties of a five-dimensional black hole. The latter was employed to perform a calculation that, via an approximate duality, provided a result in QCD: namely, the shear-viscosity-to-entropy-density ratio of the plasma, which could not be obtained in the theory of QCD describing the plasma. Thus a five-dimensional black hole is used to describe, at least approximately, an entirely different (four-dimensional!) empirical situation.

4 Empirical Equivalence of Dual Theories

In this section, I will apply the syntactic and semantic criteria of empirical equivalence, from Sect. 2, to cases of duality. The application will involve a judicious reading of these criteria.

My 'judicious reading' is liberal but also straightforward, and it has two motivations. First, it captures an important scientific practice of using dualities to construct new theories that describe empirically equivalent situations, as in the RHIC experiment, discussed at the end of the previous Section. Second, the judicious reading is independently motivated by a historical-critical analysis of van Fraassen's semantic criterion of empirical equivalence. And I will claim that it casts light on some of the alleged differences between the semantic and syntactic views of theories.

On the syntactic criterion, two theories are empirically equivalent if they imply the same observational sentences. Since a duality is an isomorphism, $d : T \to T'$, between two theories whose domains of phenomena can be very different, dual theories imply different observational sentences, and are in general *not* empirically equivalent in this sense. On their ordinary interpretations,[13] QCD and the five-dimensional gravitational theory make different predictions, even though the numerical values of their quantities agree. And under Kramers-Wannier duality, a *high-temperature* lattice maps to a *low-temperature* lattice (alternatively, strong coupling K is mapped to weak coupling K^*, according to Eq. (2)). Thus their observational sentences differ.

On the semantic criterion, two theories are empirically equivalent *if the empirical substructures of their models are isomorphic to each other* (cf. the end of Sect. 2).

Notice that it is 'isomorphism' of the empirical substructures of the theory's models, rather than 'identity', that counts here—and I will argue in Sect. 4.1 that this literal reading of van Fraassen is indeed correct. Thus let me take van Fraassen's quote from Sect. 2 literally, and look for a suitable *isomorphism* between the empirical substructures of the models that we consider. Since we are dealing with dualities, the suggestion is that the duality map gives a natural—though surprising—new candidate for such an isomorphism: which I will dub the 'induced duality map'.

Thus consider the case in which the dual theories' domains of phenomena are distinct but isomorphic, according to an 'induced duality map', $\tilde{d} : D_1 \to D_2$. The commuting diagram in Fig. 2 will not always close (the condition for its closure is that $i_2 \circ d = \tilde{d} \circ i_1$; cf. De Haro [6, Sect. 2.2.3]. But if it does, then the two theories are clearly *empirically equivalent* on van Fraassen's conception taken literally.

For Kramers-Wannier duality, the map $\tilde{d}$ replaces a lattice by its dual, and translates the value of the temperature from one to the other. In the case of the RHIC experiments, the calculations are done in the five-dimensional theory, and then they are translated (using the induced duality map) into predictions about the four-dimensional plasma.

[13] Dieks et al. [10] and De Haro [4, 5] call this an 'external' interpretation, where the meaning of the terms is fixed from outside the theory. The reinterpretations discussed below are 'internal' interpretations, which take the duality as a starting point for establishing the meanings of the terms.

$$\begin{array}{ccc} T_1 & \overset{d}{\longleftrightarrow} & T_2 \\ \downarrow{\scriptstyle i_1} & & \downarrow{\scriptstyle i_2} \\ D_1 & \overset{\tilde{d}}{\longleftrightarrow} & D_2 \end{array}$$

Fig. 2 Empirical equivalence. There is an induced duality map, $\tilde{d}$, between the domains

Thus on this literal reading of van Fraassen, the dualities that we have discussed *do* in fact (and surprisingly!) relate empirically equivalent theories, by a reinterpretation of their domains of phenomena. What *prima facie* looks like a five-dimensional black hole can, through the induced duality map, be reinterpreted as a four-dimensional plasma.

The same holds of course if we consider Newtonian mechanics with different standards of rest (an example that van Fraassen himself considers). According to Newton and Clarke, different standards of rest give different empirical situations: but they are *isomorphic* situations, because any standard of rest can be mapped to any other by a Galilei transformation. And so, different standards of rest give empirically equivalent situations on van Fraassen's criterion (as in Leibniz).[14]

Thus van Fraassen's semantic criterion—because it involves isomorphism, rather than identity, of substructures—seems more liberal than the syntactic criterion. By Quine's criterion, the examples of dualities are not cases of under-determination; by van Fraassen's, they are. This prompts two questions, the first of which leads to the second:

(A) Is my interpretation of van Fraassen's notion of empirical equivalence correct? Does his notion allow for isomorphisms that involve dualities?
(B) Can the syntactic notion of empirical equivalence also allow the same latitude?

4.1 Van Fraassen on Empirical Equivalence

As to question (A): the evidence that my interpretation is straightforward is from van Fraassen's [32, 33]. First, in his famous example of the seven point geometry [32, p. 43] that motivates the semantic notion of a 'model', he writes that 'the seven-point structure can be *embedded* in a Euclidean space. We say that one structure can be embedded in another, if the first is isomorphic to a part (substructure) of the second'. In van Fraassen [33, pp. 219–220], he adds: 'This relation [of isomorphism] is important because it is also the exact relation a phenomenon bears to some model of a theory, if that theory is empirically adequate.' Why does van Fraassen use—even emphasis—'embedding' and 'isomorphism',[15] rather than just saying that 'the

[14] For a discussion of this example as a case of duality, see [6, 9].

[15] Van Fraassen [32, 33] is constant in his use of 'isomorphism' in connection with empirical equivalence.

seven-point structure... is *equal to*, or is *the same as*, a part of Euclidean space'? Like isomorphism, an embedding is a mathematical notion: an embedding essentially comes down to an injective map, such that when restricted to the image set of the map, the map thus obtained is an isomorphism.[16] His use of both 'isomorphism' and 'embedding', rather than the more straightforward 'equality', or 'sameness', implies that he *means* isomorphism and embedding: so that the literal reading is correct—and his second quote clarifies that isomorphism is important not only because this is how models relate to one another, but also with their empirical adequacy.

Second, van Fraassen [32, pp. 45–50] discusses the empirical adequacy of Newton's theory of mechanics and gravitation, and the *empirical equivalence* of the alternatives to this theory with the additional postulate that the centre of gravity of the solar system has constant absolute velocity v (he denotes these theories by 'TN(v)'):

> When Newton claims empirical adequacy for his theory, he is claiming that his theory has some model such that *all actual appearances are identifiable with (isomorphic to) motions* in that model' (p. 45, his italics).

The use of 'isomorphism' is here essential:

> all the theories TN(v) are empirically equivalent exactly *if all the motions in a model of* TN(v) *are isomorphic to motions in a model of* TN$(v + w)$ (p. 46, *his italics*).

There can be no doubt that it is the *isomorphism* of the motions between TN(v) and TN$(v + w)$ that makes the theories *empirically equivalent*, regardless of the value of w. And in so far as one member of the family is empirically adequate (cf. p. 47), the whole family is empirically adequate. (He also argues that Maxwell's theories with different absolute velocities are empirically equivalent.)

But if all *these* theories count as empirically equivalent, then we should count other theories that are related by a similar isomorphism, of the state-spaces and quantities (i.e. dualities), as empirically equivalent.

But there is more evidence in support of my reading. For the semantic view's use of 'isomorphism' in its conception of empirical adequacy underpins van Fraassen's [32, pp. 53–56] preference for the semantic over the syntactic approach. Van Fraassen claims that the syntactic approach judges the theories TN(v) with different values of the velocity v to be *empirically inequivalent*, i.e. the syntactic approach distinguishes between the empirical consequences of theories for which there should be no such distinction:

> [On the syntactic approach,] TN(0) is no longer empirically equivalent to the other theories TN(v)' (p. 55).

Even though van Fraassen here targets an untenable version of the syntactic view (namely, the old logical positivist view that theoretical sentences are reduced to observational sentences), and the syntactic view may well have its own resources to distinguish between versions of Newtonian theory: I agree that the syntactic notion

[16] For a typical definition in the context of topology, see for example Munkres [24, p.105].

of empirical equivalence is *prima facie* stronger, because of the difference between identity and isomorphism—and coincides with my own interpretation of Quine's notion, at the beginning of this section.[17] Thus van Fraassen seems to be making the same distinction that I made earlier in this section. (I disagree that this is a reason to regard either criterion as superior, as I will explain below.)

Finally, let me add a word about why duality (and induced duality) are the right kinds of isomorphism here, i.e. why I am allowed to generalise van Fraassen's relatively simple isomorphisms to dualities. To see this, one needs to specify what is the relevant 'isomorphism': for isomorphism is a notion that is well-defined only relative to a given structure. For scientific theories, the natural isomorphisms to consider are those that relate their structures. And since van Fraassen [32, 33] presents scientific theories in terms of states and quantities, with their corresponding interpretation maps (see also 1970: pp. 329, 334–335), the relevant isomorphisms between scientific theories should relate these structures. But this is precisely what the duality and induced duality maps do: they map the states and the quantities, the values of the quantities. In other words, once one has agreed that a theory is formulated as a structured set of states, quantities, and dynamics, a principled definition of an isomorphism of theories (and models) should preserve *those* structures—and duality is such a principled isomorphism. In this light, it is no coincidence that van Fraassen's and the Schema's verdicts about the empirical equivalence of Newtonian theory coincide. Thus the claim that dual theories can be taken to be empirically equivalent is a natural (although perhaps unexpected!) application of van Fraassen's semantic criterion.

I have discussed [32] in some detail because of its wide influence, and to show how his notion of empirical equivalence applies to dualities. I submit that the textual evidence leaves no doubt that my interpretation of dual theories as being empirically equivalent, through an induced duality map $\tilde{d}$, is a straightforward application not just of van Fraassen's general notion of empirical equivalence, but indeed of his main motivation for developing a semantic account: namely, that he thinks that such an account *must* make isomorphic theories empirically equivalent—as his discussion of Newton's and Maxwell's theories, and his criticisms of the syntactic conception, show.

[17] van Fraassen [34], in his reply to Halvorson [15], has recently emphasised the importance of interpretation for questions of equivalence: he always took interpretative notions to be properly accounted for in the semantic conceptions of theories. As such, this does not contradict any of the above, which explicitly takes interpretation into account. Notice, furthermore, that van Fraassen nowhere says that 'empirically equivalent interpretations should have the same truth values' or something of the sort. Indeed, he explicitly says that 'if we believe of a family of theories that are all empirically adequate, but each goes beyond the phenomena, then we are still free to believe that each is false' (1980: p. 47). Thus the empirical adequacy (and hence equivalence) of theories is independent of their truth. Furthermore, van Fraassen [34] is concerned with Halvorson's discussion of theoretical equivalence more than with empirical equivalence.

Fig. 3 Translation, t, between observational conditionals , O_1 and O_2, on the syntactic view

$$\begin{array}{ccc} T_1 & \overset{d}{\longleftrightarrow} & T_2 \\ \big\downarrow{\scriptstyle j_1} & & \big\downarrow{\scriptstyle j_2} \\ O_1 & \overset{t}{\longleftrightarrow} & O_2 \end{array}$$

4.2 Applying the Syntactic Notion of Empirical Equivalence

As to question (B) above: I claim that the syntactic criterion of empirical equivalence can similarly be applied in a more liberal way, like the semantic criterion. And it will not even be necessary to *modify* Quine's criterion of empirical equivalence from Sect. 2; all we need is to apply it to a *new pair of theories*, where a new theory is generated through a *non-standard interpretation* of the bare theory.

Let me motivate why we would want to do this, using the example of the quark-gluon plasma, discussed at the end of Sect. 3. There, the five-dimensional black hole was used to answer empirical questions about a plasma that could not be answered using QCD. The five-dimensional black hole, plus a suitable "translation", made this possible. And this scientific practice is justified by my analysis of empirical equivalence. If dual theories could not be taken to be empirically equivalent, dualities would be of little interest for the practicing physicist. It is precisely the fact that one of the dual theories can be used in a different context to make a prediction that is otherwise unattainable, that makes dualities scientifically valuable.

This kind of isomorphism can be introduced on the syntactic conception just as it is on the semantic conception, by a suitable translation between the sets of observational conditionals, O_1 and O_2, of the two theories (see Fig. 3). If one finds a translation map t that, added to the syntactic interpretation map $j_1 : T_1 \to O_2$, makes true the very same sentences as $j_2 : T_2 \to O_2$, so that $t \circ j_1 = j_2 \circ d$, then the two theories are rendered empirically equivalent. In other words, T_1, with the *non-standard interpretation* $t \circ j_1$, is empirically equivalent to T_2. Thus we do not need to change Quine's criterion: we rather change the theory, by giving it a new (and innovative!) interpretation.[18]

But also, these reinterpretations are *not forbidden*: for notice that nobody said that we had to stick to a single interpretation in order for a theory to make empirical predictions. Although bare theories may have intended interpretations, assigned to

[18]My proposed use of Quine's notion of empirical equivalence, allowing ourselves to generate a theory with a new interpretation from the consideration of the induced duality map, is in the same spirit in which Lutz claims that the syntactic and semantic conceptions do not differ so much after all. For one of the main problems of the syntactic view is that it seemed highly language-dependent (and van Fraassen aimed to solve this problem by moving to a language-independent semantic view). Now Lutz [20, pp. 324–326] follows Glymour and others in looking for more liberal criteria, such as definitional equivalence, that can make different theories 'mutually interpretable'. Thus my proposal is that one may consider this to hold not just for the notion of theoretical equivalence, but also for empirical equivalence.

them by history and convenience, nothing in the Quinean notion of empirical equivalence prevents us from generating new theories by reinterpreting the old ones, thus extending the predictive power of a bare theory. Indeed, the exercise is not motivated by philosophical speculation, but by scientific practice. The use of this flexibility in explaining heavy-ion collisions illustrates the scientific importance of the procedure.

Notice that, as on the semantic view, the interpretation thus obtained is non-trivial, and it need not always exist: we are using the theory T_1 to produce the observational sentences O_2 of theory T_2, by giving a non-standard interpretation to T_1.

Let me briefly discuss the obvious objection: *how can a hot lattice be empirically equivalent to a cold lattice?* A thermometer surely ought to tell the difference?

But this objection can only be made if we are allowed to do measurements on the system "from the outside", i.e. measurements that are not modelled by the theory. If the measurements are modelled by the theory, then empirical equivalence by definition translates the measurements, which are ordinary physical interactions (see the end of Sect. 2). Namely, the answer to the objection is the one given to the sceptic of length contraction in special relativity, who asks: *if two observers in relative motion make different predictions about the length of a body, surely a measuring stick can say who of them is right?* But of course the measuring sticks are *themselves* contracted under a Lorentz transformation, and so seemingly irreconcilable claims turn out to be empirically equivalent after all. For a discussion of this argument for dualities, see Dieks et al. [10, pp. 209–210].

Agreed: it is of course *not mandatory* to take dual theories to be empirically equivalent. This is clearest in the syntactic conception, where a theory and its dual can always be interpreted according to their ordinary interpretations, rather than non-standard ones. And the duals then simply *disagree* about empirical matters. But also on the semantic view this is possible, by adopting non-isomorphic interpretations. This parallels the distinction between external versus internal interpretations in Dieks et al. [10] and De Haro [4, 5]: namely, interpretations that "start from the duality" vs. interpretations that are "independent of the duality". For more on when it is more appropriate to adopt one sort of interpretation or the other, see De Haro [4].

Let me briefly compare my discussion of empirical equivalence to other authors who write about dualities. While the recent philosophical discussion of dualities has focussed on the analysis of theoretical equivalence,[19] it has for the most part *assumed* that dual theories are empirically equivalent without an explicit analysis of the conception of empirical equivalence used.[20] My analysis has aimed to make explicit that the verdict of empirical equivalence of dual theories, on the standard conceptions of empirical equivalence, is not automatic: for it required, in the syntactic view, adopting a non-standard interpretation of a bare theory. And in the semantic view, it involved an unexpected application of the isomorphism criterion, which is widened from the familiar cases to theories whose structures look very different. Furthermore, we have seen that such empirical equivalence is not mandatory.

[19]For example, Matsubara [23, p. 487] and Read [28, p. 213] take empirical equivalence as a necessary condition for two theories to be dual.

[20]Read and Møller-Nielsen [29, Sect. 3.1] cite van Fraassen's notion of empirical equivalence.

5 Conclusion

Both the semantic and the syntactic views allow for special applications that render dual theories empirically equivalent: although the way in which they do this is slightly different.

In van Fraassen's semantic conception, duals are rendered empirically equivalent because his notion of empirical equivalence has 'isomorphism' built into it—and duality is a natural notion of isomorphism between scientific theories. In this way, two dual bare theories are rendered empirically equivalent under their standard interpretations. This notion of empirical equivalence is of course faithless to meanings, as van Fraassen admits—but this allows him to conclude that different versions of Newtonian mechanics are empirically equivalent, even though they cannot all be true. And this of course articulates an important practice in science.

On the syntactic view, two dual bare theories are usually empirically *in*equivalent under their standard interpretations, but are rendered empirically equivalent if one of the theories is given a *non-standard interpretation*, i.e. using a translation map induced from the duality. This does not require changing Quine's criterion of empirical equivalence: it only requires endowing the same bare theory with a new interpretation. In this way, the same bare theory can be put to different uses, including in other domains of phenomena than it was originally developed for.

And *this* move is motivated by the use of dualities in the predictions for the RHIC experiments on quark-gluon plasma: and by many other such uses of dualities by physicists in solving problems in current theoretical physics.

Acknowledgements I thank Silvia De Bianchi and Claus Kiefer for their invitation to contribute to this volume, and for their comments on the paper. I also thank Jeremy Butterfield, Nick Huggett, and James Weatherall for conversations about the contents of this paper. I also thank John Norton for a discussion of duality and under-determination. This work was supported by the Tarner scholarship in Philosophy of Science and History of Ideas, held at Trinity College, Cambridge.

References

1. R.J. Baxter, *Solved Models in Statistical Mechanics* (Academic Press, 1982)
2. J. Butterfield, On dualities and equivalences between physical theories, Forthcoming in *Philosophy Beyond Spacetime*, ed. by C. Wüthrich, N. Huggett, B. Le Bihan (Oxford University Press, Oxford, 2018)
3. R. Dawid, String dualities and empirical equivalence. Stud. Hist. Philos. Modern Phys. **59**, 21–29 (2017)
4. S. De Haro, Spacetime and physical equivalence, Forthcoming in *Beyond Spacetime. The Foundations of Quantum Gravity*, ed. by N. Huggett, C. Wüthrich, K. Matsubara (Cambridge University Press, Cambridge, 2020), http://philsci-archive.pitt.edu/13243
5. S. De Haro, Dualities and emergent gravity: gauge/gravity duality. Stud. Hist. Philos. Modern Phys. **59**, 109–125 (2017)
6. S. De Haro, Theoretical equivalence and duality, in *Synthese, Topical Collection on Symmetries*, Vol. 2019, ed. by M. Frisch, R. Dardashti, G. Valente (2019), pp. 1–39

7. S. De Haro, The empirical under-determination argument against scientific realism for dual theories, Forthcoming in *Erkenntnis* (2020)
8. S. De Haro, J.N. Butterfield, A schema for duality, illustrated by bosonization, in *Foundations of Mathematics and Physics One Century after Hilbert*, ed. by J. Kouneiher (Springer, 2018)
9. S. De Haro, J.N. Butterfield, On symmetry and duality, *Synthese, Topical Collection on 'Symmetries and Asymmetries in Physics*, ed. by M. Frisch, R. Dardashti, G. Valente (2019), pp. 1–41
10. D. Dieks, J. Dongen, S. de van Haro, Emergence in holographic scenarios for gravity, Stud. Hist. Philos. Modern Phys. **52**(B), 203–216 (2015)
11. S. French, A model-theoretic account of representation (Or, I Don't Know Much About Art... But I Know It Involves Isomorphism). Philos. Sci. **70**(5), 1472–1483 (2003)
12. S. French, J. Ladyman, Reinflating the semantic approach. Int. Stud. Philos. Sci. **13**(2), 103–121 (1999)
13. R. Frigg, J. Nguyen, *Scientific Representation*. Stanford Encyclopedia of Philosophy (2016), https://plato.stanford.edu/entries/scientific-representation
14. C. Glymour, The epistemology of geometry. *Noûs*, 227-251 (1977)
15. H. Halvorson, What scientific theories could not be. Philos. Sci. **79**, 183–206 (2012)
16. N. Huggett, Target space $\neq$ space. Stud. Hist. Philos. Modern Phys. **59**, 81–88 (2017). https://doi.org/10.1016/j.shpsb.2015.08.007
17. D. Fraser, Formal and physical equivalence in two cases in contemporary quantum physics. Stud. Hist. Philos. Modern Phys. **59**, 30–43 (2017). https://doi.org/10.1016/j.shpsb.2015.07.005
18. B. Le Bihan, J. Read, Duality and ontology. Philos. Compass **13**(e12555), 1–15 (2018)
19. V.F. Lenzen, Procedures of empirical science, in *International Encyclopedia of Unified Science*, Vol. I, ed. by O. Neurath, N. Bohr, J. Dewey, B. Russell, R. Carnap, C.W. Morris (1955), pp. 280–339
20. S. Lutz, What was the syntax-Semantics debate in the philosophy of science about? Philos. Phenomenol. Res. XCV **2**, 319–352 (2017)
21. D. Malament, Observationally indistinguishable space-times: comments on Glymour's paper, in *Foundations of Space-Time Theories*, Vol. VIII, ed. by J. Earman, C. Glymour, J. Stachel. Minnesota Studies in the Philosophy of Science (University of Minnesota Press, Minneapolis, 1977)
22. J.B. Manchak, Can we know the global structure of spacetime? Stud. Hist. Philos. Modern Phys. **40**, 53–56 (2009)
23. K. Matsubara, Realism, underdetermination and string theory dualities. Synthese **190**, 471–489 (2013)
24. J.R. Munkres, *Topology*, 2nd edn. (Prentice-Hall, 2000)
25. W.V. Quine, On the reasons for indeterminacy of translation. J. Philos. **67**(6), 178–183 (1970)
26. W.V. Quine, On empirically equivalent systems of the world. Erkenntnis **9**(3), 313–328 (1975)
27. W.V. Quine, Epistemology naturalized. A reprint of Quine's, paper, in *Knowledge and Inquiry*, ed. by K. Brad Wray (Readings in Epistemology, 1971) (2002)
28. J. Read, The interpretation of string-theoretic dualities. Found. Phys. **46**(2), 209–235 (2016)
29. J. Read, T. Møller-Nielsen, Motivating dualities, Forthcoming in *Synthese* (2018)
30. D. Rickles, Dual theories: 'Same but different' or 'different but same'? Stud. Hist. Philos. Modern Phys. **59**, 62–67 (2017)
31. R. Savit, Duality in field theory and statistical systems. Rev. Modern Phys. **52**(2, I), 453–487 (1980)
32. B.C. van Fraassen, *The Scientific Image* (Clarendon Press, Oxford, 1980)
33. B.C. van Fraassen, *Laws and Symmetry* (Clarendon Press, Oxford, 1989)
34. B.C. van Fraassen, One or two gentle remarks about Hans Halvorson's critique of the semantic view. Philos. Sci. **81**, 276–283 (2014)
35. J.O. Weatherall, Are Newtonian gravitation and geometrized Newtonian gravitation theoretically equivalent? Erkenntnis **81**(5), 1073–1091 (2016)

36. J.O. Weatherall, Equivalence and duality in electromagnetism (2019), Preprint, arXiv:1906.09699 [physics.hist-ph]. PhilSci 16149
37. H. Weyl, *Mind and Nature*. Reprinted in: P. Pesic (Ed.), *Hermann Weyl, Mind and Nature*, Selected Writings on Philosophy, Mathematics, and Physics (2009) (Princeton University Press, Princeton and Oxford, 1934)

Gauge Is More Than Mathematical Redundancy

Carlo Rovelli

Abstract Physical systems may couple to other systems through variables that are *not* gauge invariant. When we split a gauge system into two subsystems, the gauge-invariant variables of the two subsystems have *less* information than the gauge-invariant variables of the original system; the missing information regards degrees of freedom that express relations between the subsystems. All this shows that gauge invariance is a formalization of the relational nature of physical degrees of freedom. The recent developments on boundary variables and boundary charges are clarified by this observation.

Gauge invariance is often described as convenient mathematical redundancy. This description is misleading, because it hides the reason for which the world appears to be well described by gauge theories, such as Yang-Mills theory and general relativity [1–7].

The best way to appreciate the physical meaning of gauge is to consider splitting a system into components. There are various ways in which this could be done.

As a first example, consider a field theory defined by various interacting fields. We can separate one field and consider it in isolation, neglecting its interaction with the other fields. Consider for instance QED. The corresponding classical field theory is defined by the action

$$S[A, \psi] = \int \frac{1}{4} F[A]^2 + \bar{\psi} \not{D}[A] \psi \tag{1}$$

written in terms of the Dirac field ψ and the Maxwell potential A. Here $F[A]$ and $D[A]$ are curvature and covariant derivative of A. If we neglect the Dirac field, the electromagnetic field alone is described by the first term of (1) alone, the theory is invariant under the gauge

$$A \to A + d\lambda \tag{2}$$

C. Rovelli (✉)
Aix Marseille Université CNRS CPT, UMR 7332, 13288 Marseille, France
e-mail: rovelli@cpt.univ-mrs.fr

Université de Toulon CNRS CPT, UMR 7332, 83957 La Garde, France

S. De Bianchi and C. Kiefer (eds.), *One Hundred Years of Gauge Theory*,
Fundamental Theories of Physics 199, https://doi.org/10.1007/978-3-030-51197-5_4

and (if spacetime is topologically trivial) the electric and magnetic fields capture all the gauge invariant degrees of freedom. But the way electromagnetism couples to the Dirac field in (1) is via the local interaction term of the Lagrangian density

$$I = \bar{\psi} \not{A} \psi \tag{3}$$

which depends on the *non* gauge-invariant variable A. Therefore A in this formalism is not mathematical redundancy: it is the variable of electromagnetism that can couple locally to the Dirac field.

As a second example, consider a non abelian Yang-Mills theory, in a truncation where we split spacetime into discrete cells. We can represent every cell c of spacetime as a node of a lattice and the theory can be formulated using Wilson's formalism of lattice gauge theory. The theory can be discretised in terms of group variables $U_{cc'}$ associated to every couple of adjacent cells c and c'. The discrete version of the Yang-Mills gauge is the transformation

$$U_{cc'} \to \Lambda_c U_{cc'} \Lambda_c^{-1} \tag{4}$$

where Λ_c are arbitrary group elements. The gauge invariant variables are the Wilson loops

$$U_{cc'} U_{c'c''} U_{c''c'''} \dots U_{c^n c}. \tag{5}$$

Now let us split a large spacetime region Σ formed by many cells into two subregions Σ_1 and Σ_2. The variables $U_{cc'}$ split into *three* groups: those where c and c' belong (i) both to the first subregion, (ii) both to the second subregion, and (iii) to distinct subregions. Let us call these variables respectively U^1, U^2 and U^{12}. The first and the second region are described by the variables U^1 and U^2 respectively, whose gauge invariant functions are the loop variables entirely in the first or in the second region. These variables are *not* sufficient to describe the degrees of freedom of the full region Σ, because the U^{12} variables are missing. To couple the two regions we need these extra variables; notice that the variables U^{12} are not gauge invariant. They express the change of internal frame from one region to the other. They are the handles through which the two subsystems couple. Once again, non-gauge-invariant variables are ways through which a system can couple to another system, a spacetime region to another spacetime region. (On this, see also [6, 8–14].)

As a third example, consider a non rotating black hole in pure general relativity.[1] As well known, there is a single solution up to gauges that describes such a black hole in the theory. This is given, in one coordinate system, by the Schwarzschild metric. We do not associate a position or a velocity (as we do for a particle in Minkowski space) to a black hole solution in general relativity, because a Schwarzschild black hole whose coordinate position is changing in time in a given coordinate system is a solution of the Einstein's equation that is gauge equivalent to a static one. Therefore the position and velocity of a black hole are pure gauge in this sense.

[1] I thank Laurent Freidel for pointing out this example.

But does this mean that astrophysical black holes have no position or velocity? Of course not: astrophysical black holes have positions and velocities. How come so, if there is only a single solution of Einstein equations describing a non rotating stationary black hole? The answer is of course that in the universe there are other physical components than those entering he Schwarzschild solution alone, and in the coupled system formed by all the various components of the universe the position and the velocity of a black hole can be appropriately defined with respect to other objects: say for instance with respect to its its host galaxy. There is nothing particularly deep in all that, but notice something: the position and the velocity of the black hole that were *gauge variables* when considering the hole alone, become physically meaningful variables when the black hole is coupled with something else. Again, we see that gauge variables are handles through which a system couples with something else.

In all these cases, we see what is behind gauge invariance: the fact that physical degrees of freedom are often not attached to specific entities or locations, but bridge between these. Gauge-invariant quantities can be defined by coupling gauge non-invariant quantities from different systems.

Notice that this implies that in some sense one can measure a non gauge-invariant quantity of a system as long as it is relative to another system. When the measuring apparatus couples to a physical system to measure something, the coupling may be gauge invariant under a common gauge transformation on the system and the apparatus, but *the measured quantity pertaining to the system can be a non gauge-invariant variable of the system alone.* For nice examples of applications of this perspective, see [15, 16].

In recent years there has been a flourishing of interest on boundary charges, and boundary degrees of freedom and in particular asymptotic charges and asymptotic degrees of freedom [17–21]. The discussion above clarifies the physics underlying this phenomenology. To see this, consider again the case of lattice gauge theory on a region Σ split into two subregions Σ_1 and Σ_2. Suppose we study the two regions Σ_1 and Σ_2 *separately.* If we consider only the gauge invariant variables of them, and we neglect the U^{12} variables, we are clearly missing something relevant for physics. Hence we must consider these variables as well. But they sit on the boundary and are non-gauge invariant variables when considering one of the two regions alone, with its boundary. Hence, if we do not want to miss relevant physics, we must not neglect boundary variables even if they are not gauge invariant.

What do they represent? They capture aspects of the way the region can interact with whatever is on its boundary. In particular, any measure on its boundary is an interaction with the region. The measuring apparatus may couple with the region at the boundary, the coupling may be gauge invariant under a common gauge transformation on the system and the apparatus, but *the measured quantity pertaining to the bounded region can be a non gauge-invariant variable of the field in that region.*

The same is true for asymptotic symmetries and asymptotic observables: non gauge-invariant asymptotic variables are physically measurable because in a physical measurement an observer can interact with these variables. Once again: the interaction is gauge invariant, but the observed variable is not. In the above example of the black hole, the black hole can be given a position and a velocity by shifting or boosting the solution at infinity.

In summary, considering gauge invariance as mathematical redundancy obscures its physical significance: many physical quantities express relations between distinct systems.

References

1. C. Rovelli, Why gauge? Found. Phys. **44**, 91–104 (2014)
2. N. Teh, Some remarks on Rovelli's 'Why Gauge?'. http://philsci-archive.pitt.edu/10050/
3. M.M. Amaral, Some remarks on relational nature of Gauge symmetry. http://philsci-archive.pitt.edu/10995/
4. H. Gomes, Holism as the significance of Gauge symmetries. arXiv:1910.05330
5. H. Gomes, A. Riello, The quasilocal degrees of freedom in gauge theories. arXiv:1906.00992
6. L. Freidel, F. Girelli, B. Shoshany, 2+1D loop quantum gravity on the edge. Phys. Rev. D **99**, 046003 (2019)
7. A. Vanrietvelde, P.A. Hoehn, F. Giacomini, Switching quantum reference frames in the N-body problem and the absence of global relational perspectives. arXiv:1809.05093
8. J.-L. Gervais, B. Sakita, S. Wadia, The surface term in Gauge theories. Phys. Lett. B **63**, 55–58 (1976)
9. A.P. Balachandran, S. Vaidya, Spontaneous Lorentz violation in Gauge theories. Eur. Phys. J. Plus **128**, 1–9 (2013)
10. W. Donnelly, L. Freidel, Local subsystems in gauge theory and gravity. J. High Energy Phys. **09**, 102 (2016)
11. M. Geiller, Lorentz-diffeomorphism edge modes in 3d gravity. J. High Energy Phys. **02**, 029 (2018)
12. A.J. Speranza, Local phase space and edge modes for diffeomorphism-invariant theories. J. High Energy Phys. **02**, 021 (2018)
13. L. Freidel, A. Perez, D. Pranzetti, Loop gravity string. Phys. Rev. D **95**, 106002 (2017)
14. L. Freidel, D. Pranzetti, Electromagnetic duality and central charge. Phys. Rev. D **98**, 116008 (2018)
15. E. Leader, C. Lorcé, The angular momentum controversy: what's it all about and does it matter? Phys. Rep. **541**, 163–248 (2014)
16. R. Alkofer, G. Eichmann, C.S. Fischer, M. Hopfer, M. Vujinovic, R. Williams, A. Windisch, On propagators and three-point functions in Landau gauge QCD and QCD-like theories, in *Proceedings of Science*, vol. 193 (2014) (PoS (QCD-TNT-III) 003)
17. A. Strominger, Asymptotic symmetries of Yang-Mills theory. J. High Energy Phys. **07**, 151 (2014)
18. D. Kapec, M. Pate, A. Strominger, New symmetries of QED. Adv. Theor. Math. Phys. **21**, 1769–1785 (2017)
19. D. Kapec, M. Perry, A.M. Raclariu, A. Strominger, Infrared divergences in QED, revisited. Phys. Rev. D **96**, 085002 (2017)
20. M. Campiglia, R. Eyheralde, Asymptotic U(1) charges at spatial infinity. J. High Energy Phys. **11**, 168 (2017)
21. A. Nande, M. Pate, A. Strominger, Soft factorization in QED from 2D Kac-Moody symmetry. J. High Energy Phys. **02**, 079 (2018)

Homotopic Identities and the Limits of the Interpretation of Gauge Symmetries as Descriptive Redundancy

Gabriel Catren

Abstract The aim of this article is to place gauge theories in a wider horizon of philosophical and mathematical developments concerning the notion of *equality*. The first landmark in this recontextualization of gauge theories is given by the debates around Leibniz's principle of the identity of indiscernibles. The second one is given by the far-reaching groupoid-theoretical reconceptualization of the notions of equality developed in the framework of category theory and homotopy type theory. We shall argue that gauge symmetries—far from being a mere mathematical "*surplus structure*" resulting from the existence of different coordinate descriptions of each unique physical state—encode homotopic structure. We shall reinterpret the condition of gauge invariance used to define the physical observables in terms of a compatibility condition between two fundamental ontological categories, namely the category of *identity* (related to the *symmetries* of the entity) and the category of *determinacy* (related to the invariant *properties* it can support).

The aim of this article is to recontextualize gauge theories in a broader retrospective and prospective historical horizon of philosophical and mathematical problems associated to the notion of mathematical *equality*. Looking backwards, the most important cardinal point of this recontextualization is provided by the philosophical debates around Leibniz's *principle of the identity of indiscernibles* (PII in what follows). We shall argue that clinging to this principle—in spite of the fact that it is not endowed with any form of rational necessity—has constituted a sort of "*epistemological obstacle*" (Bachelard [8]) both in philosophy of mathematics (notably with respect to the philosophical interpretation of equality statements of the form $a = b$) and in philosophy of physics (notably with respect to the interpretation of gauge symmetries). We shall argue that (what we shall call) the *nominalistic interpreta-*

G. Catren (✉)
Laboratoire SPHERE (UMR 7219), Université de Paris - CNRS, 5 rue Thomas Mann, 5205 Paris Cedex 13, France
e-mail: gabriel.catren@univ-paris-diderot.fr

S. De Bianchi and C. Kiefer (eds.), *One Hundred Years of Gauge Theory*, Fundamental Theories of Physics 199, https://doi.org/10.1007/978-3-030-51197-5_5

tions of equality statements $a = b$ (where a and b are understood as different names denoting the same entity) and the epistemic interpretations of gauge symmetries as mere coordinate transformations are symptoms of the same resistance to accept that the PII is not valid in physics and mathematics.

We shall then argue that releasing mathematics from the PII naturally leads to the groupoid-theoretical (or homotopic) reconceptualization of the notion of equality. We shall call "homotopic paradigm" this far-reaching refoundation of mathematics which was mainly developed in (higher) *category theory* [5, 7] and, more recently, in *homotopy type theory* [33] and *derived geometry* [2]. We could say that the homotopic paradigm relies on a *groupoid-theoretical* and *constructive* reinterpretation of equality statements of the form $a = b$ as *types of proofs* (i.e. as types whose tokens are the concrete identifications between a and b) rather than as mere *truth values* [3, 33]. In this framework, a and b are understood as *numerical different* and *qualitative identical* elements related by (possibly multiple) concrete identifications (like for instance *isomorphisms* if a and b are objects in a category or *paths* if they are points in a space). A family of objects related by relations of identifications (including at least trivial self-identifications) defines a *groupoid*, that is a category in which every morphism is reversible (i.e. an isomorphism). Differently from *equivalence relations* (where two objects are either equivalent or not), two objects a and b in a groupoid might be identical in many different ways. In type-theoretic terms, there might be many proofs-tokens of the propositional type $a = b$.

As we shall see, this groupoid-theoretical understanding of equalities allows us to establish a link between the mathematical notion of *group* and the ontological notion of *identity*. Briefly, a group G defines a groupoid $\mathcal{B}G$ with a unique object a where each $g \in G$ defines an automorphism $g : a \to a$, i.e. a particular proof of the identity principle $a = a$. We can thus claim that the mathematical notion of group conveys a *detrivialization* of the identity principle $a = a$, where each automorphism (or symmetry) of a can be understood as a different non-trivial and entity-dependent proof of this principle. Far from being a mere contentless tautology (as it was predominantly understood in the history of philosophy), we shall argue that the identity principle becomes an *entity- and context-dependent* a priori *synthetic proposition*.

The recontextualization of gauge theories in this larger horizon will allow us to acknowledge the limits of what we shall call the *invariantist stance*, that is the thesis according to which the only thing that really matters in a gauge theory (the only features of a gauge theory that carry physical information) are the *invariants* under gauge transformations. The invariantist stance has provided up to know the main conceptual key to understand gauge theories and relies on the distinction between objective or intrinsic features (the *invariants*) and coordinate- or gauge-dependent quantities (that we could call the *variants*). While the invariants are part of the *ontological* (or physically meaningful) content of the theory, the variants are understood as mere *epistemic* (or coordinate-dependent) features deprived of any physical counterpart. According to the invariantist stance, gauge theories have a *surplus* or *excess* structure that can be removed (at least in principle) by passing to a coordinate-independent or intrinsic description (like for instance the description associated to the *reduced phase space* in the theory of *constrained Hamiltonian systems* [31]). This purely epistemic understanding of gauge symmetries has its roots in Kretschmann's objection against the physical scope of the principle of general covariance in general relativity [48–50] and was endorsed by some of the main actors in the history of gauge theories.[1]

The main objection against a purely epistemic reading of gauge symmetries result from the fact that gauge symmetries do have non-trivial physical consequences, like for instance the relation between *local gauge symmetries* and *fundamental interactions* encoded in the so-called *gauge argument* [14, 44, 56, 65], the fact that the *topological solutions* (e.g. instantons) in Yang-Mills theories require gauge transformations to define non-trivial bundles, and the relation between gauge symmetries and *renormalization* [66]. Different avenues to provide a more intrinsic understanding of gauge symmetries have been explored by Rovelli [59] , Greaves and Wallace [28], and Schreiber (see for instance [62, p. 18] and [47, gauge group]). In particular, Schreiber has been one of the first to claim that gauge symmetries—far from being a mere "*surplus structure*"—encode homotopic structure carrying relevant physical information. Schreiber argues that the removal of the gauge-dependent structure is analogous to the operation by means of which the information about the homotopy type of a topological space is truncated.

If it is indeed true that "*homotopy theory is gauged mathematics*" [47] (or, we could rather say, that *gauge theories are homotopic physics*), then it is of a paramount importance to revisit gauge theories—both conceptually and formally—from the standpoint provided by the homotopic paradigm. In this article, we shall start to analyze from a philosophical perspective some of the fundamental ideas of the homotopic paradigm and revisit some aspects of the foundations of gauge theories under this new

[1] For instance, Dirac maintains that the presence of unphysical degrees of freedom in a gauge theory is a consequence of the fact that "*we are using a mathematical framework containing arbitrary features, for example, a coordinate system which we can choose in some arbitrary way* [...]" [16, p. 17]. The understanding of gauge symmetries as mere mathematical "surplus structure" continues to provide until today the main interpretational framework. For instance, Witten writes in a recent publication that "*gauge symmetries are redundancies in the mathematical description of a physical system rather than properties of the system itself*" [70]. In their seminal monography, Henneaux and Teitelboim write that the gauge structure merely "*make*[s] *the description more transparent*", being the gauge symmetries (by means of which one can "*extract the relevant physical content*") the price that one must pay for this transparency [31, p. xxiii]. The epistemic interpretations of gauge theories are discussed in Refs. [11, 24, 25, 29, 30, 37, 42, 57] among others.

light. In particular, we shall criticize the invariantist stance by providing some clues for an ontological interpretations of both the invariants and the gauge-dependent quantities.

1 Leibniz's Principle of Identity of Indiscernibles

We shall claim that the resistance to accept that the theoretical gauge symmetries might have a physical/ontological content is a particular instance of a more general "*epistemological obstacle*" in the history of philosophy (to use Bachelard's notion [8]) associated to the debates around Leibniz's *principle of identity of indiscernibles*. Leibniz's principle provides a negative answer to the question *can two numerically different entities be qualitatively identical?*[2] The very formulation of this question relies on the distinction between two forms of difference, namely the *numerical difference* given by the fact that two things are *two* things and the *qualitative difference* given by the fact that two things might have different properties or qualities. Leibniz's principle states that differences "*solo numero*"—that is, entities that differ numerically but not qualitatively—are not possible (at least in metaphysics), i.e. that every *numerical difference* has to be grounded on a *qualitative difference*. Briefly, the PII states that perfect copies or clones of a given entity are forbidden.

Accepting the PII might seem to be an appealing option for several reasons: it allows to eliminate superfluous meta-physical entities with a parsimony worthy of Occam (as in Leibniz arguments against Newton's absolute space [1, p. 26]), it allows us to endorse a clear ontology of things *qua* bundles of properties without an underlying "substantial" support (like in Russell [60, Chap. V] and Ayer [6]), and it permits to define the notion of identity (rather than considering it as a primitive notion) along the lines of Hilbert and Bernays' *Grundlagen der Mathematik* [32] (see also [15], [61, p. 291]) and [53, pp. 61–64]). Now, in spite of its strong appeal, the PII is not endowed with any form of *a priori* logical necessity. The fact that Leibniz justified the supposed validity of the PII on different basis—e.g. as an *evident axiom*, as an *empirical fact*, as a consequence of the *principle of sufficient reason*, and as a consequence of his *doctrine of complete notions*—can be seen as a symptom of the fact that none of these justifications by itself was entirely satisfactory. Moreover, Leibniz himself maintained that the PII is a *metaphysical principle* only valid for substances, but not for mathematical entities. Whereas substances are defined—according to Leibniz—by complete notions, mathematical entities are incomplete notions that result from operations of abstraction. The interesting point is that Leibniz himself claims that incomplete and abstract notions (which are obtained by bracketing

[2]For an analysis of the different formulations and arguments proposed by Leibniz see Ref. [58] and references therein.

certain features of the entities at stake) can have instantiations which are "*perfectly similar*"). In Leibniz's own words,

> "[...] *in nature, there cannot be two individual things that differ in number alone*. For it certainly must be possible to explain why they are different, and that explanation must derive from some difference they contain. [...] never do we find two eggs or two leaves or two blades of grass in a garden that are perfectly similar. And thus, perfect similarity is found only in incomplete and abstract notions, where things are considered [*in rations veniunt*] only in a certain respect, but not in every way, as, for example, when we consider shapes alone, and neglect the matter that has shape. And so it is justifiable to consider two similar triangles in geometry, even though two perfectly similar material triangles are nowhere found" [40, p. 32]

2 On the Epistemic Interpretations of Equality Statements

In spite of the fact that the PII seems to be deprived of any *a priori* rational necessity—and that even Leibniz refused its validity in mathematics—, there seems to be a strong resistance in philosophy of mathematics with respect to the possibility of rejecting the PII. In particular, the problem of interpreting equality statements of the form $a = b$ provides an important example of the resistance to accept indiscernible entities in philosophy of mathematics.[3] Propositions of this general form are in a sense paradoxical, since they affirm an equality between terms a and b that are ostensibly different. One possible strategy to dissolve this paradox is to claim that a and b denote different *cognitive processes* referring to the same entity or different *names* of the same entity. According to these interpretations, equality statements do not engage numerically different things, which means that the validity of the PII is preserved. Let us consider some examples of this kind of interpretations.

Kant for instance reduced the paradoxical scope of equality statements by means of a cognitive interpretation. According to Kant, $2 + 2$ and 2×2 are *objectively* the same concept but *subjectively* different.[4] Later on, Frege and Quine would interpret

---

[3] Another interesting example of this resistance to reject the PII can be found in philosophy of mathematics in the debates about Keränen's "*identity problem for realist structuralism*" [35, 39, 41, 64].

[4] In a letter to Schultz, Kant presents this interpretation of equalities in the following terms:

> "I can form a concept of one and the same magnitude by means of several different kinds of composition and separation [...]. Objectively, the concept I form is indeed identical (as in every equation). But subjectively, depending on the type of composition that I think, in order to arrive at that concept, the concepts are very different. [...] Thus I can arrive at a single determination of a magnitude = 8 by means of $3 + 5$, or $12 - 4$, or 2×4, or 2^3, namely 8. But my thought '$3 + 5$' did not include the thought '2×4'. Just as little did it include the concept '8,' which is equal in value to both of these." [34, p. 283].

equality statements by substituting the cognitive dimension of Kant's interpretation with a linguistic one. For both philosophers, a statement of the form $a = b$ means that a and b are just different names that refer to the same entity.[5] In this way, equality statements—far from using names to speak about the entities denoted by these names—convey meta-linguistic claims about the names themselves. The notion of equality is understood as a mere artifact by means of which one can control the excess (we could say the *linguistic surplus structure*) of names or notations over things or subject-matters, thereby being deprived of any ontological scope. It is worth noting that we find similar interpretations of equality statements in philosophers of physics like French ("*The statement that a is identical to b, written symbolically* $a = b$*, means informally that there are not in reality two distinct items at all, but only one which may be referred to indifferently as either a or b.*" [23, p. 142]) and Saunders ("[...] *the identity sign, as it figures in extant physical theories, signifies only the equality or identity of mathematical expressions, not of physical objects.*" [61, p. 290]).

All in all, none of these interpretations (i.e. Kant's cognitive interpretation or Frege's and Quine's nominalistic interpretations) consider expressions of the form $a = b$ at face value, that is as stating that two numerically different things are qualitatively identical. In turn, this kind of epistemic interpretations of equality statements completely trivializes the identity principle $a = a$. If we understand equality statements as claims according to which two different names refer to the same entity, they become trivial when the two names are just the same name. The identity principle

[5]For instance, Frege writes in his *Begriffsschrift*:

> "Identity of content differs from conditionality and negation by relating to names, not to contents. Although symbols are usually only representatives of their contents—so that each combination [of symbols usually] expresses only a relation between their contents—they at once appear *in propria persona* as soon as they are combined by the symbol of identity of content, for this signifies the circumstance that the two names have the same content." [21, Sect. 8, p.124]

In turn, Quine writes in his *Methods of Logic*:

> "Identity is such a simple and fundamental idea that it is hard to explain otherwise than through mere synonyms. To say that x and y are identical is to say that they are the same thing. [...] For truth of a statement of identity it is necessary only that '=' appear between names of the same object; the names may, and in useful cases will, themselves be different. For it is not the names that are affirmed to be identical, it is the things named. [...] Still, since the useful statements of identity are those in which the named objects are the same and the names are different, it is only because of a peculiarity of language that the notion of identity is needed." [55, Sect. 35, p.208]

$a = a$ is then understood as a mere tautology. These two thesis about equality and identity statements have been summarized by Wittgenstein in the following terms:

> "Roughly speaking, to say of two things that they are identical is nonsense, and to say of one thing that it is identical with itself is to say nothing at all" [71, 5.5303, p. 63].

3 On Equivalence Relations as Contextual Equalities

An important step toward the clarification of equality statements was given by the interpretation of such statements in terms of *equivalences* (endorsed in particular by Whitehead in *A Treatise on Universal Algebra* [68] and *The Principle of Relativity* [69]). This interpretation states

1. that equality statements presuppose that a and b are different entities, both numerically and qualitatively,
2. and that the equality denotes a partial similarity that is only valid with respect to a particular regime of abstraction defined by an equivalence relation.

Since by definition qualitatively different entities are not identical in the strict sense of the term, equality statements merely assert that these entities can be considered equivalent *only in a certain respect*. In this conceptual framework, equality statements—far from being mere epistemic claims about the corresponding subjective cognitive processes or about theirs names—do convey a claim about the entities at stake, namely that they share some *properties* or that they instantiate some general concept.[6] The sign $=$ just denotes the fact that—under the condition of a particular process of abstraction by means of which one decides to leave aside certain differences between the two things at stake—these things share some common properties or belong to the extension of some concept. In this sense, the understanding of equalities *qua* equivalence relations is a first important step toward an ontological understanding of equalities. For instance, the equality statement $2 + 2 = 3 + 1$ means that the formulas $2 + 2$ and $3 + 1$ are equivalent when we make abstraction from their computational content in order to uniquely retain their numerical value.

[6] For instance, Whitehead writes:

> "So $b = b'$ can be interpreted as symbolizing the fact that the two individual things b and b' are two individual cases of the same general conception B. For instance if b stand for $2 + 3$ and b' for $3 + 2$, both b and b' are individual instances of the general conception of a group of five things. The sign $=$ as used in a calculus must be discriminated from the logical copula 'is'. Two things b and b' are connected in a calculus by the sign $=$, so that $b = b'$ when both b and b' possess the attribute B. But we may not translate this into the standard logical form, b is b'." [68, p. 6].

In other terms, $2 + 2$ and $3 + 1$ are *equi-valent* (they have the same numerical value) even if they are different *qua* formulas, that is even if they encode a different computational content. Strictly speaking, the equality $2 + 2 = 3 + 1$ should be written as

$$val(2 + 2) = val(3 + 1),$$

where $val : F \to \mathbb{N}$ is the valuation function that computes the number associated to each formula. The strict equality in $\mathbb{N}$ can be used to define an *equivalence relation* between formulas in F given by $f \sim f'$ if and only if $val(f) = val(f')$. Following Frege's *On Sense and Reference* [22], we could say that the elements in a fiber $val^{-1}(n)$ provide different *senses* or *modes of presentation* of the same *referent*, that is (in this particular case) different *computational modes of presentation* of the natural number n (i.e. f is a mode of presentation of n if $val(f) = n$).[7]

In a first approximation, this analysis seems to show that equivalence relations do not provide an understanding of equalities released from a commitment to the PII. Indeed, equivalence relations relate in general objects that are both numerically and qualitatively different, thereby preserving the validity of the PII. Now, this last statement presupposes a sort of *ground ontological domain* in the framework of which all entities are qualitatively different (and hence the PII is valid) and in which the equality is strict (i.e. each object is uniquely equal to itself). By forgetting certain differences between the entities in this supposed ground ontological domain, the notion of equivalence relation would allow us to put into focus certain partial similarities between them.

Now, rather than presupposing a sort of ground ontological domain endowed with a strict equality and preserving the PII, we can use equivalence relation to

---

[7] A similar Fregean analysis was proposed by Girard in the following terms:

> "This equality [$27 \times 37 = 999$] makes sense in the mainstream of mathematics by saying that the two sides *denote* the same integer [...] This is the denotational aspect, which is undoubtedly correct, but it misses the essential point: There is a finite *computation* process which shows that the denotations are equal. It is an abuse [...] to say that 27×37 *equals* 999, since if the two things we have were *the same* then we would never feel the need to state their equality. Concretely we ask a *question*, 27×37, and get an *answer*, 999. The two expressions have different *senses* and we must *do* something (make a proof or a calculation, or at least look in an encyclopedia) to show that these two *senses* have the same *denotation*." [26, pp. 1–2]

By explicitly referring to Frege, Martin-Löf writes:

> "As for Frege, elements a, b may have different meanings, or be different methods, but have the same value. For instance, we certainly have $2^2 = 2 + 2 \in \mathbb{N}$, but not $2^2 \equiv 2 + 2$." [45, p. 60]

define *ontological contexts* equipped with their own notion of equality. Equivalence relations can be used to define contextual equalities since they are defined by the main axioms that capture the notion of equality, namely reflexivity, symmetry, and transitivity. When we introduce an equivalence relation $\sim$ on a set X, the elements of X are different as elements of X. However, they might be considered qualitatively identical as elements of the so-called *setoid* $(X, \sim)$ (i.e. a set endowed with an equivalence relation). The equivalence relation $\sim$ on a set X defines a *contextual notion of equality* that can be denoted $=_X$. In the framework of Bishop's form of constructive mathematics [10], setoids—rather than being understood as sets endowed with further structure—provide the correct constructive definition of what a set is.[8] This definition presupposes that two numerically different elements of a set X might be considered equal (if they are related by the equivalence relation) with respect to the contextual equality $=_X$. Rather than defining a sort of weak equality encoding certain partial similarities between entities in a ground ontological domain, an equivalence relation is here understood as a proper equality in the corresponding contextual ontology. The only difference between the equalities defined via equivalence relations and strict equalities is that the former might relate numerically different entities (in fact, strict equalities are particular cases of equivalence relations in which every entity is uniquely equivalent to itself). Hence, this redefinition of equalities as equivalence relations does not longer cling to the validity of the PII.

4 From Equivalence Relations to Groupoids

We have seen in the previous section that equivalence relations can be used to define contextual notions of equality for which the PII is no longer valid, that is notions of equality that might relate numerically different objects. However, the notion of contextual equality given by equivalence relations is not entirely satisfactory given what we could characterize as its *ideal character* (where *ideal* should be understood as opposed to *constructive*). This ideal character results from the fact that equivalence relations encode the contextual equalities between objects *without caring about the proof of such propositions*. Now, according to Bishop's constructive characterization of sets,"[t]*o define a set we prescribe* [...] *what we must do to show that two elements*

[8]In Bishop's own terms, the notion of *setoid* derives from the fact that

> "A set is not an entity which has an ideal existence: a set exists only when it has been defined. To define a set we prescribe, at least implicitly, what we (the constructing intelligence) must do in order to construct an element of the set, and what we must do to show that two elements of the set are equal" [10, p. 2]

of the set are equal" [10, p. 2]. But how can we show that two entities are equal? In order to address this question, let us go back to the example of geometric figures proposed by Leibniz in the text quoted before. What does it mean to say that two figures a and b in the Euclidean space are equal? Remarkably enough, the answer can already be found in Euclid's *Elements*. The Notion 4 of the *Elements* states: "*Things which coincide with one another are equal to one another*" ([18, p. 155]). This "axiom of congruence" encodes the idea according to which two geometric figures are equal if they can be superposed (see the commentary of this notion in [18, pp. 224–231] and [27]). In other terms, the equality between two figures can be proved by superposing them by means of a concrete motion in the space.[9]

This definition of geometric equality is *constructive* in the sense that it specifies what has to be done in order to prove an equality statement. Whereas in (what we could call) *ideal mathematics* propositions are *truth values* or *mere propositions* (i.e. statements that are either true or false), in constructive mathematics a proposition is understood as the *type of its proofs* [3, 33]. This means that a proposition is understood as the abstract statement of which each of its proofs provides a concrete realization or instantiation. We could say in the jargon of homotopy type theory that each proof provides a particular *witness* of the fact that the proposition is true [33, 63]. According to this constructive stance, the proofs of a proposition are not an epistemic "surplus structure" that becomes irrelevant once we know that the proposition is true (like Wittgenstein's ladder which can be thrown away after one has climbed up it [71, 6.54, p. 89]), but rather the very intrinsic (homotopic in homotopy type theory [33]) structure of the proposition *qua* type of its proofs. The important point is that the (homotopic) structure of inequivalent proofs is truncated when a proposition is ideally understood as a mere truth value. We could say in a Hegelian manner that in constructive mathematics the result (the truth value) cannot be severed from the process (the proofs). Cutting the result from the process removes essential information and might lead to pathologies.[10] Hence, the constructive redefinition of propositions cannot be understood as a mere remainder of the fact that mathematics is a concrete practice in which there are no truths without proofs. More radically,

---

[9]More recently, this definition of geometric equality was commented by E. Cartan in the following terms:

> "If indeed one tries to clarify the notion of equality, which is introduced right at the beginning of Geometry, one is led to say that two figures are equal when one can go from one to the other by a specific geometric operation, called a motion." (quoted in [43, p. 19]).

[10]Mathematical examples of such pathologies are the "bad quotients" associated to group actions with non-trivial isotropies or the problems encountered in algebraic geometry to define *moduli spaces* classifying geometric objects with automorphisms. These pathologies are treated in the theory of stacks (where smooth quotient spaces and smooth classifying spaces can be defined) by keeping track of the identifications and the automorphisms respectively [46].

the constructive stance takes into consideration an intrinsic structure of the objects at stake thanks to which it becomes possible to regularize what we could call the *pathologies of abstraction*.

In order to understand the homotopic nature of the structure of proofs of a proposition, let's consider again a geometric example. As we said before, every "motion" that transports one figure a on the top of another figure b can be understood as a concrete proof or witness of the fact that the proposition $a = b$ is true. In particular, two points a and b in a space M are qualitatively equal (albeit numerically different) if there is a path $\gamma : a \to b$ between them (i.e. a map $\gamma : [0, 1] \to M$ with $\gamma(0) = a$ and $\gamma(1) = b$). In this way, each path γ between two points a and b in a space M provides a proof of the proposition $a = b$. The existence of such a path means that a and b are in the same connected component of the space M. If a and b were in different connected components of M, then the property of belonging to a certain connected component would define an individuating predicate that would distinguish the points (i.e. two points in different connected components are not only numerically different but also qualitatively different).

Differently from equivalence relations in which any two terms are either equal or not, concrete identifications between two terms might be multiple. For instance, two points in a space might be connected by multiple paths. We can in turn consider the set of identifications-paths between two points a and b and endow it with a contextual criterium of identity. In the geometric framework, this can be done by applying once again Euclid's axiom of congruence to the geometric figures given by the paths themselves: two paths with the same endpoints are equal if there is a "motion" (a continuous deformation) by means of which one of them can be transported to the other one. This kind of motion between paths is formalized by the notion of *homotopy*.[11] Homotopies $h : \gamma_1 \Rightarrow \gamma_2$ can be understood as 2-identifications between 1-identifications (paths) between points and can be represented by diagrams of the form:

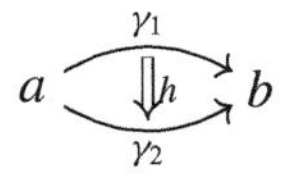

Paths are not homotopic when the space has holes that obstruct the existence of an homotopy between them. Such non-homotopic paths provide inequivalent proofs of the corresponding equality statement $a = b$. This means that two points a and b can be equal (i.e. qualitatively identical) in different inequivalent manners.

[11] Two paths $\gamma_1, \gamma_2 : a \to b$ are homotopic if there exists a smooth map (called *homotopy*)

$$h(s, t) : [0, 1] \times [0, 1] \to M$$

such that $h(0, t) = \gamma_1$ and $h(1, t) = \gamma_2$ (while the parameter s spans all the paths that interpolate between γ_1 and γ_2, the parameter t runs along each path).

In the general framework provided by category theory, two objects a and b in a category are qualitatively identical if they are *isomorphic*, that is if they can be concretely identified by means of a reversible morphism $f : a \to b$ (where the reversibility condition means that there exists another morphism $f^{-1} : b \to a$ such that $f^{-1} \circ f = id_a$ and $f \circ f^{-1} = id_b$). It is worth noting that the term *isomorphism* keeps track of the geometric origin of this notion of equality: two things are isomorphic if—by having the same "form"—they can be superposed by means of a "motion" given by the reversible morphism. Once again, two numerically different objects in a category can be isomorphic in different ways, where each isomorphism provides a concrete proof of their qualitative identity. It is also worth noting that this definition of identity is clearly contextual since the notion of isomorphism depends on the corresponding category. In this way, we arrive to the categorical notion of *groupoid*, that is a category given by a multiplicity of objects related by means of (possibly multiple) identifications.

The statement according to which objects in a category "*can be isomorphic but not* [strictly or numerically] *equal*" [7, p. 7] can be understood as a negation of the PII. By paraphrasing Baez and Dolan, we could say that the basic philosophy of the homotopic paradigm is simple: *never mistake qualitative indiscernibility (isomorphism) for numerical identity (strict equality)* [7, p. 46]. The principle according to which "*isomorphic objects are* [qualitatively] *identical*" was called by Awodey the *Principle of Structuralism* [4].

5 Revisiting the Identity Principle

Maybe the most surprising philosophical consequence of the constructive understanding of propositions is the impact that it has on our understanding of the *identity principle*

$$a = a. \tag{1}$$

Let's consider first what we could call the classical (or pre-homotopic) understanding of this principle. First of all, we could say that the identity principle is valid *a priori*, which means that the proposition (1) is always true independently of the entity a that we are considering. In this sense, we can claim with Quine that there is "*no entity without identity*" [54, p. 23], that is that there is no entity a for which the proposition (1) is not true. Since the notion of identity applies to any entity whatsoever we can maintain that identity is an ontological category, that is a category that informs the very definition of what it means to be an entity. Second, the identity principle is an analytical principle (in the Kantian sense of the term) in the sense that it carries no synthetic information about a. The knowledge that a

is identical with itself does not teach us anything about a. In Wittgenstein terms, "[...] *to say of one thing that it is identical with itself is to say nothing at all.*" [71, Sect. 5.5303, p. 63]. Since the classical identity principle is valid for any a (*aprioricity*) and it carries no information about a (*analyticity*), this principle does not allow us to make the difference between entities. This point has been stressed by Quine in the following terms "*identity theory knows no preference*" in the sense that "*it treats of all objects impartially*" [53, p. 62].[12] Third, the identity principle is *trivially* valid, which means that the truth of the proposition (1) does not depend on any demonstration.[13] Fourth, the classical notion of identity is characterized by its *unicity*, which means that any entity a is identical to itself in a unique (trivial) manner. Finally, the classical notion of identity is *context-independent*, which means that the proposition $a = a$ is not indexed by (it does not depend on) any reference to an ontological/linguistic/conceptual context or domain. We can then summarize the classical understanding of the identity principle by saying that this principle is a *context-independent* proposition which is *apriorically*, *analytically*, and *trivially* true and which states that any entity a is identical to itself in a unique trivial manner.

Let's consider now the impact of the constructive understanding of propositions on this classical notion of identity. The constructive prescription according to which an equality $a = b$ has to be understood as the type of the concrete identifications between a and b implies in particular that the *identity principle* $a = a$ might admit inequivalent proofs depending on a. Let's consider for instance a point a in a space M. According to what we said before, any path in M going from a to a is a proof of the statement $a = a$. The homotopy classes of loops based at a defines the *fundamental group* $\pi_1(M, a)$ of M at a. Since there is always the trivial loop $id_a : a \to a$ given by the constant path $id_a(t) = a$ for all $t \in [0, 1]$—whose equivalence class plays the role of the neutral element in the fundamental group $\pi_1(M, a)$—, every point is necessarily identical to itself. In this sense, the identity principle $a = a$ is (also in this new constructive context) an *a priori* proposition, i.e. a proposition that is necessarily true for every a. However, the identity principle might no longer be trivial since the proposition $a = a$ might admit—depending on the topology of M—

[12]This point was also stressed by Frege in the following terms: "*Identity is a relation given to us in such a specific form that it is inconceivable that various forms of it should occur.*" [Die Identität ist eine so bestimmt gegebene Beziehung, dass nicht abzusehen ist, wie bei ihr verschiedene Arten vorkommen können.] [20, p. 254].

[13]This point was expressed by Fichte in the following terms:

> "The proposition A *is* A (or $A = A$, since that is the meaning of the logical copula) is accepted by everyone and that without a moment's thought: it is admitted to be perfectly certain and established. Yet if anyone were to demand a proof of this proposition, we should certainly not embark on anything of the kind, but would insist that it is *absolutely* certain, that is, *without any other ground* [...]" [19, p. 94].

other proofs given by the homotopy classes of non-trivial loops based at a. In this way, the identity principle $a = a$ for a point a of a space M—far from not conveying any synthetic information about a—becomes an entity-dependent proposition that encodes homotopic information about the space M that contains the point a. By borrowing Kantian jargon we could say that the identity principle $a = a$ is no longer an *analytic a priori* proposition, but rather a *synthetic a priori* proposition, that is a proposition that

(1) is necessarily true independently of a (*a priori*),
(2) carries non-trivial information about the entity a (synthetic).

In category theory, the *a priori* nature of the identity principle is attested by the fact that—according to the very definition of a category—every object is endowed with an identity morphism which acts as a neutral element for compositions. Now, an object a also has a non-trivial identity if it has automorphisms $f : a \to a$ other than the trivial identity id_a. Since in that case there exists a morphism f^{-1} such that $f \circ f^{-1} = id_a$, we can say that each automorphism f provides a particular *factorization* of the trivial identity id_a. The set of automorphisms $a \to a$ have the structure of a group, the group $Aut(a)$ of automorphisms of a (where the composition of morphisms and the trivial identity id_a define the group operation and the neutral element respectively). Reciprocally, any group G defines a groupoid $\mathcal{B}G$ with a unique object a. Each group element defines an automorphism $g : a \to a$ and the group operation defines the composition of morphisms. In this presentation, each automorphism (or symmetry) of a can be understood as a particular proof of the identity principle $a = a$. We could then say that a *symmetric object carries a non-trivial identity*. Moreover, the non-trivial identity of a carries synthetic information about a, namely information about its symmetries. On the contrary, an object without symmetries is a rigid object whose self-identity is only witnessed by the trivial automorphism id_a. The consequences of the categorical presentation of the notion of group with respect to the understanding of the notion of identity has been clearly phrased by Badiou in the following terms:

> "[...] the categorical definition of group G makes G appear as the set of the different ways in which object-letter a is identical to itself. [...] This means that the real purpose of a group is to set the plurality of identity. [...] Among the different ways of being identical to oneself, there is 'inert' identity, that is, the null action id_a. What a group indicates is that this identity is but the degree zero of identity, its immobile figure and at rest. The other arrows of the group are dynamic identities. They are the active ways of being identical to oneself. [...] Let us say that a group is the minimal presentation of the intelligible in terms of the otherness of the same." [9, p. 149]

The Platonic undertones of Badiou's quote (coming from his explicit reference to "*the dialectic of the Same and the Other*" in Plato's *Sophist*) can also be justified by referring to Plato's *Timaeus*. In this work, Plato gets close to the idea according to which self-identity might be non-trivial and entity-dependent. Speaking about spheres, Plato writes, "*there is no shape more perfect and none more similar to itself.*" [51, 33b]. Plato's statement seems to presuppose that there are degrees according to which a shape can be similar to itself, that is that different entities might be characterized by different degrees of self-similarity.

6 On Truncations and Resolutions

Let us summarize the ideas introduced thus far. The nominalistic interpretation of equality statements $a = b$ preserves the validity of the PII by maintaining that a and b are just different names of the same entity. On the contrary, we can interpret equality statements in an ontological manner (i.e. as statements about things and not about their names) by openly assuming—in the wake of Leibniz himself—that the PII is not valid in mathematics. The content of the statement $a = b$ is that the numerically different entities denoted a and b are qualitatively identical (where the term entity is relative to the corresponding ontological context). Moreover, we can understand equality statements of the form $a = b$ in a constructive manner by taking into account that a and b might be identified by means of a multiplicity of concrete identifications. In this way, we naturally arrive to the categorical notion of *groupoid*, i.e. of a category defined by a multiplicity of objects uniquely related (at most) by means of isomorphisms. In turn, by passing to higher category theory, the isomorphisms in a groupoid might also be identified by means of 2-isomorphisms (reversible 2-morphisms) and so on and so forth all the way up. By unfolding (or *resolving*) this structure of higher identifications we arrive to the notion of ∞-*groupoid* [52]. In the case of the example given by the points of a space as objects and the paths as isomorphisms, the corresponding ∞-groupoid encodes the *homotopy type* of the space (i.e. the space up to homotopy equivalence).

Now, this infinite structure of identifications, identifications between identifications and so forth can be *truncated* at any level. This can be done by making abstraction from the concrete identifications between two entities (e.g. points, paths, objects in a category) and only retaining their equivalence classes. That is, we could consider in an ideal manner that the structure of identifications is a dispensable "*surplus structure*" and that the only thing that matters is the truth value of the equality propositions. In philosophical terms, such truncations amount to force the validity of the PII: the numerical multiplicities of qualitatively identical entities are flattened on a set of equivalence classes where the only identifications are the trivial self-identifications associated to the elements of the set. In other terms, each numerical multiplicity of qualitatively identical objects is amalgamated into a single object, the equivalence class. In the particular case given by the identifications between two points a and b in a space, we could consider that the only thing that matters is whether the propositions

$a = b$ is true or not. By making abstraction from the concrete identifications between the points, we obtain a set of equivalence classes of identical points called 0-*homotopy set* and denoted $\pi_0(M)$ (where each equivalence class in the set corresponds to a connected component of M). Less radically, we can preserve the numerical difference between identical points as well as the corresponding identifications/paths, but make abstraction from the homotopies between the latter. This amounts to uniquely retain the equivalence classes of homotopic paths. By doing so, we obtain the 1-groupoid of the space (and by choosing a point $a \in M$ we obtain the fundamental group $\pi_1(M, a)$ of the space). The operation by means of which the numerical multiplicities of qualitatively identical objects are substituted by single equivalence classes is called *truncation* in homotopy type theory [33] and *decategorification* in category theory [7].

7 Toward a Homotopic Interpretation of Gauge Theories

Equipped with the homotopic notion of identity, we can now revisit the problem of symmetries in gauge theories. According to the standard interpretation of gauge theories, the gauge transformations are understood as passive transformations of the corresponding coordinate systems and only the invariants are supposed to carry the physical information. In the general framework provided by the theory of constrained Hamiltonian systems, the invariants are given by the Dirac observables, i.e. by the functions on the reduced phase space obtained by quotiening the constraint surface by the action of the gauge group [16, 31]. In this conceptual framework, the condition of gauge invariance simply encodes the fact that the objective physical predictions of the theory cannot depend on the "subjective" arbitrary choice of a coordinate system. In this way, gauge symmetries would be nothing but a mere mechanism by means of which we can get rid of the "*surplus structure*" [57] or the mathematical redundancy associated to the different re-coordinatizations of a unique physical (i.e. coordinate-independent) situation described by gauge invariant quantities. In Kantian terms, we could say that gauge invariance can be understood as a sort of transcendental principle dealing with the relation between the subjective and the objective components of physical theories, i.e. the arbitrary coordinate systems and the intrinsic invariants respectively. Gauge symmetries should not be attributed to "nature in itself", but rather to the symbolic means by means of which we constitute an objective description of nature. As we have mentioned in the introduction, the main obstructions to this kind of epistemic interpretation of gauge symmetries are given by the non-trivial physical consequences of local gauge invariance, notably the heuristic relation—encoded in the so-called *gauge argument*—between local gauge invariance and the notion of physical interaction [14, 44, 56, 65] and the fact that the *topological solutions* (e.g. instantons) in Yang-Mills theories require gauge transformations to define non-trivial bundles. Now, the homotopic paradigm provides an alternative conceptual framework that might lead to a better comprehension of the non-trivial physical consequences of gauge symmetries on many different levels. We shall now briefly describe some of these repercussions of the homotopic paradigm on the foundations of gauge theories.

First, we shall stress that the underlying geometric structure of Yang-Mills theories provides an example of a mathematical situation that is not compatible with the PII. The geometric setting of a Yang-Mills theory is given by a bundle $E \overset{\pi}{\to} M$ of copies S_x—parameterized by spacetime M—of a *standard fiber* S endowed with a left G-action $G \times S \to S$ (see the notion of *G-bundle* in Ref. [36, Definition 10.1, p.86]). Since a G-action can be defined as a group homomorphism $\varphi : G \to Aut(S)$, each group element defines an automorphism (or a symmetry) of S. In Leibnizian terms, we can say that a G-bundle $E \overset{\pi}{\to} M$ is a numerical multiplicity of symmetric objects S_x that are qualitatively identical. Given an open covering $\{U_i\}$ of M, such a bundle can be presented by means of a *fiber bundle atlas*

$$(U_i, \psi_i : \pi^{-1}(U_i) \overset{\simeq}{\to} U_i \times S)$$

defined by a family $\{\psi_i\}$ of *local trivializations*. The *transition functions*

$$\psi_{ij} : U_i \cap U_j \to G \tag{2}$$

associated to this atlas are defined by the expressions

$$\begin{aligned} f_{ij} = \psi_i \circ \psi_j^{-1} : U_j \times S|_{U_{ij}} &\overset{\simeq}{\to} U_i \times S|_{U_{ij}} \\ (x, s) &\mapsto (x, \psi_{ij}(x)s) \end{aligned} \tag{3}$$

and satisfy the *cocycle conditions* $\psi_{ii}(x) = id_G$ (for each $x \in U_i$) and $\psi_{ij}(x)\psi_{jk}(x) = \psi_{ik}(x)$ (for each $x \in U_i \cap U_j \cap U_k$). These expressions encode the fact that it is thanks to the G-action associated to the symmetries of each fiber that it is possible to define fiber bundles with non-trivial topologies. Indeed, the local trivializations are not glued in the intersections $U_i \cap U_j$ by means of strict equalities, but rather by means of the different isomorphisms (3). In turn, the different isomorphisms are defined by means of the G-valued transition functions (2). All in all, it is thanks to the G-action on S that it is possible to introduce topological twists in the non-empty intersections $U_i \cap U_j$. In conceptual terms, the topological richness of Yang-Mills theories relies on the fact that there exist many possible identifications between qualitatively identical copies of a symmetric object S. This point has been described in the *nlab* in the following terms:

"[...] all fibers (over all points of X) are equivalent. But the point is that any F may be equivalent to itself in more than one way (it may have 'automorphisms'), and this allows non-trivial global structure even though all fibers look alike" [47, fiber bundles in physics].

Second, let us revisit the epistemic interpretation of gauge symmetries. Each fiber S_x of the G-bundle $E \xrightarrow{\pi} M$ can be identified with the standard fiber S by means of different non-canonical isomorphisms $p : S \to S_x$, each of which defines what we shall call a *framing* of S_x (we follow here the original article by Ehresmann [17]). To "frame" a fiber S_x means to choose a concrete identification between S_x and the standard fiber S. The fact that S has non-trivial automorphisms implies that there is not a canonical isomorphism between S and its copies S_x. Therefore, the multiplicity of local "frames" at a point $x \in M$ is a consequence of the intrinsic geometric nature of the notion of G-bundle, more precisely of the fact that the fibers are symmetric objects. In fact, the G-action on the standard fiber S induces a right action on the set of frames $Iso(S, Sx)$ on each x given by precomposition

$$\begin{aligned} Iso(S, S_x) \times G &\to Iso(S, S_x) \\ (S \xrightarrow{p} S_x, g) &\mapsto (pg : S \xrightarrow{g} S \xrightarrow{p} S_x). \end{aligned} \tag{4}$$

In this way, the freedom to change the local frames $p : S \to S_x$ is just an "epistemic" consequence of the intrinsic automorphisms of the standard fiber S. It follows that the epistemic invariance of a gauge theory under local gauge transformations results from the intrinsic fact that we are dealing with families of identical copies S_x of a structure S with symmetries (or, in homotopic terms, with a non-trivial identity). But this epistemic consequence of the presence of symmetries should not overshadow the intrinsic or ontological scope of the latter *qua* non-trivial identities of the fibers.

As we have argued elsewhere [12, 13], the gauge symmetries of both general relativity (local Lorentz invariance plus diffeomorphisms) and Yang-Mills theories (group of vertical automorphisms of the corresponding G-principal fiber bundle) result from the fact that general solutions of the theories are obtained by "connecting" identical symmetric building blocks (Klein geometries in general relativity and G-torsors in Yang-Mills theories) in a curved manner.[14] The local gauge symmetries of these theories are in the last instance a consequence of the fact that these building blocks are symmetric objects. The fact that local gauge transformations do not have any observable effect might seem to support the thesis according to which they are a "surplus structure" lacking empirical content. However, we could argue in the opposite direction. For instance, the fact that the rotation of a sphere does not have any observable effect is a consequence of an intrinsic property of the sphere, namely

[14] Whereas in a Yang-Mills theory the connection is an *Ehresmann connection*, in the gauge-theoretic formulation of general relativity the connection is a *Cartan connection* [13].

its spherical symmetry. In Yang-Mills theories, the inobservability of local gauge symmetries is a consequence of an intrinsic feature of the underlying geometry, namely that every possible solution can be understood as the result of "connecting" in a particular (possibly curved) manner identical symmetric objects.[15]

Strictly speaking, the previous comments about topological solutions and local gauge symmetries do not rely on the formal apparatus of the homotopic paradigm. They rely however on what we consider here to be the fundamental conceptual bedrock of this paradigm, namely a) the rejection of the PII and b) the fact that there might be different identifications between identical entities. We want now to stress that the homotopic paradigm might also play a central role in the further development of gauge theories. In this respect, we could say the main claim of the homotopic interpretation of gauge symmetries started (as far as the author knows) by Schreiber is that the interpretation of the gauge structure (i.e. the gauge transformations, the transformations between gauge transformations, and so on) as mere "*surplus structure*" truncates the homotopic structure of the corresponding spaces of states. In fact, the homotopic interpretation of gauge symmetries provides a concrete pragmatic prescription: rather than quotiening out the gauge transformations, we should unfold them all the way up (to transformations between gauge transformations and so on and so forth). Interestingly enough, this is exactly what the most sophisticated formalism for dealing with gauge symmetries in quantum field theory already does (at least at an infinitesimal level), namely the BRST formalism [31, 38]. The BRST formalism is a method for working *equivariantly* on the reduced phase space of a gauge theory. To do so, the BRST formalism provides a (co)homological description of the two operations associated to the reduction of a Hamiltonian theory with constraints, namely the *restriction* to the constraint surface defined by the (first-class) constraints and the *quotient* of this surface to the orbit space defined by the group action generated by the constraints. Rather than removing the gauge symmetries, the BRST formalism unfolds them all the way up by introducing the so-called *ghosts* (that encode the gauge transformations), the *ghosts of ghosts* (that encode the transformations between gauge transformations) and so on and so forth.[16] In Henneaux and Teitelboim's terms:

[15]It is worth noting that the idea according to which the "elementary" building blocks of a gauge theory are given by symmetric entities can also be traced back to Plato's *Timaeus* [51, 55c-56a]. In this work, Plato argues that the four *elements* (earth, air, fire, and water) can be associated to symmetric objects given by the cube, the octahedron, the icosahedron, and the tetrahedron.

[16]Technically, the BRST formalism yields a *derived ∞-Lie algebroid* that encodes the infinitesimal structure of the ∞-groupoid defined by the gauge transformations as morphisms and the higher gauge transformations as higher morphisms [47, field (physics)]. An analysis of the BRST formalism in the framework of derived geometry can be found in Ref. [2, Sect. 3.5].

> "It is a remarkable occurrence that the road to progress has invariably been toward enlarging the number of variables and introducing a more powerful symmetry rather than conversely aiming at reducing the number of variables and eliminating the symmetry." [31, p. xxiii]

The historical fact described by Henneaux and Teitelboim acquires now a clear interpretation: eliminating the gauge symmetries and reducing the number of variables—far from being a mere epistemic reduction that would remove the "*surplus structure*" for the benefit of the physically relevant information—amounts to truncate relevant homotopic information. In the terms of the *nlab*:

> "[...] the idea that symmetry is *just* a redundancy is a mistake of decategorification: passing from a groupoid of configurations—where different configurations are related by morphisms called gauge transformations—to the quotient space of configurations modulo gauge transformations is the decategorification of the groupoid. More technically speaking, it is the 0-truncation. It computes the 0-th homotopy group and forgets all the higher homotopy groups." [47, gauge group]

In this way, the "ideal" truncation of the gauge structure for the sole benefit of the corresponding equivalence classes might remove intrinsic homotopic information. In particular, passing to the quotient of the corresponding constraint surface by the action of the gauge group without taking into account the possible isotropy groups might lead to a pathological (non-smooth) reduced phase space. According to the modern methods to define smooth quotient spaces (e.g. stack theory [2, 46]), smoothness can be recovered by preserving the numerical differences between identical states (i.e. states connected by the group action) as well as the corresponding identifications.

In this way, the main claim of the homotopic interpretation of gauge theories is that the gauge structure is not a dispensable "*surplus structure*", but rather an intrinsic homotopic structure. The formal and conceptual understanding of the precise physical meaning of the different levels of this homotopic structure is a central challenge for the forthcoming philosophy, physics, and mathematics of gauge theories.

8 On Transformations and Invariants

In order to conclude, we shall address the relation between

- the homotopic idea according to which a symmetry $g \in Aut(a)$ of an entity a can be understood as a generalized identity, i.e. as a non-trivial proof of the identity principle $a = a$
- and the fact that the determinacy of an entity (its intrinsic properties) are given by the *invariants* under symmetry transformations.

The key point to understand the relation between the notion of symmetry *qua* generalized identity and the notion of invariant is given by the fact that a non-trivial identity imposes a constraint on the qualitative (or observable) properties that the entity can support, namely the constraint of being compatible with its identity. This means that only the properties that are invariant under the identity transformations define intrinsic properties of the system. Indeed, a self-identification that transforms the entity into itself cannot modify the properties that characterize the entity as such. Now, the gauge invariance condition is nothing but a particular case of this compatibility condition. In his analysis of the notion of general covariance in the framework of homotopy type theory, Shulman described this compatibility condition in the following terms:

> "[...] we say that there is one Minkowski spacetime, namely $\mathbb{R}^4$, and that it can be identified with itself in many ways, such as translations, $3D$ rotations, and Lorentz boosts. These extra added identifications force everything we say about 'Minkowski spacetimes' to be invariant under their action." [63, p. 54]

.

In this way, the transformations and the invariants define the two correlative sides of an ontology in which the entity's generalized *identity* (i.e. its automorphisms) and the entity's *determinacy* (i.e. its properties) are entangled by the compatibility condition of gauge invariance. While the epistemic interpretation of gauge theories associates the variations to an epistemic dimension and the invariants to an ontological one, the homotopic interpretations stresses the ontological scope of both the variations (associated to the entity's *identity*) and the invariants (associated to the entity's *determinacy*). By paraphrasing Benoît Timmermans we could say that it might be utterly misleading to conceal the pluralism of gauge transformations (by considering them as mere mathematical redundancies associated to the symbolic constitution of objective knowledge) for the sole benefit of the realism of invariants [67, p. 292]. In his own terms,

> "There is not on one side the transformations that would give nothing but the appereance, the superficial part of the things, and on the other side the invariants, the laws or the immutable relations so to speak that are inaccessible to the transformations."[17]

References

1. H. Alexander (ed.), *The Leibniz-Clarke Correspondence. Together with Extracts from Newton's Principia and Opticks* (Manchester University Press, Manchester and New York, 1955)

---

[17] "Il n'y a pas d'un côté les transformations qui ne livreraient que l'apparence, la part superficielle des choses, et de l'autre côté les invariants, lois ou rapports immuables pour ainsi dire inaccessibles aux transformations." [67, p. 86].

2. M. Anel, The geometry of ambiguity: an introduction to the ideas of derived geometry, in *New Spaces in Mathematics. Formal and Conceptual Reflections.* Ed. by M. Anel and G. Catren (Cambridge: Cambridge University Press, 2020)
3. S. Awodey, A proposition is the (homotopy) type of its proofs (2016). ArXiv:1701.02024 [math.LO]
4. S. Awodey, Structuralism, invariance, and univalence. Philosophia Mathematica **22**(1), 1–11 (2014)
5. S. Awodey, *Category Theory* (Oxford University Press, New York, 2006)
6. A.J. Ayer, *The Identity of Indiscernibles.* In *Philosophical Essays* (Palgrave Macmillan, London, 1972)
7. J.C. Baez, J. Dolan, Categorification, in *Higher Category Theory.* Ed. by E. Getzler and M. Kapranov, 1–36. Contemp. Math.; 230 (American Mathematical Society, Providence. Rhode Island, 1998)
8. G. Bachelard, *The Formation of the Scientific Mind.* Trans. by M. McAllester. (Clinamen Press, Manchester, 2002)
9. A. Badiou, *Briefings on Existence: A Short Treatise on Transitory Ontology.* Trans. by N. Madarasz. Albany: SUNY Press (2006)
10. E. Bishop, *Foundations of Constructive Analysis* (McGraw-Hill Inc, USA, 1967)
11. K. Brading, H. Brown, Are gauge symmetry transformations observable? Brit. J. Philos. Sci. **55**, 645–665 (2004)
12. G. Catren, Klein-Weyl's program and the ontology of gauge and quantum systems. Stud. History Philos. Modern Phys. **61**, 25–40 (2018)
13. G. Catren, Geometric foundations of Cartan gauge gravity. Int. J. Geom. Methods Mod. Phys. **12**(4), 1530002-1–1530002-33 (2015)
14. G. Catren, Geometric foundations of classical Yang-Mills theory. Stud. History Philos. Modern Phys. **39**, 511–531 (2008)
15. A. Caulton, J. Butterfield, On kinds of indiscernibility in logic and metaphysics. Brit. J. Philos. Sci. **63**(1), 27–84 (2011)
16. P.A.M. Dirac, *Lectures on Quantum Mechanics* (Yeshiva University, New York, 1964)
17. C. Ehresmann, Les connexions infinitésimales dans un espace fibré différentiable. Séminaire N. Bourbaki **24**, 153–168 (1950)
18. Euclid, *The Thirteen Books of Euclid's Elements.* Trans. from the text of Heiberg with introduction and commentary by T.L. Heath, Vol.1, Introduction and Books I and II. (Cambridge University Press, Cambridge, 1968)
19. J.G. Fichte (1794), *Science of Knowledge.* Trans. by P. Heath and J. Lachs. (Cambridge University Press, Cambridge, 1991)
20. G. Frege. *Grundgesetze der Arithmetik.* Hildesheim (Georg Olms, 1903)
21. G. Frege (1879), Conceptual notation: A formula language of pure thought modeled upon the formula language of arithmetic [Begriffsschrift], in *Conceptual Notation and Related Articles.* Trans. by T. W. Bynum. (Clarendon Press, Oxford, 1972)
22. G. Frege (1892), On sense and reference, in *Translations from the Philosophical Writings of Gottlob Frege.* Ed. and Trans. by P. Geach and M. Black (Basil Blackwell, Oxford, 1960)
23. S. French, Why the principle of the identity of indiscernibles is not contingently true either. Synthese **78**(2), 141–166 (1989)
24. S. Friederich, Symmetries and the identity of physical states. EPSA15 Selected Papers **5**, 153–166 (2016)
25. S. Friederich, Symmetry, empirical equivalence, and identity. Brit. J. Philos. Sci. **66**(3), 537–559 (2015)
26. J.-Y. Girard, *Proofs and Types* (Cambridge University Press, New York, 1990)

27. M. Goldstein, The historical development of group theoretical ideas in Connexion with Euclid's axiom of congruence. Notre Dame J. Formal Logic **XIII**, N°3 (1972)
28. H. Greaves, D. Wallace, Empirical consequences of symmetries. Brit. J. Philos. Sci. **65**, 59–89 (2014)
29. R. Healey, Perfect symmetries. Brit. J. Philos. Sci. **60**(4), 697–720 (2009)
30. R. Healey, *Gauging What's Real. The Conceptual Foundations of Contemporary Gauge Theories* (Oxford University Press, Oxford, 2007)
31. M. Henneaux, C. Teitelboim, *Quantization of Gauge Systems* (Princeton University Press, Princeton, NJ, 1994)
32. D. Hilbert, P. Bernays, *Grundlagen der Mathematik*, vol. 1 (Springer, Berlin, 1934)
33. The Univalent Foundations Program, *Homotopy Type Theory: Univalent Foundations of Mathematics*. Institute for Advanced Study (2013)
34. I. Kant, Letter to Johann Schultz, November 25, 1788, in *Correspondence*. Trans. by A. Zweig (Cambridge University Press, New York, 1999)
35. J. Keränen, The identity problem for realist structuralism. Philosophia Mathematica **9**(3), 308–330 (2001)
36. I. Kolář, P.W. Michor, J. Slovák, *Natural Operations in Differential Geometry* (Springer, Berlin, Heidelberg, 1993)
37. P. Kosso, The empirical status of symmetries in physics. Brit. J. Philos. Sci. **51**(1), 81–98 (2000)
38. B. Kostant, S. Sternberg, Symplectic reduction, BRS cohomology, and infinite dimensional clifford algebras. Ann. Phys. **176**, 49–113 (1987)
39. J. Ladyman, Mathematical structuralism and the Identity of Indiscernibles. Analysis **65**(3), 218–221 (2005)
40. G.W. Leibniz, *Philosophical Essays*. Ed. and Trans. by R. Ariew and D. Garber (Hackett Publishing Company, Indiana, 1989)
41. H. Leitgeb, J. Ladyman, Criteria of identity and structuralist ontology. Philosophia Mathematica **16**(3), 388–396 (2008)
42. H. Lyre, The principles of gauging. Philos. Sci. **68**(3), 371–381 (2001)
43. J.-P. Marquis, *From a Geometrical Point of View* (Springer, A Study of the History and Philosophy of Category Theory, 2009)
44. C.A. Martin, Gauge principles, gauge arguments and the logic of nature. Philos. Sci. **69**, 221–234 (2002)
45. P. Martin-Löf, *Intuitionistic Type Theory* (Bibliopolis, Napoli, 1984)
46. N. Mestrano, C. Simpson, Stacks, in *New Spaces in Mathematics. Formal and Conceptual Reflections*. Ed. by M. Anel and G. Catren (Cambridge University Press, Cambridge, 2020)
47. nlab. https://ncatlab.org/nlab/show/HomePage
48. J. Norton, General covariance, gauge theories, and the Kretschmann objection, in *Symmetries in Physics. Philosophical Reflections*. Ed. by K. Brading and E. Castellani (Cambridge University Press, Cambridge, 2003)
49. J. Norton, General covariance and the foundations of general relativity: Eight decades of dispute. Reports Progress Phys. **56**, 791–858 (1993)
50. J. Norton, The physical content of general covariance, in *Studies in the History of General Relativity*, ed. by J. Eisenstaedt, A.J. Kox (The Center for Einstein Studies, Boston University, Boston, 1992)
51. Plato, *Timaeus*. Trans. by D.J. Zeyl (Hackett Publishing Company, Indianapolis, 2000)
52. T. Porter, Spaces as ∞-groupoids, in *New Spaces in Mathematics. Formal and Conceptual Reflections*, ed. by M. Anel and G. Catren (Cambridge University Press, Cambridge, 2018)
53. W.V. Quine, *Philosophy of Logic* (Cambridge: Harvard University Press, (1986)
54. W.V. Quine, *Ontological Relativity and Other Essays* (Columbia University Press, New York, 1969)

55. W.V. Quine, *Methods of Logic* (Holt, Rinehart, and Winston, Inc, USA (1966)
56. L. O'Raifeartaigh, *The Dawning of Gauge Theory* (Princeton University Press, Princeton, NJ, 1995)
57. M. Redhead, The interpretation of gauge symmetry, in *Symmetries in Physics. Philosophical Reflections*. Ed. by K. Brading and E. Castellani, 124–139 (Cambridge University Press, Cambridge, UK, 2003)
58. G. Rodriguez-Pereyra, *Leibniz's Principle of Identity of Indiscernibles* (Oxford University Press, Oxford, 2014)
59. C. Rovelli, Why gauge? Found. Phys. **44**, 91–104 (2014)
60. Russell, B., *Principles of Mathematics* (Abingdon, Oxon: Routledge, (1903) [2010])
61. S. Saunders, Physics and Leibniz's Principles, in *Symmetries in Physics. Philosophical Reflections*. Ed. by K. Brading and E. Castellani (Cambridge University Press, Cambridge, 2003), pp. 289–308
62. U. Schreiber, Differential cohomology in a cohesive ∞-topos (2013). ArXiv:1310.7930 [math-ph]
63. M. Schulman, Homotopy type theory: A synthetic approach to higher equalities, in *Categories for the Working Philosopher*. Ed. by E. Landry (Oxford University Press, New York, 2017), pp. 36–57
64. S. Shapiro, Identity, indiscernibility, and ante rem structuralism: the tale of i and $-i$. Philosophia Mathematica **16**(3), 285–309 (2008)
65. P.R. Teller, The gauge argument. Philos. Sci. **67**, S466–S481 (2000)
66. G. t'Hooft, M. Veltman, Regularization and renormalization of gauge fields. Nucl. Phys. B **44**(1), 189–213 (1972)
67. B. Timmermans, *Histoire philosophique de l' algèbre moderne. Les origines romantiques de la pensée abstraite* (Classiques Garnier, Paris, 2012)
68. A.N. Whitehead, *A Treatise on Universal Algebra* (Cambridge University Press, New York, (1898) [2009])
69. A.N. Whitehead, *The Principle of Relativity with Application to the Physical Science* (Cambridge University Press, Cambridge, 1922)
70. E. Witten, Symmetry and emergence. Nature Phys. **14**, 116–119 (2018)
71. L. Wittgenstein, *Tractatus Logico-Philosophicus*. Trans. by D.F. Pears and B.F (McGuinness, NY: Routledge, (1918) [1974])

Weyl's *Raum-Zeit-Materie* and Its Philosophical Underpinning

Reichenbach, Weyl, Philosophy and Gauge

Dennis Dieks

Abstract Hermann Weyl connected his epoch-making work on general relativity and gauge theory to his Husserlian views about the phenomenological essence of space and time. This philosophical stance of Weyl's has received considerable attention in recent years and has been favorably compared and contrasted with the "logical-empiricist" approach of Reichenbach, Weyl's contemporary who wrote extensively about relativity and the philosophy of space and time. We will argue, however, that Weyl's use of phenomenology should be seen as a case of personal heuristics rather than as a systematic and viable philosophy of physics. We will explain and defend Reichenbach's sophisticated empiricism, which in our opinion has often been misunderstood, and argue that it is better suited as a general philosophical framework for the natural sciences than Weyl's phenomenology.

1 Introduction: Philosophy and Early 20th Century Physics

It is an acknowledged historical fact that much of early-20th century philosophy of physics grew on the soil of Kantian philosophy. Leading authors of the period like Weyl, Reichenbach, Schlick and Cassirer thought it essential to define their positions with respect to Kant, and many were raised in the Kantian tradition. It is typical of Kantian philosophy to emphasize the complicated nature of observation and empirical data: the external world is not simply empirically given to us, but is at least partly *constructed* by our "sensibility" (*reine Anschauung*) and understanding. A central Kantian motif in the elaboration of this idea is the thought that constitutive concepts and principles are needed to make scientific experience possible at all; only when perceptions are subsumed under concepts, which themselves possess an *a priori* character, do they become experiences.

In the second half of the nineteenth and the beginning of the twentieth century Kant's original version of this "*synthetic a priori*" doctrine came under increasing

---

D. Dieks (✉)
History and Philosophy of Science, Utrecht University, Utrecht, Netherlands
e-mail: d.dieks@uu.nl

S. De Bianchi and C. Kiefer (eds.), *One Hundred Years of Gauge Theory*, Fundamental Theories of Physics 199, https://doi.org/10.1007/978-3-030-51197-5_6

pressure due to developments in mathematics and physics. An important example is provided by debates that arose about the status of Euclidean geometry: according to Kant the validity of this geometrical system was a necessary precondition for the possibility of spatial experience, but this was first made less than obvious by the development of non-Euclidean mathematical geometries and was then problematized in physics by relativity theory. Similarly, the intuitive conception of absolute time was called into question by the theory of relativity.

For many philosophers the natural response was to adapt their Kantian views to the new scientific reality by embracing a more flexible form of the synthetic *a priori* doctrine, in which the constitutive role of *a priori* concepts was maintained, but in which the notion that these concepts were given once and for all was abandoned. Instead, it was proposed that constitutive concepts and principles may change as a consequence of the evolution of science, so that experience itself may evolve, by being structured in terms of new concepts. This gave rise to the notion of the *relativized a priori*: constitutive principles and concepts used to make scientific sense of observations are informed by previously acquired scientific knowledge and therefore are historically contingent—their applicability is relative to a certain stage in the development of science.

Hans Reichenbach, one of the main figures of the twentieth century philosophy of physics was among those who started following this neo-Kantian path. However, Reichenbach was swayed by Schlick, in 1920, to refrain from using Kantian terminology and to replace talk about "relativized constitutive *a priori* principles" by the use of the terms *conventions* or *definitions*, conventionally chosen rules and principles that connect a mathematical formalism to the empirical world. According to a now standard story, this episode ushered in *logical empiricism*, in which arbitrarily chosen and precisely formulated "coordinative definitions" (Reichenbach) or "rules of correspondence" (Carnap) took the place of the Kantian *a priori* and injected physical content into mathematically formulated theories.

Among those who did not move into the direction of this brand of empiricism but tried to maintain a position closer to Kant's transcendental idealism, Hermann Weyl stands out because of his ground-breaking work in mathematics and theoretical physics. Weyl positioned himself within Husserlian phenomenology, according to which it is possible to arrive at the essence of our intuitive concepts by an analysis in which all accidental details are removed (epoché). Weyl argued that our pure intuitions of space and time, once all non-essential elements in them have been eliminated ("bracketed"), constitute a precondition for our spatio-temporal thinking and are therefore normative for theory construction in physics. As Weyl made clear in his celebrated *Raum, Zeit, Materie* (*Space, Time, Matter*) and other writings, this philosophical stance played a key role in his thinking about space and time and led to his 1918 proposal for a gauge theory meant to generalize the general theory of relativity (Ryckman, 2004, provides an extensive analysis). So Weyl's enormously influential and important gauge ideas appear as fruits of his phenomenological starting points.

However, we will argue that Weyl's phenomenological approach, although undoubtedly of great value for Weyl himself, cannot bear the burden of providing a general and fundamental conceptual basis for physical theories. We also argue that

Reichenbach's empiricism, stripped of oversimplifications and misunderstandings that surround many of its presentations, is better suited to serve as a framework for the analysis of empirical science. In fact, it is very plausible to arrive at "Weylian" ideas about the merely local validity of our usual geometrical notions ("*Nahegeometrie*") and gauge freedom on the basis of empiricist principles, as we shall see. Weyl's philosophy is therefore not needed for arriving at Weyl's physics. Moreover, Weyl's actual scientific reasoning indicates a strong reliance on empirical and mathematical arguments, with his phenomenology playing a heuristic rather than foundational role.

2 Reichenbach's Struggle with Kant

When Hans Reichenbach wrote his 1916 dissertation *The Concept of Probability in the Mathematical Representation of Reality* [13], his aim was to show how the original Kantian system, with minor "updates", could deal with modern science. Kant had claimed that the assumption of strict causality, i.e. the presence of deterministic cause-effect relations, was a necessary precondition for the possibility of knowledge in the natural sciences. However, Reichenbach argued, Kant's argument should be viewed in the historical context of his time, in which the dominant Newtonian paradigm made it virtually impossible to think of probability as a fundamental scientific concept. If Kant's transcendental proof for the necessity of causality is scrutinized it turns out, Reichenbach argued, that it actually shows only the necessity of positing the existence of *well-defined probability functions* behind the statistical distributions of results that we find in experiments [2]. So Kant's *a priori* scheme had to be adapted, but only in the sense of a harmless minor correction.

Three years later, in 1919, Reichenbach was among the five students attending Einstein's first course on relativity in Berlin. The confrontation with the new physics had a great impact on Reichenbach and led him to rethink the relation between Kantian philosophy and science. In his 1920 book *Relativity Theory and Apriori Knowledge* (*Relativitätstheorie und Erkenntnis A Priori*) Reichenbach comes to the conclusion that Kant's system cannot accommodate relativity, and moreover cannot be saved by small adaptations and completions; a more radical step is needed. According to Reichenbach, the original Kantian system with, for example, its *a priori* imposition of absolute temporal order and Euclidean spatial geometry has actually become an obstacle to the development of science. Although it is possible to adopt strategies aimed at preserving a number of Kant's *a priori* requirements, for instance by assuming a fixed Euclidean background even in general relativity, it turns out that one then has to introduce "unKantian" changes at other places in the theory in order to salvage the theory's empirical adequacy—so that the theory's complete conceptual package must necessarily violate *some* Kantian *a priori* requirements. In other words, the totality of Kant's *a priori* principles is incompatible with the new physics, and this constitutes a refutation of Kant's system.

Nevertheless, for Reichenbach this does not imply the downfall of the Kantian approach. As he sees it, at the core of that approach is the notion that constitutive principles and concepts are needed in order to create physical meaning in abstract mathematical theories, and thus to make empirical science possible at all. This idea remains valid, Reichenbach argues, even if the specific collection of *a priori* principles posited by Kant himself turns out to be untenable. As Reichenbach wrote to Schlick, in reply to Schlick's reproach that he had not abandoned Kantianism altogether in his book [21, 29 November]:

> I believe like you that my criticism implies a break with a very central principle of Kant's (compare p. 89 of my book). If it nevertheless seemed to me that my views could be seen as a new and further development of those of Kant, this is due to the circumstance that the emphasis on the constitutive character of the concept of an object always appeared to me the most essential aspect of Kant's work—perhaps merely because I personally have first learned this way of thinking from Kant.[1]

The view put forward by Reichenbach in *Relativity Theory and Apriori Knowledge* is that we cannot expect that interpretative frameworks for scientific theories are given once and for all: the contribution of human reason to empirical knowledge consists in defeasible posits made by us, in specific historical circumstances. Kant's doctrine of the *a priori* may accordingly be split into a constitutive and an "apodictic" part: the constitutive *a priori* has to be retained, whereas apodicticity, i.e. the permanent and absolute validity of our structuring concepts, should be rejected. The principles that we use to interpret experimental results and to link them to the abstract formalisms of mathematical physics have a restricted scope: they are valid relative to a certain stage in the development of physics.

Interestingly, in his book Reichenbach goes into some detail about how conceptual changes actually come about. He argues that this is a process of "continuous extension" (stetige Erweiterung): our conceptual network will usually be adapted in a piecemeal fashion, in such a way that central components that are well-embedded in our previous practices are likely to remain unchanged, whereas more peripheral parts will be candidates for revision. So even though all our concepts are open to revision *in principle*, from a logical point of view, some of them will actually be almost immutable. For example, everyday interpretative schemes used in describing daily observations can hardly be expected to be abandoned. The resulting piecemeal extension of conceptual frameworks leads to a continuity of descriptions without which it would become hard to compare and contrast how new and old theories deal with experimental results.

After the appearance of *Relativity Theory and Apriori Knowledge*, Reichenbach sent a copy to Moritz Schlick. Schlick responded positively, but urged Reichenbach to be more consistent in his criticism and rejection of Kant.[2] Schlick objected that it is

[1]The original reads: "Dass meine Kritik einen Bruch mit einem sehr tiefen Prinzip Kants bedeutet, glaube ich auch (vergl. S. 89 meines Buches). Wenn es mir trotzdem schien, dass meine Auffassung als eine neuere Fortführung der Kant'schen angesehen werden kann, liegt das wohl daran, dass mir die Betonung des konstitutiven Characters im Objektbegriff immer als das Wesentlichste bei Kant erschienen ist—vielleicht nur deshalb, weil ich persönlich diese Gedanken zuerst durch Kant gelernt habe.".

[2]The exchange that followed has received considerable attention in the philosophical literature (e.g. [5, 12]).

exactly the *combination* of the apodictic and the constitutive that defines Kant's philosophy; separating these two aspects in Reichenbach's manner leads to *empiricism* rather than to a position that can still be called Kantian. Schlick therefore admonished Reichenbach to stay away from Kantian terminology and to use the term "convention", à la Poincaré, instead of speaking about "*a priori* constitutive principles". As Schlick wrote to Reichenbach on 26 November 1920 [3]:

> The central point of my letter is that I cannot find out what the difference really is between your a priori principles and conventions, so that it seems that we agree on the essential issue. What has amazed me most in your manuscript is that you dismiss Poincaré's conventionality doctrine with only so few words.[3]

In his extensive reply, Reichenbach [21, 29 November] retorts that he thinks that the term "convention" is misleading, because even though it is true that individual principles in a theory could in principle be chosen differently, this would require compensating changes in other principles in order to preserve the empirical adequacy of the theory. *Combinations* of "constitutive principles" are therefore not conventional at all but possess factual content—this is precisely one of the main arguments of *Relativity Theory and Apriori Knowledge*. Moreover, Reichenbach expresses doubt about whether the choices actually made in physics (for example, in general relativity, the choice for non-Euclidean geometry combined with general covariance, instead of the combination of Euclidean geometry and privileged coordinate systems) are mere conventions, adopted for the sake of simplicity, as Schlick and Poincaré would maintain. In fact, he reports feeling an "instinctive repulsion" against this conventionalist view ("Ich habe gegen diese Deutung eine instinktive Abneigung"); but having no compelling analysis immediately available to decide which choice is to be preferred over others he defers discussion of this problem, at least for the time being ("Nach welcher Rangordnung hier entschieden wird weiss ich vorläufig einfach nicht").

3 Definitions and Conventions

In his reply Schlick dismissed Reichenbach's qualms by commenting that Poincaré already had dealt with arguments of this sort, and declared that he and Reichenbach now agreed on all relevant points. After this exchange Reichenbach indeed changed his terminology in the direction recommended by Schlick. In particular, in his famous and influential 1928 book *The Philosophy of Space and Time* (*Philosophie der Raum-Zeit-Lehre*), Reichenbach discusses the link between the uninterpreted mathematical formalisms of physical theories and observational results as being given by "coordinative *definitions*", which "like all definitions" are *arbitrary*. Essentially identical statements can be found in his much later *The rise of scientific philosophy* [20].

[3] German original: "Es ist der Kernpunkt meines Briefes, dass ich nicht herauszufinden vermag, worin sich Ihre Sätze a priori von den Konventionen eigentlich unterscheiden so dass wir also im wichtigsten Punkte einer Meinung waren. Dass Sie über die Poincaresche Konventionslehre mit so wenigen Worten hinweggehen, hat mich an Ihrer Schrift am meisten gewundert.".

So it would appear that Reichenbach in 1920 fully converted to Schlick's conventionalism and early version of logical positivism. It is therefore understandable that Reichenbach's work has shared the fate of logical positivism: it is often criticized for its linguistic orientation, its theory of meaning that rests on a strict division between analytic statements and synthetic statements, and its conventionalism (see, e.g., Ryckman 22, Giovanelli 6).

However, closer examination of Reichenbach's work after 1920 shows that it is not so clear whether a radical break with the essential ideas of his revamped Kantianism ever took place: many of the motifs that we find in his 1920 book can still be identified in his later work (see also 2, 3). It is true that Reichenbach switched over to a terminology that may well suggest otherwise, but we submit that this terminology often served the purposes of pedagogy and ease of exposition. In our opinion, the message that Reichenbach aimed to convey is more subtle and complicated than *prima facie* consideration suggests.

A case in point are Reichenbach's many discussions of the nature of physical geometry. The starting point is always the observation that geometry as dealt with by pure mathematics does nor refer to anything outside of mathematics; what is needed to supply physical content is a link between on the one hand the mathematical formalism, in which concepts like "point", "line" and "distance" are only implicitly defined by their place in the network of abstract concepts, and on the other hand the physical world. A standard way to provide this link is by associating a *mathematically defined* distance function with numbers resulting from a *physical procedure*, for example the number of times a measuring rod can be laid down between two physical points. This is an example of what Reichenbach calls a *coordinative definition*: a physical object or procedure is coordinated to a mathematically defined concept or quantity. Geometry only becomes physical after such coordinative definitions have been put in place, and only then the question of which geometrical system applies to reality starts making sense. Reichenbach explained the status of these definitions in his 1924 *Axiomatization of the Theory of Relativity* (*Axiomatik der relativistischen Raum-Zeit-Lehre*), where he first made systematic use of them.

In §2 of this book, *The Logical Status of the Definitions*, Reichenbach begins by stating that quite in general definitions are arbitrary stipulations that are neither true nor false. However, coordinative definitions are different from mathematical definitions because they consist in the coordination of a "piece of physical reality" to an already mathematically defined concept. This involves a difficulty, because any verdict of what is "physically real" already presupposes a theoretical framework:

> the physical thing that is coordinated is not an immediate perceptual experience but must be constructed from such an experience by means of interpretation. ... We eliminate this difficulty [in the following way]: for elementary interpretations we use coordinative definitions whose degree of precision is not important and which, in particular, do not make use of relativistic definitions. This device leads to a restriction in the arbitrariness of coordinative definitions: they must not contradict elementary definitions.

New definitions should consequently respect, and make use of, basic meanings and interpretations already in use in classical physics or daily life—for example, concerning notions like "measuring rods" or "clocks". This is the same thought that we already encountered in Reichenbach's 1920 explanation of the method of "continuous extension" (*stetige Erweiterung*).

Reichenbach also makes it clear that his definitions should not be seen as unique "suppliers of meaning": a concept may acquire physical content in more than one way. Indeed, Reichenbach [16] explains (§3) that "it is customary to use rigid bodies for spatial measurements, natural clocks for temporal measurements, and light signals for simultaneity only", but adds that he is going to use a different procedure in this book:

> It should be noted that for the purpose of employing a minimum number of axioms in our presentation, the space-time metric of the special theory of relativity is defined by light signals alone, and therefore light signals are the basis not only for simultaneity but also for uniformity and equality of spatial distances.

Reichenbach here switches easily from using material rods for the determination of length to the use of light rays for the same purpose. In fact, later in the book his claim that light alone is sufficient to determine the geometry proves too optimistic: light signals only determine the metric up to a scale factor, so that use of something material is still needed in order to define a unit—for example a material measuring rod. Reichenbach [16, p. 154] adds, however, that other options are available: "the path of a force-free mass point may take the place of a material thing, as H. Weyl has shown." Apparently, completely different but equivalent procedures (using rods, light rays or particle trajectories) may be used to "define" the metric.

Typically, Reichenbach uses the terms "measurement" and "definition" alternately in this context (as in the above quotation). Given the standard logical-positivist reading of Reichenbach, one would rather respect a strict separation of these terms. Indeed, one would expect that a *definition bestows physical meaning* on a concept, whereas a *measurement* assigns a numerical value to an already meaningful quantity. Given such an already defined quantity, there will generally be alternative ways of measuring it. By contrast, associating different procedures or different physical objects to a mathematical quantity by means of different *coordinative definitions* should result in different and mutually exclusive physical meanings of a concept. If "length" is *defined* via the use of measuring sticks, statements about the number of times such a device can be laid down become *analytic*, whereas statements about the time needed by light to traverse the interval in question are *synthetic*; and *vice versa*. But when Reichenbach analyzes concrete cases like these (as opposed to his general and schematic philosophical explanations of the need for a connection between mathematical formalism and physical reality) he is flexible between such alternatives.

In *The Philosophy of Space and Time* [18, §27] Reichenbach emphasizes again that his "light geometry" constitutes only one recipe to determine space-time distances: he could have started with measuring rods instead of light signals. As he writes, "since there are also measuring instruments other than light, namely measuring rods and clocks, it is important to ask how these objects behave in relation to the light geometry." He explains that the relation between light signals and rods is given by an empirically justified "axiom": measurements by light signals and by material rods lead to exactly the same geometry. So here we have *two* coordinative *definitions* of length, and an *axiom* saying that they are equivalent. It seems clear that in this final situation the question of whether one or the other definition leads to analytic statements has been dissolved.

That Reichenbach did not take his "definitions" as inflexible determinants of meaning and of what counts as analytic and synthetic, but rather used this term as a simple expository device for explaining the general difference between uninterpreted mathematical formalisms and physical theories, is also made clear by other examples. One of them is Reichenbach's discussion of the "relativity of geometry" in §8 of *The Philosophy of Space and Time* . Reichenbach there explains that when we define spatial distances by means of measuring rods, we should not do so by pointing to actual objects but should keep in mind that these have to be corrected for distorting influences. In other words, the coordinative definition for length refers to a theoretically determined, ideal object that can only be approximated by actual rods. But the needed approximation, via corrections for deformations, is not without complications: "differential" forces deform rods made of different materials to different degrees and can therefore be detected experimentally via comparisons; but the presence of "universal forces", which act in a way that is insensitive to the rods' chemical composition, cannot be found out this way. Since only the *combination* $G + F$ of geometry plus physics is empirically testable, this means that there are infinitely many different geometries that, when combined with suitable physical force fields, lead to the same empirical predictions. These different geometries G deliver different verdicts about which distances are equal to each other, even though they all agree about the coordinative "definition" of length, namely that the unit of length is realized by an ideal, undeformed measuring rod. Consequently, even after a definition has been given there is still room for different empirically equivalent theories.[4]

In short, Reichenbach's empiricism is more sophisticated than what is suggested by the slogan "concepts are completely fixed by coordinative definitions that like all definitions are arbitrary". In line with this assessment, in the foreword to the English edition of *Axiomatik der relativistischen Raum-Zeit-Lehre* Wesley Salmon observes that Reichenbach did not intend his axiomatization to be a mathematically elegant deductive systematization of relativity theory, but rather as a presentation highlighting the physical facts that necessitated changes to the classical conceptual framework. This assessment is confirmed by Reichenbach's bitter reaction to Weyl's negative review of his work[5] [27]. Weyl had written that the major part of the *Axiomatik* had a

[4]Grünbaum [7] therefore considered Reichenbach's introduction of universal forces to be a severe philosophical mistake based on a misunderstanding of what it means to lay down a coordinative definition. Grünbaum argued that once a coordinative definition of length has been specified length simply is what it has been defined to be, in terms of some measuring procedure, so that there can be no place for undetectable deforming forces.

[5]According to Rynasiewicz [23], part of Weyl's displeasure with Reichenbach's book may have related to the fact that Weyl had alerted Reichenbach, in earlier correspondence, to the possibilities but also technical limitations of the light geometrical method for measuring the metric—Reichenbach had not taken full account of these problems nor had he acknowledged Weyl's suggestions.

formal mathematical/logical nature and that the book should be judged accordingly, after which he characterized it "as far from satisfactory, awkward and opaque (wenig befriedigend, umständlich und undurchsichtig)." Reichenbach [17] retorted:

> The plan of my research project has been dictated by the intention to lay bare the results of physical experience as clearly as possible and to extract from each new empirical proposition as many derivable consequences as can be done. If one works with a minimum of concepts, many steps will become more awkward than if from the start one begins with the totality of all available resources.[6]

And he continues:

> I consider it very regrettable, though, if a mathematician of Weyl's rank misunderstands the aim of such a epistemological-logical investigation to such an extent, and with his authority tries to suppress the attempt to give the theory of relativity, which has been so fruitfully developed mathematically and physically, now finally also the logical foundation which at the end of the day alone can secure its validity.[7]

In *The Philosophy of Space and Time* Reichenbach famously proposed to solve the problem of how to choose between different empirically equivalent geometrical systems (the issue about which he was hesitant in his 1920 correspondence about conventionality with Schlick) by the methodological principle to set universal forces equal to zero ($F = 0$). Over the years, Reichenbach often returned to this topic. In *The rise of scientific philosophy* [20, pp. 136–137], written against the end of his career, he contrasted his own views with Poincaré's conventionalism in a way that is virtually identical to what we find in his defense of neo-Kantianism in his 1920 correspondence with Schlick:

> Assume that empirical observations are compatible with the following two descriptions [a and b]: Class I. a) The geometry is Euclidean, but there are universal forces distorting light rays and measuring rods. b) The geometry is non-Euclidean, and there are no universal forces. ...
>
> Now assume that in a different world, or in a different part of our world, empirical observations were made which are compatible with the following two descriptions: Class II. a) The geometry is Euclidean, and there are no universal forces. b) The geometry is non-Euclidean, but there are universal forces distorting light rays and measuring rods.
>
> [As before] Poincaré is right when he argues that these two descriptions [a and b] are both true; they are equivalent descriptions.
>
> But Poincaré would be mistaken if he were to argue that the two worlds I and II were the same. They are objectively different. Although for each world there is a class of equivalent

[6]Der Plan meiner Untersuchung ist durch die Absicht diktiert, die Resultate der physikalischen Erfahrung möglichst deutlich aufzudecken und aus jedem neuen Erfahrungssatz so viel an ableitbaren Folgerungen herauszuholen, als irgend angeht. Wenn man mit einem Minimum von Begriffen arbeitet, wird dabei mancher Schritt umständlicher werden, als wenn man von vornherein mit der Gesamtheit aller verfügbaren Hilfsmittel beginnt.

[7]Ich halte es aber für sehr bedauerlich, wenn ein Mathematiker von Herrn Weyls Rang den Zweck einer solchen erkenntnistheoretisch-logischen Untersuchung derart verkennt und mit seiner Autorität den Versuch zu unterdrücken sucht, der mathematisch und physikalisch so fruchtbar ausgebauten Relativitätstheorie jetzt endlich auch den logischen Unterbau zu geben, der letzten Endes allein die Gewähr ihrer Gültigkeit tragen kann.

> descriptions, the different *classes* are not of equal truth value. Only one class can be true for a given kind of world; which class it is, only empirical observation can tell. Conventionalism sees only the equivalence of descriptions within one class, but stops short of recognizing the differences between the classes. The theory of equivalent descriptions, however, enables us to describe the world objectively by assigning empirical truth to only one class of descriptions, although within each class all descriptions are of equal truth value.
>
> Instead of using classes of descriptions, it is convenient to single out, in each class, one description as the *normal system* and use it as a representative of the whole class. In this sense, we can choose the description in which universal forces vanish as the normal system, calling it *natural geometry*.

Although Reichenbach here emphasizes the existence of many empirically equivalent descriptions, as a general philosophical point, he singles out one of them as the *normal* description, leading to the *natural geometry*. In his correspondence with Schlick, Reichenbach had appealed to "physical intuition" as the justification of this choice. But it seems clear from the above that the method of continuous extension is here at work: the use of measuring devices, be they material measuring rods or light beams, that are only corrected for *differential* distortions is the traditional and standard way of proceeding in physical geometry. Abandoning this practice, although possible in abstract principle, would be a break with the practice of physics and would obscure the relation with earlier space-time theories.

Summarizing all this: Even though Reichenbach often explains the general ideas of his empiricism in language that smacks of a rigid logical positivism and Schlick-style conventionalism, his actual analyses are much more flexible and retain the general outlook of the revised and modernized Kantianism of his 1920 book. With hindsight, this position of Reichenbach's can be said to anticipate a good deal of the later criticism of logical positivism, for example concerning the analytic-synthetic distinction.[8]

4 The Rejection of Kantianism as Emancipation from Intuition

If in "arbitrary choice of definitions" neither *arbitrary choice* nor *definition* should be given the significance that many have taken for granted, why did Reichenbach find it important to make use of this terminology at all? I submit that the answer to this question does relate to a change of mind by Reichenbach concerning Kantianism, but in a way that is slightly different from the standard account of this episode. The core of Reichenbach's change was not that he embraced the details of Schlick's conventionalism and converted to a strict logical-postivist philosophy of science, but rather that he distanced himself even more radically from Kant's apodictic *a priori* and became convinced by Schlick that it was unwise to keep on using terminology that could remind of Kant's doctrine of apodicticity. The emphasis on the "arbitrary"

---

[8]Reichenbach's sophisticated empiricism as construed here comes close to the position attributed to Einstein by Hentschel [8] and Howard [9].

(*willkürlich*, literally: as we like to choose it) character of the relation between theoretical concepts and measurement results served the purpose of making it clear that if we want to make progress in empirical science no concept should be assumed to be unalterable in principle. Science, in particular fundamental physics, is only able to evolve because even cherished and ingrained concepts (like classical simultaneity) may be abandoned.

Reichenbach wrote frequently about the lessons philosophy should draw from the revolutions in physics, in particular from Einstein's theories of relativity. Typical are his 1922 publication *La signification philosophique de la théorie de la relativité* [15][9] and his contribution to the 1949 Einstein-Schilpp volume, entitled *The philosophical significance of the theory of relativity* [19]. In spite of the more than two decades between these two essays, they cover much of the same ground and their conclusions overlap.

In his 1922 article Reichenbach endeavors to drive home the point that it is absolutely wrong to object that the theory of relativity is "not intuitive" or "unintelligible". Such qualifications show only one thing, says Reichenbach, namely that the objector accords an absolute meaning to representations derived from tradition.[10] In particular, Reichenbach discusses the objection that relativistic simultaneity, though a logically flawless conceptual construct, contradicts an "immanent intuition of reason". As he explains, this objection presupposes a special faculty of reason; "some call this with Kant *reine Anschauung* or an *a priori faculty*; others speak about *phenomenological experience*".[11] It will come as no surprise that Reichenbach rejects these transcendental notions in the strongest terms.

It is relevant here to pay attention to Reichenbach's empiricist analysis of time, which he also sets against the Kantian reliance on *a priori* intuition. He says:

> Precisely because the theory of relativity considers any definition of time as admissible, we have the possibility to choose a certain definition of time and to demand that all observers will make use of it. For example, a moving observer may define time in such a way that it is identical to the time of a resting observer, so that the principle of the constancy of the speed of light will not hold for him though he still succeeds in describing all phenomena unambiguously. But it would be meaningless to call a time of this kind "absolute", since any other definition would serve the latter purpose as well.[12]

---

[9]This is a French translation of a German manuscript of Reichenbach's (translated by Léon Bloch). As far as I am aware the German original is no longer extant.

[10]c'est qu'on accorde une prédominance absolue aux représentations tirées de la tradition.

[11]Examinons alors une autre objection : bien que le temps des relativistes ne soit pas contradictoire logiquement, il contredit pourtant une intuition immanente de la raison... D'après cette objection, il existerait donc au-dessus de notre faculté de déduction logique une puissance particulière de la raison qui édicterait des prescriptions particulières en matière de simultanéité. Certains l'appellent avec Kant "intuition pure" ou "faculté a priori"; d'autres parlent d' "expérience phénoménologique".

[12]Précisément parce que la théorie de la relativité considère toute définition du temps comme admissible, nous avons la faculté de choisir une certaine définition du temps et de demander que tous les observateurs se servent de ce temps. Par exemple, un observateur en mouvement peut définir le temps de façon qu'il soit identique avec le temps d'un observateur immobile, alors le principe de constance de la vitesse de la lumière cesse d'être valable pour lui et pourtant il parvient à décrire

As becomes clear from this passage (and others of the same kind), Reichenbach employs the "arbitrariness" of the definition of time to argue that one is free to introduce a notion of time that is different from the one suggested by everyday intuition and that is more adapted to the requirements of physics. It is not an academic exercise in the logic of definitions that motivates Reichenbach's emphasis on "arbitrariness". Rather, it is the concrete opportunity for conceptual change in science, through the possibility of breaking away from traditional intuition, that lead him to stress our liberty in establishing new relations between theoretical concepts and empirical data.

Summarizing the philosophical lesson taught by relativity, at the end of his 1922 essay, Reichenbach mentions as the decisive point that the apodictic status of Kant's "forms of intuition" has been refuted because relativity theory has shown the possibility of empirical knowledge under other preconditions than the Kantian ones. To salvage Kant one should have to show that it is *necessary* for obtaining knowledge to presuppose the Kantian categories and forms of pure intuition; but relativity theory has demonstrated beyond any doubt that the Kantian system is not needed for a consistent and potentially valid description of nature. Even though a stubborn Kantian could hold fast to a non-relative notion of simultaneity without an immediate conflict with experiment, it has become impossible to believe that such a notion of time constitutes a *precondition* of experience, as there are other viable definitions of time that in fact are used by physicists.[13] This is exactly where the non-uniqueness of coordinative definitions comes in.[14]

In his contribution to *Albert Einstein: Philosopher-Scientist* [19], entitled *The philosophical significance of the theory of relativity*—virtually the same title but 27 years later—Reichenbach again starts out by contrasting Einstein's theories to Kant's transcendental philosophy and "reine Anschauung". As he announces as his final conclusion already in the beginning of the article, Kant's doctrine has become untenable in view of relativity. Elaborating, Reichenbach makes the diagnosis that the philosophical significance of the theory of relativity lies in the discovery that many statements that were once regarded as demonstrable truths or falsities turn out to be "mere definitions". Such definitions are arbitrary, we hear again, and this leads to a plurality of empirically equivalent descriptive systems, with the same empirical content. For example, observers in different inertial systems are entitled, because of this "arbitrariness" of the definition of simultaneity, to define different classes of events to be simultaneous, and they can do so without coming into conflict with

---

sans ambiguïté l'ensemble des phénomènes. Mais il serait dépourvu de sens d'appeler "absolu" un temps de ce genre, car toute autre définition pourrait rendre le même office.

[13] Il est seulement impossible de croire que ce temps absolu soit une condition de l'expérience.

[14] It is to be noted that in this 1922 essay, dated after his exchange with Schlick and his supposed conversion to logical positivism, Reichenbach still emphasizes that the rejection of apodicticity does not imply that we do not need *constitutive* principles; quite the opposite, physics does need such principles, although they must be considered as changing and never final. "Nous ne pouvons affirmer qu'une chose, c'est que dans l'état présent de notre savoir, on emploie tels et tels principes pour la définition des objets, et nous devons essentiellement admettre que non seulement les résultats de chaque science mais encore le concept de la chose physique, du réel et de sa détermination, sont soumis à une évolution continuelle.".

measurement results. This possibility, of which relativity theory takes advantage, implies a dissolution of Kant's *synthetic a priori* of absolute time.

Indeed, Reichenbach [19, p. 307] writes, "It is this process of a dissolution of the *synthetic a priori* into which we must incorporate the theory of relativity, when we desire to judge it from the viewpoint of the history of philosophy." As he explains [19, p. 309]:

> Kant believed himself to possess a proof for his assertion that his synthetic a priori principles were necessary truths: According to him these principles were necessary conditions for knowledge. ... What has happened, then, in Einstein's theory is a proof that knowledge within the framework of Kantian principles is not possible. For a Kantian, such a result could only signify a breakdown of science. It is a fortunate fact that the scientist was not a Kantian and, instead of abandoning his attempts of constructing knowledge, looked for ways of changing the so-called *a priori* principles. Through his ability of dealing with space-time relations essentially different from the traditional frame of knowledge, Einstein has shown the way to a philosophy superior to the philosophy of the *synthetic a priori.*

Reichenbach's argument here faithfully echoes that of his 1922 essay and his 1920 book: The philosophical importance of the theory of relativity is its break with principles that once seemed unquestionably valid. Einstein's theory refutes the apodicticity of the Kantian principles, and this is made possible by replacing so-called necessary principles of thought and intuition with "arbitrary definitions".[15]

5 Weyl's Phenomenology of Space and Time

According to Hermann Weyl philosophy had an even more important role to play in science than judged by his contemporaries, like Reichenbach and Schlick. In the Preface to the first edition (1918) of his famous *Raum, Zeit, Materie* (*Space, Time, Matter*) Weyl [25] wrote: "I wanted to illustrate with this great subject how philosophical, mathematical and physical thought permeate each other, a topic that is dear to my heart."[16] This statement neatly illustrates Weyl's view that philosophy is not only a tool to be used on a meta-level, with the sole purpose of analyzing existing scientific theories, but also has a direct significance for ongoing scientific research. Philosophical analysis is relevant for the formation of scientific concepts and the development of new theory.

---

[15]Even in its details it is striking how Reichenbach's later essay follows his work from the 1920 s; for example, his earlier relativized *a priori* returns as "changing a priori principles". It is also true that Reichenbach uses typical logical-positivist language in some places, for instance by briefly referring to the verifiability theory of meaning. In our opinion these are philosophical adornments that do not affect the core of Reichenbach's message, which is that there exists a conceptual freedom in science that was denied by traditional philosophy.

[16]Zugleich wollte ich an diesem großen Thema ein Beispiel geben für die gegenseitige Durchdringung philosophischen, mathematischen und physikalischen Denkens, die mir sehr am Herzen liegt.

A large part of the Introduction of *Raum, Zeit, Materie* is taken up by a summary of the philosophical ideas Weyl has in mind, and differences with empiricism soon become clear. Weyl declares: "the real world, and every one of its constituents with their accompanying characteristics, are, and can only be given as, intentional objects of consciousness." And, a bit later: "*Pure consciousness* is the seat of that what is philosophically *a priori*." It follows that if we want to understand with what concepts we should describe physical reality, we have to start with an investigation of what can be found in consciousness. This applies to concepts used in order to describe objects, but also to space and time. Indeed, Weyl states, time is the form of the stream of our consciousness, and space is the form by which we order and represent external material reality. Both are forms of our perception.

According to Weyl, a philosophical analysis of concepts that we use for ordering our perceptions should try to penetrate to the essential core contained in those concepts. The method to achieve this consists in mentally removing aspects that are only contingent, by comparing different perceptions of an object, under different conditions. In this way we can distinguish characteristics that vary from case to case from the *essence*, which is always present, under all circumstances.

It is not difficult to recognize in this Husserl's method of "bracketing" (*ausklammern*) non-essential elements by means of the mental operation of "epoché", mentioned in the Introduction. Indeed, Weyl adds the note: "The detailed development of these ideas follows very closely the lines of Husserl in his *Ideen zu einer reinen Phänomenologie und phänomenologischen Philosophie*." The Introduction of *Raum, Zeit, Materie* can in fact be seen as a quick overview of Husserl's brand of neo-Kantian transcendental idealism, *phenomenology* [10], as emphasized and further explained by Ryckman [22]. The phenomenological point of view is not restricted to Weyl's Introduction, but returns repeatedly in the book's later, technical chapters, in particular in the context of Weyl's proposal for a generalization of general relativity (see the next section). It can also be found in other publications by Weyl.

The stress laid by Weyl on the indispensable role of consciousness and intuition, and their ineliminable contribution to concept formation, may suggest that introspection paired with bracketing can in principle lead to definitive verdicts about the essential content of our concepts. That impression would be wrong, however. It is not Weyl's intention to propose a completely rationalist *a priori* philosophy of science. Since new perceptions will be realized continuously, more and more aspects of concepts will be found to be contingent, so that we can come ever closer to the true essences—although we may never actually attain them. The concepts that we use to structure experience are therefore always tentative and approximate, and subject to revision under the pressure of observation. As Weyl writes in his Introduction [25, English translation by Brose, p. 5][17]:

[17]The German original differs slightly in its nuances: "Es liegt im Wesen eines wirklichen Dinges, ein Unerschöpfliches zu sein an Inhalt, dem wir uns nur durch immer neue, zum Teil sich widersprechende Erfahrungen und deren Abgleich unbegrenzt nähern können. In diesem Sinne ist das wirkliche Ding eine Grenzidee. Darauf beruht der empirische Charakter aller Wirklichkeitserkenntnis.".

> It is the nature of a real thing to be inexhaustible in content; we can get an ever deeper insight into this content by the continual addition of new experiences, partly in apparent contradiction, by bringing them into harmony with each other. In this interpretation, things of the real world are approximate ideas. From this arises the empirical character of all our knowledge of reality.

This acknowledgment of the importance of empirical data, and the possibility of conceptual revision, brings Weyl closer to the philosophy of the relativized *a priori* and (sophisticated) empiricism. There remains an important difference, though. According to Weyl, there exists a rock-bottom limit of our representational means, linked to the possibilities of our intuition and consciousness; whereas for Reichenbach, as we have seen, concepts are "completely arbitrary definitions", where "arbitrary" should be interpreted as "in principle free from limitations imposed by intuition".

In *Raum, Zeit, Materie* Weyl uses his phenomenology as the backdrop against which he formulates and explains relativity theory and his own "gauge" extension of it. In the context of space-time theories, like general relativity, it is important to ask how an ideal perceiving subject, the "transcendental ego", constitutes the essential geometrical forms of representation of space and time. In line with what was just explained, to answer this question Weyl investigates which elements of all geometrical perceptions are immediately given to us and evident, and therefore not contingent. He argues that these essential aspects pertain only to geometrical relations in infinitely small regions, centered around the perceiving ego. The reason is that when we start to imagine geometrical objects at some distance from each other and from ourselves, their mutual relations become less evident for our geometrical intuition; a feeling of uncertainty creeps in. For example, it is not immediately clear in pure perception whether geometrical figures at a finite distance from each other are congruent, or whether distant vectors point in the same direction. Therefore, Weyl claims, space-time in fundamental physics should be characterized by a geometry that can be constructed from the geometries of infinitesimal regions, i.e., the immediate infinitesimal neighborhoods of space-time points (22, p. 148, 29):

> Only the spatio-temporally coinciding and the immediate spatial-temporal neighborhood have a directly clear meaning exhibited in intuition. . . . The philosophers may have been correct that our space of intuition bears a Euclidean structure, regardless of what physical experience says. I only insist, though, that to this space of intuition belongs the ego-center and that the coincidences, the relations of the space of intuition to that of physics, becomes vaguer the further one distances oneself from the ego-center.

According to Weyl infinitesimal geometry is the only geometry that remains when we remove all geometrical notions that are contingent and not necessary for our spatio-temporal intuition and thinking. This geometry is essential for our representation of space and time. Infinitesimal geometry should consequently constitute the *a priori* basis of our space-time description, also in theoretical physics: it defines what space-time *is* for us, when stripped from all non-essential characteristics. It is then left to experience to decide which of the many possible space-time theories that can be theoretically constructed by pasting infinitesimal regions together, with different geometrical connections between these regions, is empirically adequate [26].

This infinitesimal grounding of the representation of space-time is the motivating idea behind Weyl's proposed generalization of general relativity, in which there is no globally defined standard of length and in which the laws of physics are required to be invariant under local scale transformations.

6 Weyl's Gauge Theory

As we have seen, according to Weyl pure intuition informs us about the most general representation of space-time: it is an infinitesimal geometry built up from infinitesimal regions, without *a priori* imposed connections between them. For intuition it only makes sense to compare quantities at infinitesimally close points, and when we move further away "from the ego-center" our ability to immediately compare geometric objects disappears. As Weyl sees it, this should be reflected in the most general possible form our space-time theories may assume: in this most general form theories should not contain an absolute standard of length, but rather should be invariant under local changes of scale. In other words, there should be a "relativity of magnitude". As Weyl [25, p. 283] states: "The same intuitive certainty that characterizes the relativity of motion accompanies the principle of the relativity of magnitude."

This thought led Weyl [25, §16] to consider metrical spaces in which the metric at each point can be calibrated arbitrarily. That is, compared to a Riemannian-Einsteinian space with a position-dependent metrical tensor $G(x)$, Weyl considered spaces in which this tensor can be multiplied by an arbitrary position dependent calibration scalar $c(x)$, so that $G'(x) = c(x)G(x)$. In an arbitrary calibration it will be the case that if l is a distance at a point P, and $l + dl$ the measure of this distance when congruently displaced to the infinitesimally near point P', we will find $dl = -ld\varphi$, in which the infinitesimal factor $d\varphi$ is independent of l. When a different calibration is chosen (i.e. a different unit of length at each position) with calibration ratio λ we will have $\bar{l} = \lambda l$, $d\bar{l} = -\bar{l}d\bar{\varphi}$, with $d\bar{\varphi} = d\varphi - d\lambda/\lambda$. The necessary and sufficient condition that an appropriate value of λ can make $d\bar{\varphi}$ vanish at P for infinitesimal displacements (as required, for we can directly, intuitively and thus objectively compare distances in infinitesimal regions) is that $d\varphi$ is a differential form: $d\varphi = \varphi_i dx_i$.

So the metrical properties of the manifolds considered by Weyl are characterized by two forms: a quadratic metrical form, just as in standard general relativity, plus a linear form $d\varphi = \varphi_i dx_i$ characterizing how lengths change under congruent transport. As just stipulated, it is always possible to choose a calibration (gauge) so that l remains constant under infinitesimal displacements. However, such a choice does not exclude that transporting a length along two different *finite* paths to a new position will yield different results: there may be a global path-dependence of length under congruent transport, just as there is a global path-dependence of *direction* under parallel transport in standard general relativity (depending on the curvature of space-time).

When we calculate this length difference for the case of congruent transport along an infinitesimal parallelogram with sides dx_i, δx_i, we find $\Delta l = -l\Delta\varphi$, with $\Delta\varphi = f_{ik}dx_i\delta x_k$ and $f_{ik} = \partial\varphi_i/\partial x_k - \partial\varphi_k/\partial x_i$.[18]

So f_{ik} can be interpreted as a measure of "distance curvature"; its vanishing is necessary and sufficient for the possibility of transporting any distance unambiguously to any other place (without path-dependence). This case of vanishing distance curvature returns us to standard general relativity. The formalism of Weyl spaces thus suggests a generalization of general relativity in which the distance curvature does *not* vanish. Weyl famously proposed that this distance curvature is dynamically determined, analogously to the dynamical determination of Riemann curvature in the general theory of relativity. The Riemann curvature depends on energy and momentum, and Weyl's analogous proposal is that the distance curvature is determined by electric charges and currents. Or, viewing it from the opposite direction, the electromagnetic field tensor f_{ik} in Weyl's gauge theory is a manifestation of the distance curvature and so of the geometrical structure of the world.

Weyl [25, p. 283] accordingly wrote:

> it immediately suggests itself to us ... to identify the coefficients of the linear groundform $\varphi_i dx_i$ with the electromagnetic potentials. The electromagnetic field and the electromagnetic forces are then derived from the metrical structure of the world.

7 Non-Weylian Routes to Infinitesimal Geometry and Gauge Freedom

The core idea of Weyl's approach is that space-time theories in physics should start from our most basic and essential spatio-temporal intuitions. This led him to the general conceptual framework in which infinitesimal space-time regions, accessible to pure consciousness, build up an infinitesimal geometry (*Nahegeometrie*). The details of the physical relations *between* these infinitesimal space-time regions, for example the affine connection governing parallel transport and the distance curvature governing length comparisons, are not part of the *a priori* framework but must be left to empirical determination.

Weyl's appeal to phenomenological foundations is certainly atypical within modern physical theorizing, but the notion of infinitesimal geometry itself is not at all an uncommon ingredient of theory construction in mathematical physics. Weyl's analysis is clearly not *necessary* to arrive at the ideas of *Nahegeometrie* and gauge freedom. Indeed, in Reichenbach's work we find ideas that are very close to those of Weyl. As Reichenbach argued in several places, for example in *The Philosophy of Space and Time*, Chapter I, in physical geometry the unit of length can in principle be chosen independently in each spatial point. For instance, as he argues, we could stipulate that a given tiny measuring rod represents one unit at P, two units at P',

[18] The tensor with components f_{ik} satisfies the equation $\partial f_{kl}/\partial x_i + \partial f_{li}/\partial x_k + \partial f_{ik}/\partial x_l = 0$, with suggests a relation with Maxwell's equations—this suggestion is taken up by Weyl.

three units at P'', and so on. Different choices of this kind lead to different verdicts about the applicable geometry. In practice we do not make use of such outlandish measures of length, as it would complicate matters unnecessarily and, above all, would spoil the relation with what we understand by "equal length" in daily life. Accordingly, we use the "natural geometry" that results from using units that are considered to preserve their lengths under transport. This enables us, in principle, to define a unit of length at each position by transporting one ideal standard rod.[19] Now there is an important *observational fact* that makes this standard length definition with the help of transported rods unambiguous: two measuring rods shown to be equal in length by local comparison at a certain point will be found to be equal in length by local comparison also at all other space points, whether they have been transported along the same or different paths [18, pp. 16–17]. If this factual relation did not hold, a definition of the unit of length would have to be given for every space point separately.

There is an implicit appeal to infinitesimal regions in this account, via its definition of the metric and other quantities *per point*. As Reichenbach explains, there is a physical justification for appealing to this "differential outlook". Indeed, rods, clocks and other measuring devices should be "closed systems"; that is, systems for which the quotient of external and internal physical forces becomes negligible when the size of the system decreases [16, p. 148]. Only such infinitesimal closed systems can be considered to be completely unaffected by distorting external forces. Of course, the in-principle demand that rods and clocks should possess infinitesimal dimensions is usually unimportant in practice. However, in the most general situations of vehemently accelerating systems and rapidly varying forces the limit of vanishing sizes should actually be considered.

In Reichenbach's account we find back all the conceptual ingredients that we encountered in the discussion of Weyl's gauge ideas. There is one important difference though: Weyl's motivating thought about the accessibility for pure intuition of infinitesimally small regions is in a Reichenbachian approach replaced by a *physical motivation* for focusing on the infinitesimally small and its *Nahegeometrie*. This physical motivation boils down to the argument that "rigid bodies" of finite size do not exist and that different parts of finite systems will generally be affected differently by external forces. This in turn connects to the physical fact that causal influences propagate with speeds lower than, or equal to, the speed of light (light is a "first-signal", as Reichenbach put it).

In fact, in his 1922 essay on the philosophical significance of the theory of relativity Reichenbach himself compares his ideas to those of Weyl. He writes:

> One cannot say in an absolute sense that a ruler is equal to another ruler at a distance, but one can arbitrarily stipulate a procedure for comparing them. From the viewpoint of the theory of knowledge this is the meaning of the statement that size is relative. But Weyl goes beyond this statement. He demands that under all definitions for comparing lengths the laws of physics keep the same form; this is the physical claim of the relativity of length. This claim

[19] As Reichenbach remarks elsewhere, as we have pointed out before, it is not essential to connect the concept of length to the existence of material measuring rods; we may replace measurements with rods by some other procedure, for example using light and particle geodesics [16, p. 154].

> means that the systems of physics that are possible according to the theory of knowledge, and that arise from different definitions for comparing length, are mutually equivalent. However great Weyl's ingenuity has been in mathematically developing this theory, its validity can only be decided by experience.[20]

Weyl would actually agree with the observation that only empirical data can decide about the fate of his gauge theory—it was not his intention to suggest that the complete theory can be deduced and proved valid *a priori* [26]. The real distinction between Weyl and Reichenbach is in the difference between Weyl's motivation based on a phenomenological analysis of intuition and the nature of space, and Reichenbach's argument that *physical facts* leave the choice of units at different points open.

A similar story can be told regarding the concept of *time*. One of the topics for which *The Philosophy of Space and Time* has become famous (or perhaps notorious) is Reichenbach's discussion of relativistic simultaneity, and in particular the thesis that simultaneity in special relativity is conventional [18, §19]. Reichenbach's analysis runs as follows. When we want to synchronize two clocks A and B we need to know the velocity of a signal (for example a light signal) connecting the clocks. However, any measurement of this velocity requires the use of already synchronized clocks, so that we get caught in a circle, leading to underdetermination of the theoretical framework by empirical data. This underdetermination can be expressed with the help of Reichenbach's famous ϵ formula. If we attempt to synchronize clocks A and B by sending light from A to B and back again, so that a light signal leaves at time t_1 (as indicated by A), arrives at B at time t_2 (indicated by B), and returns to A at t_3 (on A again), we can use the following criterion as a synchronization rule: $t_2 = t_1 + \epsilon(t_3 - t_1)$, with $0 < \epsilon < 1$. That is, if t_3—indicated by clock B at the moment B is hit by the light ray—satisfies this formula, we conclude that B is synchronized with A. The standard ("natural") choice is of course given by $\epsilon = 1/2$, which implies that the speeds of light to and fro between A and B are equal to each other. Other values of ϵ lead to different velocities of light in the two directions. We have empirical access to t_1, t_2 and t_3, and to the unchanging distance L between the clocks, but not independently to the one-way speeds of light. The only speed that we can directly measure is the round-trip speed $2L/(t_3 - t_1)$. It is this round-trip speed of light, as measured in inertial frames, that is always found to have the same value c.

The different time accounting systems that result from different choices for ϵ are analogous to the different geometrical systems associated with different choices of the unit of length, at different spatial positions, in our discussion above. As in the

[20] On ne peut pas dire qu'une règle soit égale au sens absolu à une autre règle placée à distance, mais on peut disposer arbitrairement du procédé de comparaison de ces règles. Tel est le sens, au point de vue de la théorie de la connaissance, de l'affirmation de la relativité de la grandeur. Mais Weyl va au delà de cette affirmation. Il exige que pour toute définition de la comparaison des grandeurs les lois physiques gardent la même forme, c'est là l'affirmation physique de la relativité de la grandeur. Elle signifie que les systèmes de physique possibles au point de vue de la théorie de la connaissance et qui prennent naissance par suite de définitions différentes de la comparaison des grandeurs sont équivalents entre eux. Quelle que soit l'ingéniosité mise par Weyl dans le développement mathématique de cette théorie, son exactitude ne peut être décidée que par l'expérience.

spatial case, the situation can be described in the language of gauge theories [1, 11]. When we denote the "natural" time ($\epsilon = 1/2$ everywhere) by t, the time that results from a position-dependent non-standard synchronization will be different and can be written as: $t' = t + \varphi(x)$. In this time system the one-way velocity of light along the unit vector $\vec{n}$ will be given by:

$$\vec{v} = (c\vec{n})/(1 + c\vec{n}.\nabla\varphi(x)), \tag{1}$$

with c the ordinary speed of light. The round-trip time needed by light to go along a closed path of length L is given by

$$\tau = L/c + \oint \vec{A}.d\vec{l}, \tag{2}$$

with $\vec{A} \equiv \nabla\varphi$ so that $\nabla \times \vec{A} = 0$. This means that the round-trip speed of light will always be c, regardless of the chosen gauge, exactly as it should be. This is analogous to what we would find in the case of a non-standard choice for the unit of length at different positions, in other words an arbitrary *length gauge*, under the condition that the result of the congruent transport of a given length should be path-independent (that is, when we describe ordinary Riemannian geometry by means of a Weyl space-time).

That the round-trip speed of light always and everywhere equals c is in this case due to the relation $\nabla \times \vec{A} = 0$. This is therefore a relation that expresses a global connectedness in the manifold that may be considered to be at odds with the idea of *Nahegeometrie*, an infinitesimally defined geometry. This then invites a mathematical/physical *hypothesis*: might the formalism be generalized, so that the round-trip velocity of light no longer remains always and everywhere the same? This can be achieved by positing $\nabla \times \vec{A} \neq 0$. When we now repeat the above calculations, we find that light traversing the same closed curve in opposite directions will generally need different amounts of time to come back to its initial position. If we take for the field $\vec{A}$ the specific form $\vec{A}(\vec{x}) = (\vec{\omega} \times \vec{x})/c^2$, we obtain for the time difference:

$$\Delta\tau = 2\oint \frac{(\vec{\omega} \times \vec{x}).d\vec{l}}{c^2}. \tag{3}$$

This formula is in fact well-known: it represents the time difference of the so-called Sagnac effect, which occurs in rotating systems. On a disk rotating with angular velocity $\vec{\omega}$ with respect to an inertial system a light signal traveling clockwise along a circle needs a different amount of time to come back than a signal going counterclockwise, and the difference is given by Eq. (3).

We could speculate further: In the above formula, $\vec{\omega}$ is a fixed ("absolute") quantity referring to rotation relative to a pre-given global inertial system. We could try to turn the equation into a dynamical one, by relating $\vec{\omega}(\vec{x})$ to sources in terms of masses and energy. This might inform us about rotation with respect to dynamically determined local inertial frames and thus might lead into the direction of a theory of gravity. So by hypothetically postulating a dynamic instead of a rigid $\vec{A}$ field we might be led into the direction of new physics, following exactly the same route as in Weyl's gauge theory or in the gauge theory of electromagnetism. But it is not our intention to explore such possibilities. Our point is simply that empirically and logically motivated ideas naturally lead to hypotheses about physical gauge theories of exactly the same sort as Weyl put forward on the basis of his Husserlian phenomenological reflections.

8 Conclusion: Physics and Intuition

The use of phenomenological intuition à la Weyl is therefore certainly not *necessary* for arriving at the uncontroversially fruitful idea of infinitesimally defined physical quantities. But neither does this phenomenological intuition appear *sufficient* for the purpose. To see this, it should first be noted that it is not self-evident that the phenomenological procedure of removing all non-essential elements (epoché) from our pure spatio-temporal representations actually leads us to infinitesimal geometry; it is even controversial whether the infinitesimally small is accessible to intuition at all. In the history of mathematics the notion of infinitesimal quantities has long been suspect for exactly this reason—our spatial and temporal intuition deals with small, but still *finite*, spatio-temporal intervals whereas the infinitesimally small has only become acceptable through abstract mathematical arguments.

But even if we were to accept that infinitesimally small space-time regions constitute the essence of our intuitive notions of space and time, what consequences should this have for the fundamental concepts to be used in theoretical physics? It is true that the phenomena that we perceive should be explainable by fundamental physics, and the same might be argued for our perceptions themselves, and *a fortiori* for what is common to them under all possible circumstances. But the direction of explanation cannot be reversed: it does not follow that physics should use what is common to all our perceptions, and in this sense is essential for perception and intuition, as conceptually fundamental. It could well be that on a truly fundamental level the world according to physics is not close to phenomenological intuition at all. In that case, the imposition of a phenomenological *a priori* would hinder the development of physics rather than providing a basis for it, just as the original Kantian "conditions for the possibility of knowledge" risked to do.

To illustrate that this possibility should be taken seriously, one has to think only of quantum mechanics. In this theory physical systems and their properties often have "holistic" features. For example, many-particle systems usually have properties that do not supervene on local properties of the composite systems. This is due to the fact that in general many-particle states are entangled, and such entangled states—famous for their role in discussions of Bell inequalities—defy a point-by-point definition in three-dimensional space.[21]

Another example is furnished by present-day research in quantum gravity: the dominant opinion in this area of physics is that space and time are not truly fundamental notions at all—instead, space and time are thought to *emerge* from a deeper layer of reality that is not spatio-temporal in character and not amenable to visualization. Such a layer of reality would be beyond the reach of phenomenological intuition (see, e.g., Dieks et al. 4).

In fact, Weyl himself seemed to be sensitive to similar points. In the motivation of his 1929 new gauge theory [28], in which electromagnetism appears as a quantum phenomenon linked to invariance under local gauge transformations of the phase of the wave function (instead of to length calibrations), we find no reference to phenomenology anymore. Weyl there states, "the electromagnetic potentials have an invariance property that formally (*in formaler Hinsicht*) resembles that of my old gauge theory: $\Psi \to \Psi.e^{i\lambda(x)}$; $\varphi_j(x) \to \varphi_j(x) - \partial\lambda(x)/\partial x_j$." There is thus only a *formal* continuity with his earlier considerations about space and time. Indeed, it is notoriously difficult to associate anything intuitive and visualizable, and in this sense non-formal, with the quantum wave function and its phase—the wave function is not even defined in ordinary space, but rather in a mathematically constructed higher-dimensional configuration space. Weyl accordingly concludes (my italics), "it seems to me that this new gauge principle, *which stems not from speculation but from experience*, compellingly shows that the electric field is a necessary consequence not of gravity but of matter, represented by Ψ." This statement seems almost a disavowal of his earlier methodological starting point, which is now referred to as speculative. Similarly, in his discussion with Einstein about the empirical viability of his first gauge theory, Weyl [24] argued famously that the transported length figuring in his gauge theory had to be considered a length living in a mathematically described abstract phenomenological space, and that it was not immediately evident to what this phenomenological length corresponded in physical applications of the scheme, in terms of the behavior of measuring rods or other material systems. This also comes close to a confession of skepticism regarding the direct physical significance of phenomenological space and intuition.

In §8 of his *Philosophy of Space and Time* Reichenbach argued that even if it were correct (which he thought not to be the case) that intuition singles out one unique geometry, this should not be taken as normative for physics, because physics is not about our possibilities of intuitive representation but rather about empirically found relations between physical objects and processes that may be far removed from intuition. It seems that Weyl's introduction of the idea of local gauge invariance in physics illustrates the correctness of this verdict rather than that it provides reasons to revise it.

---

[21] Different interpretations of quantum mechanics take different views concerning how this should be understood. But this is not relevant: the standard interpretation makes use of holistic features, and this is sufficient to show that the conceptual framework of infinitesimal geometry and physical quantities defined per point is not *indispensable* for making theoretical sense of our spatio-temporal perceptions.

References

1. R. Anderson, I. Vetharaniam, G. Stedman, Conventionality of synchronization, gauge dependence and test theories of relativity. Phys. Rep. **295**, 93–180 (1998)
2. F. Benedictus, D. Dieks, Reichenbach's transcendental probability. Erkenntnis **80**, 15–38 (2015)
3. D. Dieks, Reichenbach and the conventionality of distant simultaneity in perspective, in *The Present Situation in the Philosophy of Science*, ed. by F. Stadler, et al. (Springer, Dordrecht, 2010), pp. 315–333
4. D. Dieks, J. van Dongen, S. de Haro, Emergence in holographic scenarios for gravity. Stud. Hist. Philos. Mod. Phys. **52**, 203–216 (2015)
5. M. Friedman, *Reconsidering Logical Positivism* (Cambridge University Press, Cambridge, UK, 1999)
6. M. Giovanelli, Talking at cross-purposes: how Einstein and the logical empiricists never agreed on what they were disagreeing about. Synthese **190**, 3819–3863 (2013)
7. A. Grünbaum, *Philosophical Problems of Space and Time* (Reidel, Dordrecht, 1963)
8. K. Hentschel, Einstein, Neokantianismus und Theorienholismus. Kantstudien **78**, 459–470 (1987)
9. D. Howard, Einstein and the development of twentieth-century philosophy of science, in *The Cambridge Companion to Einstein*, ed. by M. Janssen, C. Lehner (Cambridge University Press, Cambridge, 2009), pp. 354–376
10. E. Husserl, Ideen zu einer reinen Phänomenologie und phänomenologischen Philosophie. Jahrbuch für Philosophie und phänomenologische Forschung **1**, 1–323 (1913)
11. E. Minguzzi, On the conventionality of simultaneity. Found. Phys. Lett. **15**, 153–169 (2002)
12. F. Padovani, *Probability and Causality in the Early Works of Hans Reichenbach.* Faculty of Letters, Dissertation at the University of Geneva, Geneva (2008)
13. H. Reichenbach, *Der Begriff der Wahrscheinlichkeit für die mathematische Darstellung der Wirklichkeit.* Leipzig: Johann Ambrosius Barth. English translation by F. Eberhardt and C. Glymour (2008), *The Concept of Probability in the Mathematical Representation of Reality.* Open Court, Chicago and La Salle, Illinois (1916)
14. H. Reichenbach, Relativitätstheorie und Erkenntnis Apriori. Berlin: Julius Springer; English translation by M. Reichenbach, *The Theory of Relativity and A Priori Knowledge* (University of California Press, Berkeley, 1965) (1920)
15. H. Reichenbach, La signification philosophique de la théorie de la relativité (traduit par L. Bloch). Revue philosophique de la France et de l'Étranger **93**, 5–61 (1922)
16. H. Reichenbach, Axiomatik der relativistischen Raum-Zeit-Lehre. Braunschweig: Friedrich Vieweg und Sohn. English translation by M. Reichenbach, *Axiomatization of the Theory of Relativity* (University of California Press, Berkeley, 1969) (1924)
17. H. Reichenbach, Über die physikalischen Konsequenzen der relativistischen Axiomatik. Zeitschrift für Physik **34**, 32–48 (1924)
18. H. Reichenbach, Philosophie der Raum-Zeit-Lehre. Berlin: Walter de Gruyter. English translation by M. Reichenbach, *The Philosophy of Space and Time* (Dover Publications, New York, 1958) (1928)
19. H. Reichenbach, The philosophical significance of the theory of relativity, in P.A. Schilpp, Ed., *Albert Einstein. Philosopher-Scientist* (Open Court, La Salle, Illinois, 1949), pp. 289–311
20. H. Reichenbach, *The Rise of Scientific Philosophy* (University of California Press, Berkely, 1951)
21. H. Reichenbach, M. Schlick, Correspondence. http://echo.mpiwg-berlin.mpg.de/content/space/space/reichenbach1920-22 (1920–1922)
22. T. Ryckman, *The Reign of Relativity-Philosophy in Physics 1915–1925* (Oxford University Press, Oxford, 2004)
23. R. Rynasiewicz, Weyl vs. Reichenbach on Lichtgeometrie, in A.J. Kox, J. Eisenstaedt (Eds.), *The Universe of General Relativity* (Birkhäuser, Boston, 2005), pp. 137–156
24. H. Weyl, *Erwiderung auf Einsteins Nachtrag zu H. Weyl, Gravitation und Elektrizität.* Sitzungsberichte der Preussischen Akademie der Wissenschaften (1918), pp. 478–480

25. H. Weyl, *Raum, Zeit, Materie. Vorlesungen über allgemeine Relativitätstheorie, vierte Auflage.* Berlin: Julius Springer. English translation by Henry L. Brose: *Space-Time-Matter* (Dover Publications, New York, 1952) (1921a)
26. H. Weyl, Über die physikalischen Grundlagen der erweiterten Relativitätstheorie. Physikalische Zeitschrift **22**, 473–480 (1921b)
27. H. Weyl, Rezension von Hans Reichenbach, Axiomatik der relativistischen Raum-Zeit-Lehre. Deutsche Literaturzeitung **45**, 2122–2128 (1924)
28. H. Weyl, Gravitation and the electron. Proc. Natl. Acad. Sci. USA **15**, 323–334 (1929)
29. H. Weyl, Geometrie und Physik. Die Naturwissenschaften **19**, 49–58 (1931)

Hermann Weyl, the Gauge Principle, and Symbolic Construction from the "Purely Infinitesimal"

Thomas Ryckman

Only in the infinitely small may we expect to encounter elementary uniform laws; hence the world must be comprehended through its behavior in the infinitely small.
(Weyl [32, 61, 44, 86])

Abstract The modern gauge principle stipulates that every global symmetry of a quantum field theory be replaced by a local one. It has an unusual "context of discovery": imposed in 1918 on philosophical grounds by Hermann Weyl, it led to a purely formal unification of Einstein's gravitational theory and electromagnetism. This achievement prompted Weyl's purely mathematical turn in 1925–6 to Lie theory and of course Lie groups and Lie algebras play prominent roles in the subsequent development of the gauge principle leading up to the Standard Model. I show that the gauge principle as well as Weyl's predominant interest in Lie theory stem from two complementary philosophical interests: (i) phenomenological evidential requirements imposed on differential geometric construction, and (ii) the epistemological command of "*Nahewirkungphysik*": to comprehend the world from its behavior in the infinitely small. (i) and (ii) productively meet in Weyl's notion of "symbolic construction". The gauge principle's "radical locality" is the basis of Weyl's constitution of objectivity as an intersubjectivity requiring local degrees of freedom. This leads to a necessary redundancy of physical description, a philosophical puzzle that might be elucidated by revisiting the philosophical underpinnings of the gauge principle.

1 Introduction

The gauge principle is a broad moniker about invariance properties of fundamental physical laws. It stipulates that every global symmetry of a quantum field theory be replaced by a local one; in effect, that every continuous symmetry of a quantum field

---

T. Ryckman (✉)
Stanford University, Stanford, CA, USA
e-mail: tryckman@stanford.edu

S. De Bianchi and C. Kiefer (eds.), *One Hundred Years of Gauge Theory*,
Fundamental Theories of Physics 199, https://doi.org/10.1007/978-3-030-51197-5_7

(i.e., under which the field Lagrangian transforms invariantly) be a local symmetry. The gauge principle is not quite a priori physics yet remarkably it provides an a priori framework for constructing the form of interaction between force and matter fields, a procedure that has been shown to be highly empirically successful. Invoked in 1918 on largely philosophical grounds by mathematician Hermann Weyl, it emerged in the context of classical general relativity and, in Weyl's hands, led to a purely formal unification of Einstein's gravitational theory and electromagnetism. For Weyl, the gauge principle encapsulated two desirable but complementary philosophical demands: (i) the phenomenological evidential requirements of "eidetic insight" and "eidetic analysis" imposed on differential geometric construction, and (ii) the metaphysical command of *"Nahewirkungphysik"*, that "the true lawfulness of nature is expressed in laws of nearby action, connecting only values of physical quantities at spacetime points in the immediate vicinity of one another" Weyl [32, 61, 44, 86]. The idea did not work in the context of general relativity but in 1929 Weyl himself carried the gauge principle over to quantum theory, its proper setting. Revived by Yang and Mills [51], the gauge principle's mandate of *radical locality* plays a central unifying role in the Standard Model (SM) of elementary particles, the quantum field theories describing three of the four known fundamental forces and the regnant theory of matter since 1978. The philosophical origin yet astonishing success of the gauge principle presents something of a puzzle. Moreover, a gauge symmetry is widely understood to be a redundancy of description, i.e., an unphysical symmetry merely relating mathematically different representations of the same physical state or history. One can agree with the assessment of a prominent philosopher of physics, that "the elucidation of [the gauge principle] is the most pressing problem in current philosophy of physics" [18]. Our thesis is that revisiting the philosophical underpinnings of the gauge principle can contribute to this elucidation.

2 Hermann Weyl and the Philosophical Origins of the Gauge Principle

To the philosopher of science who demurs from scientific realism but finds that the alternatives of antirealism or instrumentalism yield only an anemic understanding of the cognitive role of physical theory, the origin of the gauge principle by Hermann Weyl (1885–1955) presents an instructive case study. A preeminent mathematician of the 20th century, Weyl also made seminal contributions to the twin pillars of fundamental physical theory, general relativity and quantum mechanics. Nearly a half-century after his death, Fields medalist Sir Michael Atiyah noted the continuing extent of Weyl's influence:

> No other mathematician could claim to have initiated more of the theories that are now being exploited. His vision has stood the test of time. [1, 13]

In pure mathematics, Atiyah pointed to Weyl's work on the theory of Lie groups and algebras. In physical theory, Atiyah underscored Weyl's idea of gauge invariance that subsequently became the unifying framework of the SM. In fact, these two currents are thematically and philosophically related. And though not mentioned by Atiyah, Weyl authored a handful of philosophical works giving expression to the reflective musings of a highly innovative scientist. Weyl's philosophical orientation lies far from what logical empiricism regarded as "scientific philosophy"; closely intertwined with his scientific achievements, they are somewhat idiosyncratic. Drawing from figures and traditions largely unknown to contemporary philosophers of science (post-Kantian transcendental idealism including Fichte and Husserl, Nicholas of Cusa, even the medieval mystic Meister Eckert), Weyl's philosophical remarks are broadly stated, purposefully hesitant, and not articulated in any detail. After all,

> an epistemological conscience (*Erkenntnisgewissen*), sharpened by work in the exact sciences, does not make it easy … to find the courage for philosophical statement. One cannot get by entirely without compromise. (Weyl [46, 648])

Nevertheless, it is not difficult to identify a metaphysics of transcendental subjectivity underlying the two central accomplishments alluded to by Atiyah: the origin of the idea of gauge invariance in 1918 with its restatement in the context of quantum mechanics in 1929, and the 1925–6 purely mathematical work on Lie theory (on representations of semisimple Lie groups and Lie algebras), results that Weyl himself regarded as his greatest mathematical triumph. Both achievements are heuristically motivated by an injunction to comprehend the world from its behavior in the infinitely small, an evidential constraint implicating an intersubjective constitution of physical objectivity related to the intentional analyses of transcendental phenomenological idealism. While Weyl traced the impetus for this "purely infinitesimal" explanatory agenda back to Leibniz, he specifically associated it with Bernhard Riemann (in particular, the theory of Riemannian manifolds) and with Sophus Lie (the theory of continuous groups and their infinitesimal generators). Weyl would coin the term "Lie algebra" for the infinitesimal group structure of a Lie group; he also showed that this infinitesimal structure is a real-valued linear space, in fact, a vector space, a concept Weyl was in fact the first to define [23]. It is only fitting that Lie algebras play an important role in the contemporary gauge theories of the Standard Model.

2.1 Idealism in the Infinitesimal

Following the apt term of Bernard [2], Weyl's transcendental metaphysics is an "idealism in the infinitesimal". It is a modern descendant of Leibniz's *principle of continuity* ("*natura non facit saltus*") i.e., that all finite changes are to be comprehended as arising through infinitesimal increments acting in sequence.[1] Its modern

[1] On the principle of continuity, see Leibniz's letter to Varignon, 1702: "*Assurément je pense que ce Principe* ["*mon Principe de Continuité*"] *est general, et qu'il tient bon, non seulement dans la*

mathematical setting was provided by two titans of mathematics in the second half of the 19th century, Bernard Riemann and Sophus Lie:

> The productivity shown by the differential calculus, by contiguous action [field] physics (*Nahewirkungsphysik*), and by Riemannian geometry certainly rests upon the principle: To understand the world, according to its form and content, by its behavior in the infinitely small, clearly because *all problems can be linearized in passing to the infinitely small.*[2]

In geometry, Weyl observed, Riemann took the step Faraday and Maxwell had taken in physics, according to "the principle: to understand the world from its behavior in the infinitely small."[3] Geometrically, this is the tangent space T_P to each point P of a Riemannian manifold M. Just as in elementary differential calculus, functions differentiable at a point P on the function's graph are locally linear there, so in a Riemannian manifold the tangent space T_P to each point P is a linear space. These manifolds, according to Riemann, exhibited "planeness in their smallest parts", hence only linear relations are required.[4] The geometric concept of manifold was also the starting point of Sophus Lie's theory of continuous groups. A Lie group is a differential manifold whose points are the group elements, parameterized by continuous real variables. The points are combined by an operation (action) obeying the group axioms; they compose continuously to form a 'space'. Lie's seminal idea was to investigate the group actions on a manifold infinitesimally; indeed, the core idea of Lie theory, as characterized by Weyl, is "descent to the infinitely small".[5]

Géometrie, mais encore dans la Physique. La Géometrie n'étant que la science des limites et de la grandeur du Continu, il n'est point étonnant, que cette loi s'y observe par-tout: car d'où viendroit une subite interruption dans un subject, qui n'en admet pas en vertu de sa nature? Aussi savons-nous bien, que tout est parfaitement lie dans cette science, et qu'on ne sauroit alléguer un seul exemple, qu'une propriété quelconque y cesse subitement, ou naisse de même, sans qu'on puisse assigner le passage intermédiaire de l'une à l'autre, les points d'inflexion et de rebroussement, qui rendent le changement explicable; de manière qu'une Equation Algébraique, qui représente exactement un état, en représente virtuellement tous les autres, qui peuvent convenir au même sujet. L'universalité de ce Principe dans la Géometrie m'a bientôt fait connoitre, qu'il ne sauroit manquer d'avoir lieu aussi dans le Physique: puisque je vois, que, pour qu'il y ait de la règle et d'ordre dans la Nature, it est necessaire, que la Physique harmonie constamment avec le Géométrique; ..." "Brief von Leibniz an Varignon über das Kontinuiätsprinzip", in E. Cassirer (ed.), *G. W. Leibniz: Hauptschriften zur Grundlegung der Philosophie,* Bd. II. Leipzig: Verlag der Dürr'schen Buchhandlung, 1906, 556.

[2]Weyl [29, 9]: "*Es beruht ja die Leistungsfähigkeit des in der Differentialrechnung, der Nahewirkungsphysik und der* ***Riemann****schen Geometrie zum Durchbruch kommenden Prinzips: die Welt nach Form und Inhalt aus ihrem Verhalten im Unendlichkleinen zu verstehen, eben darauf, dass* **alle Problem durch den Rückgang aufs Unendlichkleine linearisiert werden**." See also Weyl [24, 82, 32, 61, 44].

[3]"Vorwort des Herausgebers", in Weyl (ed.) [26].

[4]A crucial feature of Einstein's theory of gravitation is that it allows (pseudo-) Riemannian geometry ("pseudo" since time is treated differently than the three space dimensions) to be the appropriate mathematical framework for the concept of "local inertial frame" and so to uphold the "infinitesimal" validity of special relativity in that theory.

[5]Weyl [29, 34]: "*Die Ersetzung der endlichen Gruppe durch die infinitesimale-- das ist wieder der 'Rückgang aufs Unendlichkleine'! - ist einer der Hauptgedanken der Lieschen Theorie.*" Original emphasis. Hawkins [7] quotes from an 1879 paper of Lie, "In the course of investigations on first-order partial differential equations, I observed that the formulas that occur in this discipline become

Though generically non-linear, Lie groups can be linearized in passing to a local (infinitesimal) group acting in the tangent space of the group identity. In the simplest case, the local Lie group acts on itself by left (or right) translations forming what Weyl [41] termed the "Lie algebra" of the Lie group. The Lie algebra is a much simpler and cognitively more accessible structure than the Lie group yet it contains most (non-topological) information about the group. This is the crucial fact on which the modern structure theory of Lie groups is based: their respective algebras yield a precise classification of Lie groups (see Sect. 3.1).

In the application of both Riemannian geometry and Lie theory to physics, Weyl's "idealism in the infinitesimal" is an epistemological mandate that *comprehensibility* of the physical world, i.e., phenomenological "sense-constitution", is to be gained by bottom-up symbolic construction starting from mathematical relations in the infinitely small. With both Riemannian manifolds and Lie groups in mind, Weyl would regard infinitesimal linear spaces as the legitimate epistemic reach of "eidetic vision" or "insight" or *"Evidenz"* (or whatever one will call it) of the cognizing, constructing subject, an "ego-center" whose secure epistemic range is mathematically understood as the "infinitesimally small" bounded linear region surrounding each point.[6] This is the entry point of transcendental phenomenological sense constitution and a core assumption of "idealism in the infinitesimal",

> Only the spatio-temporally coinciding and the immediate spatial-temporal neighborhood has a directly clear meaning exhibited in intuition. … The philosophers may have been correct that our space of intuition bears a Euclidean structure, regardless of what physical experience says. I only insist, though, that to this space of intuition belongs the ego-center [*Ich-Zentrum*] and that … the relations of the space of intuition to that of physics, becomes vaguer the further the distance from the ego-center. (Weyl [38, 49, 52])

Phenomenological requirements on evidence, on what is given to consciousness and what can be evidently constituted (in Weyl's mathematically expanded notion of *Wesensschau*) on that basis, are expressly tied to the infinitely small: only in this limited region can a cognizing consciousness impose *evident* elementary and uniform laws. Other mathematical resources may be required for manifolds as a whole; their evidential basis is accordingly less direct. The injunction to comprehend the world

amenable to a remarkable conceptual interpretation by means of the concept of an infinitesimal transformation. In particular, the so- called Poisson-Jacobi theorem is closely connected with the composition of infinitesimal transformations. By following up on this observation I arrived at the surprising result that all transformation groups of a simply extended manifold can be reduced to the linear form by a suitable choice of variables, and also that *the determination of all groups of an n-fold extended manifold can be achieved by the integration of ordinary differential equations*. This discovery … became the starting point of my many years of research on transformation groups."

[6]Husserl, e.g. [12, 141], is careful to distinguish the usual (and "countersensical") philosophical notion of evidence as the absolute criterion of truth from evidence as *"that performance on the part of intentionality which consists in the giving of something-itself* [*die intentionale Leistung der Selbstgebung*]. More precisely, it is the universal pre-eminent form of 'intentionality', of 'consciousness of something', in which there is consciousness of the intended-to objective affair in the mode itself-seized-upon, itself-seen—correlatively, in the mode: being with it itself in the manner peculiar to consciousness.

from "its behavior in the infinitely small" is a recurrent theme running through Weyl's writings from 1918 to at least 1949 (cf. Weyl [24, 82, 44, 86]).

2.2 *Transcendental Phenomenological Idealism and "Symbolic Construction"*

Weyl's injunction to understand the world from its behavior in the infinitely small is an evidential constraint upon a transcendental idealism according to which objects of knowledge (natural science) are *constituted* via a process Weyl termed "symbolic construction":

> Science concedes to idealism that its objective reality is not given but to be constructed (*nicht gegeben, sondern aufgegeben*), and that it cannot be constructed absolutely but only in relation to an arbitrarily assumed coordinate system, and in mere symbols. (Weyl [32, 83, 44, 117])

Readers of Kant's Transcendental Dialectic (A647/B675) will recognize the passage as an avowal of transcendental idealism. It is also a transcendental *phenomenological* idealism insofar as symbolic construction of the "objective reality" of the purportedly mind-independent objects of physics is, *per* Husserl, a constitution of the *sense* of such objects as having "the sense of 'existing in themselves'".[7] Weyl himself expressed this understanding of sense-constitution in a densely exposited account of the phenomenology of perception in the "Introduction" to all five editions of his masterful text on general relativity *Raum-Zeit-Materie* (*RZM*):

> Upon general in-principle grounds: The real world, in each of its components and all their determinations, is and can only be, given as intentional objects of acts of consciousness. Given, purely and simply, are the conscious experiences that I have - as I have them. Certainly, in no way do they consist, as positivists perhaps maintain, only of the mere stuff of sensation. Rather a perception, for example, an object standing bodily there before me, each experience of which is known to everyone but not more exactly describable, is taken up in a completely characteristic manner to be designated, with Brentano, through the expression "intentional object". While I am perceiving, as in seeing this chair, I am thoroughly directed to it. I "have" the perception, but only when I make this perception itself into the intentional object of a new, inner perception (of which I am capable in a free act of reflection), do I "know" something about it (and not merely about the chair) .... In this second act the intentional object is immanent like the act itself; it is an actual component of my stream of experience;

[7] Husserl, ca. [10, 382]: "My transcendental method is transcendental-phenomenological. It is the ultimate fulfillment of old intentions, especially those of English empiricist philosophy, to investigate the transcendental-phenomenological 'origins'. the origins of objectivity in transcendental subjectivity, the origin of the relative being of objects in the absolute being of consciousness." Transcendental sense constitution of objective nature is founded on one of *empathy*; e.g., Husserl [13, 92]: "a *transcendental theory of experience of the other* (*Fremderfahrung*), the so-called *empathy (Einfühlung)*" has within its scope "the founding of *a transcendental theory of the objective world* ... in particular, of objective Nature to whose existence sense (*Seinsinn*) belongs *there-for-everyone* (*Für- jedermann-da*)."

> but in the primary perceptual act, the object is *transcendent,* i.e., actually given in a conscious experience but not an actual component. The immanent is absolute, that is, exactly what it is as I have it and am able to bring its essence (*Wesen*) to givenness (*Gegebenheit*) before me in acts of reflection. ... The given-to-consciousness (*Bewußtseins-Gegebene*) is the starting point at which we must place ourselves in order to comprehend the sense and the justification of the posit of reality (*Wirklichkeitssetzung*). (Weyl [24, 3–4, 28, 3–4])

While the Husserlian resonances are unmistakable, the book's first endnote states that the "precise wording" is "closely modeled" upon Husserl's *Ideen* [11]. This passage, utterly remarkable in a monograph establishing much of the modern mathematical machinery of general relativity, is not idle philosophical window dressing. A similar declaration occurs in Weyl's 1930 Rouse Ball lecture at Cambridge, already some years after his most intense period (1918–25) of rather explicit immersion in Husserlian phenomenology,

> Reality [*Wirklichkeit*] is not a being-in-itself [*Sein an sich*] but rather is constituted *for* a consciousness.[8]

These passages set out a theme crucial to Weyl's *transcendental* philosophy of natural science: the central phenomenological distinction between "objects" beyond (*transcendent* to) experience and those immanent within experience, i.e., "intentional objects" produced in acts of reflection upon experience. The former is the realm of mind- independent objects of the physical world, the target of physical theory; the latter are the idealized mathematical symbolic surrogates of physical theory.

Around 1925 Weyl began to use the expression "symbolic construction" to underscore this distinction between mind-*transcendent* objects, and their symbolic surrogates in physical theory. The term itself originates in Weyl's intervention in the period controversy over foundations of mathematics. His first work in philosophy of mathematics, the predicative analysis of [25], drew upon Husserlian phenomenology. By the early 1920s Weyl had become an enthusiastic proponent of Brouwerian intuitionism though largely as interpreted through a phenomenological lens. Still, he was all-too-aware of the severe limitations intuitionistically acceptable methods placed on classical mathematics, writing in 1925 that "full of pain, the mathematician sees the greatest part of his towering theories dissolve into fog" (Weyl [31, 534]).

Weyl's subsequent divergence from Brouwer was prompted by Hilbert [8], who had entered the lists against Brouwer and Weyl. Declaring "in the beginning was the sign", Hilbert's idea was to begin with the intuitively given but otherwise meaningless signs of "concrete, intuitive" number theory (elementary arithmetic), a finite part of mathematics, including recursion and intuitive induction for finite existing totalities, grounded in "purely intuitive considerations" (*rein anschauliche Überlegungen*), hence acceptable to intuitionism. This finite formal part is to be supplemented by a strict axiomatic formalization of the rest of the mathematical theory (its infinitary part), including its proofs. Questions of the truth or validity of mathematical statements are to be replaced by the metamathematical demand for a consistency

[8] Weyl [38, 49]; compare this passage of Husserl [11, 121], "The existence of a Nature *cannot* be the condition for the existence of consciousness, since Nature itself turns out to be a correlate of consciousness: Nature *is* only as being constituted in regular concatenations of consciousness."

proof of the theory's axioms, to be obtained in a "formal proof theory" in which proofs are rule-governed arrays of concrete and displayable formal signs. A formal consistency proof guarantees the reliability of mathematical theory in yielding the conclusion that the permitted arrays cannot yield a contradiction, such as $0 = 1$.[9] A few years later Hilbert [9] distinguished between *finitary* and *ideal statements* in an attempt to justify appending to the finitary part of mathematics the contentious infinitary part (Cantorian set theory) by appeal to a Kantian Idea of Reason, i.e., the regulative demand to complete the concretely given in the interest of totality.[10] With this step, Hilbertian metamathematics persuaded Weyl that the evidentiary demands of symbolic construction could not be reduced to "the demands of open-eyed [*schauenden*] certainty" for each individual statement of a theory [36, 29]. The Hilbertian shift from truth of individual statements to consistency, a global requirement on theories, prompted analogy to modern physical theories. As the justification of any particular mathematical statement ultimately requires reference to the entire theory, via a metamathematical proof of that theory's consistency, in theoretical physics evidential justification of particular statements involves an often complex and indirect relation between theoretical terms and their ties to observation and experiment.[11] Following Hilbert, Weyl would take it to be a metatheoretic axiom of "symbolic construction" that *evidence* bears on a theory not statement by statement but only on the theoretical system as a whole. Even so, Weyl dismissed Hilbert's metamathematical "game of formulae" as an adequate philosophical justification of

[9]Famously, Gödel in 1931 showed that any such consistency result could only be relative, since a consistency proof could only be carried out in a stronger theory.

[10]Hilbert [9]: "The role that remains to the infinite is ... merely that of an idea—if, in accordance with Kant's words, we understand by an idea a concept of reason that transcends all experience and through which the concrete is completed so as to form a totality". Weyl commented [36, 28]: "Hilbert himself says somewhat obscurely that infinity plays the role of an idea in the Kantian sense, by which the concrete is completed in the sense of totality. I understand this to mean something like the way in which I complete what is given to me as the actual content of my consciousness, into the totality of the objective world, which certain includes much that is not present to me. The scientific formulation of this objective concept of the world occurs in physics, which avails itself of mathematics as a means of construction. However, the situation we find before us in theoretical physics in no way corresponds to Brouwer's idea of a science. That ideal postulates that every judgment has its own meaning achievable in intuition. The statements and laws of physics, nevertheless, taken one by one, have no content verifiable in experience; only the theoretical system as a whole allows itself to be confronted by experience. What is accomplished here is not the intuitive insight into singular or general contents and a description that truly renders what is given, but instead a theoretical, and ultimately purely symbolic, construction of the world."

[11]Weyl [33, 147–8]: "[The] individual assumptions and laws [of theoretical physics] have no separate fulfilling sense [that is] immediately realized in intuition (in *der Anschauung unmittelbar zu erfüllender Sinn eigen*); in principle, it is not the propositions of physics taken in isolation, but only the theoretical system as a whole that can be confronted with experience. What is achieved here is not intuitive insight [*anschauende Einsicht*] into particular or general states of affairs and a faithfully reproduced *description* of the given (*das Gegebene*), but rather theoretical, ultimately a purely symbolic, *construction* of the world." Weyl goes on to state that if Hilbert's view prevails over Brouwer's, as indeed appears to be the case, then this represents "*a decisive defeat of the philosophical attitude of pure phenomenology*, as it proves insufficient to understand creative science in the one domain of knowledge that is most rudimentary and earliest open to evidence, mathematics."

the cognitive importance of mathematics. Instead, he sought to find epistemological warrant for mathematics in its application to the symbolic constructions of physical theory. Of course, unlike mathematics, physics posits a mind-transcendent reality. Still, philosophical reflection (*Besinnung*) on the constructions of physical theory reveals that the physicist must remain content to represent this reality only in symbols (*"das Tranzendente darzustellen ...nur im Symbol"*).[12] "Fusing" mathematics with physics thus again brings out transcendental idealism's central theme, that physical objectivity is constituted through symbolic (i.e., mathematical) relations, not "given" through relations of reference and designation.

"Symbolic" signifies more than the truism that mathematics is the necessary instrument of exact natural science. More significantly, the intent of the term is to underscore the conviction that the finite human mind, rooted in "all too human ideas with which we respond to our practical surroundings in the natural attitude of our existence of strife and action" (Weyl [40, 6]) *can* attain only a *symbolic* (neither literal nor pictorial) understanding of the infinite (in mathematics[13]) or of the mind-independent real world posited by physical theory.[14] In particular, the twin revolutions of twentieth century physics, relativity and quantum mechanics, have demonstrated that symbolic representation is essential, that "here we are in contact with a sphere which is impervious to intuitive evidence; cognition necessarily becomes symbolic construction."[15] Relativity and quantum mechanics also instruct that physical objectivity is constituted structurally, as "invariance under a group of automorphisms" (Weyl [43]). Transcendental idealist limitations on the scope and character of cognition of nature are thus reformulated through the notions of *symbolic construction* and *invariance*:

> A science can only determine its domain of investigation up to an isomorphic mapping. In particular, it remains quite indifferent as to the 'essence' of its object. .... The idea of isomorphism demarcates the self-evident boundary of cognition. [32, 22, 44, 25–6]

The term "construction" also echoes Weyl's oft-expressed predilection for constructive vs. axiomatic (i.e., predicative or intuitionist vs. set-theoretic) mathematics (Weyl [48]). It reflects evidential preference for theories resting on 'visualizable', iterative or recursive basal structures, for geometry and point-set topology over those of modern abstract algebra even as Weyl insisted that mathematics requires both

---

[12]Weyl [31, 540], also [46, 645]. The latter is a lecture entitled *"Erkenntnis und Besinnung (Ein Lebensrückblick)"*. *Besinnung*, here translated 'reflection', is a technical term in Husserlian phenomenology, having the meaning of "sense-investigation"; e.g., Husserl [12, 8]: "Sense-investigation (*Besinnung*)... radically understood, is *originary sense-explication (ursprüngliche Sinnesauslegung*, orig. emphasis), transforming and above all striving to transform sense in the mode of unclear opinion into sense in the mode of full clarity or essential possibility (*Wesensmöglichkeit*)."

[13]Weyl [40, 7]: "*Mathematics is the science of the infinite,* its goal the symbolic comprehension of the infinite with human, that is finite, means."

[14]The influence of Husserl is also apparent in Weyl's use of the term "natural attitude".

[15]Weyl [40, 82]. For example, the central underlying theoretical device of quantum mechanics, densities of a complex valued, infinite-dimensional wave function, *can* only be symbolically represented. Dirac, influenced by Weyl [34] offers the same philosophical message regarding the necessity of symbolic methods in the first sections of his well-known textbook on quantum mechanics.

(Weyl [39]). Above all, it signifies that 'objectivity' in physical theory is constructed as an invariance "for a subject with its continuum of possible positions", and that it arises in step-by-step construction from a basis of what is *aufweisbar* (evident), "something to which we can point to *in concreto*" as demonstrably evident to the constituting consciousness.

> (T)he constructions of physics are only a natural prolongation of operations [the] mind performs in perception, when, e.g., the solid shape of a body constitutes itself as the common source of its various perspective views. These views are conceived as appearances, for a subject with its continuum of possible positions, of an entity on the next higher level of objectivity: the three-dimensional body. Carry on this 'constitutive' process in which one rises from level to level, and one will land at the symbolic constructions of physics. Moreover, the whole edifice rests on a foundation which makes it binding for all reasonable thinking: of our complete experience it uses only that which is unmistakably *aufweisbar.* [45, 628, 627]

The *fons et origo* of all meaning-constitution, i.e., what is given *in evidence* to 'pure consciousness' remains, but "symbolic construction" is the constitutive process wherein concrete symbols go proxy for Husserlian data of 'pure consciousness'. It is Weyl's mathematical rendering of Husserlian analysis of essences (*Wesensanalyse*) in terms of a step-by-step symbolic construction beginning from the evidentially privileged standpoint of the "purely infinitesimal". "Symbolic construction" is then Weyl's term of art for what is yielded by philosophical reflection upon the enterprise of theoretical natural science. It is his generic term for *sense-constitution* of the transcendent physical world, i.e., a physical theory is a symbolic construct that must not be conflated with that "true real world". [16]

2.3 *Transcendental-Phenomenological Origins of Gauge Invariance*

We have previously argued that [24] reformulation of Einstein's general relativity (GR) within a "purely infinitesimal geometry" was largely spurred by his philosophical orientation to transcendental phenomenological idealism [20]. The mandate of *RZM* "to comprehend the sense and the justification of the posit of reality (*Wirklichkeitssetzung*)" beginning from the starting point of the "given-to-consciousness (*Bewußtseins-Gegebene*)" quickly became an explicit recognition that in general relativity this starting point, the primary locus of sense-constitution, lay in the "purely

[16] Weyl [45, 627]: "... the words 'in reality' must be put between quotation marks; who could seriously pretend that the symbolic construct is the true real world?" The term 'symbolic construct' encompasses not merely the symbolic universe in which physical systems, states, transformations and evolutions are mathematically defined in terms of manifolds, functional spaces, algebras, etc., but also a symbolic specification of idealized procedures and experiments by which the basic physical quantities or observables of the theory are related to observation and measurement. It reflects an insistence, reinforced by quantum mechanics, that physical quantities (beginning with 'inertial mass') are not simply given, but "constructed" [37, 76, 41, 109ff].

infinitesimal" for the infinitesimally small requires only elementary linear analytic relations, essentially encapsulated in what Weyl termed a linear "connection". The "purely infinitesimal" thus bounds the immediate evidential reach of a situated ego, the constituting transcendental subject that, through step-by-step construction, invests derived mathematical structures with meaning. In this regard, Weyl's purely infinitesimal reconstruction of general relativity brings phenomenological reflection upon the levels of sedimented geometrical structures defined on general relativistic spacetime manifolds, a reflection that reveals how there is

> a surreptitious substitution of the mathematically substructed world of idealities for the only real world, the one that is actually given through perception, that is every actually experienced, and experienceable … (Husserl [14])

The injunction "to comprehend the world" (i.e., structures on the entire manifold) from "the purely infinitesimal" is then just a requirement that sense-constitution of the finite and global mathematical structures defined upon these manifolds ultimately derives from this evidential basis.

General relativity indeed provided the physical and philosophical impetus for the birth of the gauge principle. General relativity preeminently features a local symmetry, the invariance of the form of the Einstein field equations under arbitrary curvilinear coordinate transformations, i.e., the requirement of "general covariance". Einstein embraced this formal requirement as a heuristic corresponding to a supposed generalized principle of relativity, that physical laws appear the same to all observers regardless of their state of motion—accelerating, rotating, inertial—of their reference frames.[17] While the first edition of *RZM* was still in press, Weyl saw how electromagnetism (the only other known interaction in 1918) could also be presented as a manifestation of a kind of local symmetry, analogous to the local symmetry of general relativity.

The additional invariance that Weyl sought to exploit was an invariance with respect to a change of scale ("gauge", hence the name) at each point of spacetime. He did so on grounds that the Riemannian geometry of Einstein's theory was "inconsistent" with the basic thrust of the Riemannian theory of manifolds and the variable curvature of Einsteinian spacetimes. Following Riemann, Einstein allowed the direction of a vector to change as the vector is transported (by an affine connection) "parallel to itself" from point to point around a closed curve in spacetime; the angle between the initial and the returning vector at the same point is the indicator of spacetime curvature. But, of course, vectors have two properties, direction and magnitude, and the Riemann-Einstein geometry required the magnitude of the vector to remain the same while traversing a closed curve. Weyl proposed to rectify this "inconsistency" by requiring a local gauge invariance according to which a "length connection" allows scale changes to vary smoothly from point to point of spacetime. The demand that physical laws remain invariant under these local changes of scale resulted in new degrees of freedom that mathematically, as Weyl showed, brought

[17] In fact, general covariance is a purely formal requirement having nothing to do with a generalized principle of relativity.

electromagnetism in addition to gravitation into the metric of spacetime. That is to say, the metric of a [24] "purely infinitesimal" geometry of spacetime, unlike Einstein's Riemannian geometry, allowed for variation of unit of scale ("gauge") at each point; this new degree of freedom, a function φ_μ [$\mu = 1, 2, 3, 4$] of the four spacetime coordinates of the point, was shown to be mathematically identical to the vector four potential A_μ of relativistic electromagnetic theory. The end result was that Weyl's geometry yielded fundamental field equations of *both* Einstein gravitation and electromagnetism. From Weyl's "purely infinitesimal" starting point, global field laws (the "physical world") are constructed that satisfy not only, as in general relativity, general covariance (freedom to choose spacetime coordinates) but also "gauge invariance" (freedom to choose scale at each point). To be sure, Einstein immediately identified an empirical objection to [24] idea of gauge invariance. So the idea lay dormant until reformulated by Weyl as pertaining to a quantum mechanical factor of phase in 1929 when bringing the relativistic equation of the electron (the Dirac equation) into the four-dimensional spacetime context of general relativity. Indeed, contemporary physical theory pertains to so-called "internal symmetries" of quantum fields rather than to a factor of scale and so from a modern point of view, the gauge principle is misnamed.[18] On the other hand, the basic idea of gauge invariance, that it involves an arbitrary function of the spacetime coordinates, remains. To Weyl, the requirement of gauge invariance, however interpreted, has the "character of a 'general' ('*allgemeiner*') relativity".[19]

2.4 *From the "Raumproblem" to Lie Groups and Lie Algebras*

In a natural development from his 1918 "purely infinitesimal" reformulation of general relativity, Weyl turned to the new "space problem" posed by the variably curved manifolds permitted in Einstein's theory. In the late 1860s, Helmholtz had characterized the geometry of "space" by a set of conditions termed "free mobility" whereby geometrical quantities (lengths, angles, etc.) could be compared throughout the space via the free motions of a rigid body serving as a measuring instrument. Lie recast free mobility in the language of continuous transformation groups acting on a space, thereby representing the congruences of the space. Hence the Helmholtz-Lie solution to the old "space problem" presupposed a geometric space permitting free mobility of rigid bodies, i.e., a homogeneous space of constant curvature; this is made obsolete by general relativity where the metric is no longer homogeneous (and

[18]Internal symmetries refer to the fact that particles occur in multiplets, members of which can be considered as "the same" under the symmetry of the interaction. Mathematically, the multiplets are realizations of an irreducible representation of some internal symmetry group. See further below.

[19]Weyl [35, 246]. Cf. Weyl [37, 220]: "One can in fact take it as a general rule that an invariance property of the kind met in general relativity, involving an arbitrary function, gives rise to a differential conservation theory. In particular, gauge invariance is only to be understood from this standpoint."

not a positive definite quadratic form but one that is indefinite) and the medium itself is variably curved spacetime.

Weyl's new solution (1921–23) came about by drawing again from the "purely infinitesimal". He noted that the old Helmholtz-Lie solution retained its validity in the infinitely small if posed initially in terms of a point-dependent Lie congruence group G acting transitively in the homogeneous tangent space $T_P M$ centered on each point $P \in M$ (Weyl [49]). The infinitesimal linear action of the group (Weyl showed that it takes values in what today is termed the Lie algebra of the group G, see below) is comparable to rotations about points in Euclidean geometry, yielding "point congruences", generalized metrical relations within each $T_P M$. Assuming the volume of parallelepipeds formed by basis vectors at P is preserved by these "rotations", these are subgroups of the special linear group $SL(n)$. Metrical relations in the infinitesimally close neighborhood **U** of P are then given through the "congruent transport" of a vector from P to a neighboring point $P' \in \mathbf{U}$ by a linear connection that intuitively combines parallel transport ("affine connection") and the infinitesimal rotations. This is a "metric connection" A linking infinitesimal congruence relations at P with those at P'. An equivalence class of such connections characterizes the same infinitesimal congruence structures at each point differing only by point-dependent infinitesimal rotations.[20] Thus the "nature" of space at each point P is metrically the same, while the different points may have distinct "orientations". Invoking what he termed the "fundamental fact of infinitesimal geometry", that the metric connection uniquely determines the affine connection, Weyl showed that the local group of isometries must be the (pseudo-) Euclidean group. As required by general relativity, the rotation groups at the various points may have different "orientations" due to variations of matter and energy, but they all share the same infinitesimal Pythagorean (pseudo-Euclidean) metric group structure [29, 43–61].

Already in the 1880s Sophus Lie had reduced the concept of continuous group to the "germ" of infinitesimal elements that generated it. In abstract form, these *"groupes infinitésimaux"* had been extensively studied and classified in 1894 by É. Cartan in his doctoral thesis (Paris), building upon earlier results of W. Killing.[21] But Weyl was the first to explicitly recognize that the simple tools of linear algebra could be brought to bear on this infinitesimal structure. Lie's concept of "infinitesimal group" was essentially repurposed in Weyl's solution to the "Space Problem" in the light of the variably curved four-dimensional spacetimes of Einstein's theory. In his new solution to the "Space Problem", Weyl showed that this infinitesimal group structure could be axiomatically expressed in algebraic terms [29, 82]. In particular, the structure of the tangent space surrounding the point that is the group identity (where G can be considered homogeneous and linear), is a linear vector space, i.e., there is a linear algebra structure in the tangent space to the identity of a Lie group. This

[20] For a clear account of mathematical details in more modern terms, see Darrigol [3] and Scholz [21].

[21] See [5, 7]. Cartan's "structure theory" for infinitesimal Lie groups (today, Lie algebras) identifies isomorphic groups through their "structure constants". Cartan worked exclusively at the level of abstract groups; Weyl would translate Cartan's structure theory into the language of matrix groups, group representations by matrices, the language of most interest to physics.

axiomatized infinitesimal structure of a Lie group was first termed a "Lie algebra" by Weyl in lectures ten years later at the Princeton Institute of Advanced Study.[22] In short, Weyl's solution to the new "Space Problem" crucially rested upon the concept of infinitesimal group, recast in language of linear vector spaces. To Weyl this was compelling evidence that "mathematical simplicity and metaphysical primitiveness (*Ursprünglichkeit*) are narrowly bound together".[23] The purely infinitesimal solution to the new "Space Problem" in turn led to purely mathematical research on representations of semisimple Lie groups and Lie algebras.[24] And it is in the guise of linear vector spaces that the concept of Lie algebra (and its representations) appears in the contemporary gauge theories of the Standard Model (see Sect. 3.1).

3 The Gauge Principle

A symmetry of a (quantum) field theory is a group of transformations that leave the equations of motion of the field (encapsulated in its Lagrangian) unchanged in form. Symmetries may be discrete or continuous, global (applying in the same way everywhere) or local (applying differently at each point of space). The gauge principle refers more specifically to a logic or procedure, or really a *recipe*, for constructing the *form* or *template* of an interaction mechanism within a free (non-interacting) quantum field whose law of motion (Lagrangian density) transforms invariantly under a continuous *global* symmetry (e.g., a rotation in an "internal" dynamical space closed under the action of a given Lie group). A global symmetry is one that applies in the same way at each point of space(time). The form of interaction emerges from "gauging" the global symmetry, i.e., requiring the global symmetry to become a *local* symmetry so that, e.g., the rotations can be different at each point of space(time). In consequence of gauging the global symmetry, a so-called "gauge field" (manifested in particle terms as spin 1 (in units of h), "gauge bosons") appears that is required to transform in a way compensating for the change from global to local symmetry. Supplemented with additional information (appropriate kinetic and

---

[22]Weyl [42, 4]. In modern terms, a Lie algebra to a Lie group G is usually denoted by the Gothic character g and is defined by three properties: (1) the elements X, Y, etc. of g form a linear vector space; (2) the elements of g close under a commutation relation $[X,Y] = -[Y,X], \forall X,Y \in$ g; (3) the Jacobi identity $[X,[Y,Z]] + [Y,[Z,X]] + [Z,[X,Y]] = 0$ is satisfied. Using Cartan's "structure theory", the structure of a Lie algebra is completely determined by its "structure constants" c^k_{ij} that appear in the commutator of any two basis vectors $[X_i, X_j] = c^k_{ij} X_k$.

[23]Weyl [27, 329]: "auf diesem Felde mathematische Einfachheit und metaphysische Ursprünglichkeit in enger Verbindung miteinander stehen."

[24]Weyl [30]. A (non-abelian) Lie group whose Lie algebra cannot be factorized into two commuting subalgebras is called *simple*. A direct product of simple Lie groups is called *semi-simple*. To be sure, in these papers Weyl supplemented Cartan's infinitesimal abstract group viewpoint with global and topological properties of Lie groups, thus in our view, "comprehending the world beginning from the infinitesimal". Indeed, linearizing a Lie group G in the tangent space of the identity to form its Lie algebra g destroys *G's* global properties, i.e., what happens far from the identity. Hence the need for integral and topological methods.

coupling terms), the Lagrangian density of the now interacting field is invariant with respect to the new extended group of local transformations.

The so-called "gauge argument" illustrates the canonical way to "gauge" a field theory. The paradigm is set by quantum electrodynamics (QED), the first quantum field theory; in virtue of the logical pattern of the argument in QED, one can view the SM as essentially a generalization of QED. The new gauge degree of freedom here appears not as a factor of scale but as a local phase factor in wave function of electron. One begins with a free electron field $\Psi(x)$ that is determined up to a phase factor θ. The Maxwell-Dirac Lagrangian for the electron field that is the basis of QED transforms invariantly under the *global* phase transformation (applying at each point in the same way),

$$\Psi(x) \Rightarrow \Psi'(x) = e^{i\theta}\Psi(x)$$

where $e^{i\theta}$ is Euler's formula. $\Psi(x)$ is then invariant under a global U(1) internal symmetry group; the global invariance of the matter system implies, via Noether's theorem, the existence of a conserved quantity, a matter current $j\mu$.

One then "promotes" the global symmetry to a local phase symmetry; this means that an independent U(1) group is associated with *each* space-time point. Then requirement of gauge symmetry demands that the phase parameter vary as a function of space-time position *x*, and the phase invariance is *local*:

$$\Psi(x) \Rightarrow \Psi'(x) = e^{i\theta(x)}\Psi(x) \tag{1}$$

Typically, Lagrangians depend not only on the field magnitudes but also on their (at least first) derivatives. However, by imposing a local symmetry, the derivative of the field $\partial\mu\Psi(x)$ picks up an extraneous term $\partial\mu(x)$ in its transformation; as $\theta(x)$ is a function varying with spacetime position, it is not a covariant object. In order to cancel this unwanted term, the "gauge covariant derivative" is introduced,

$$\partial_\mu \Rightarrow D_\mu = \partial_\mu - ieA_\mu$$

where *e* is the charge of the electron field. The new covariant derivative transforms as

$$D_\mu \Rightarrow D'_\mu = e^{i\theta(x)}D_\mu$$

and $A_\mu = A_\mu(x)$ is an "invented vector field required to transform as

$$A_\mu(x) \Rightarrow A'_\mu(x) = A_\mu(x) - \frac{\partial_\mu\theta(x)}{e} \tag{2}$$

showing proportionality of the transformation to *e*, the electric charge. The resulting Lagrangian is then invariant under joint local transformation of $\Psi(x)$, given by (1), and of $A_\mu(x)$, given by (2), the added partial derivative in (2) *exactly compensating the extraneous position-dependent variation of the phase factor.* Moreover, in imposing the requirement of local symmetry in steps (1) and (2), the free electron field $\Psi(x)$ is coupled to the electromagnetic field represented by the Faraday tensor, obtained by taking the derivative of the four-potential $A_\mu(x)$,

$$F_{\mu\nu} = \frac{\partial A_\mu}{\partial x_\nu} - \frac{\partial A_\nu}{\partial x_\mu},$$

and the conserved current now appears in an interaction of the form $ej\mu A\mu$ in conformity with Maxwell theory. For obvious reasons, the new term (x) is now called a "gauge field". The "gauge argument" then shows how introducing local symmetries dictates the form of the interaction of matter fields (Yang [50, 20]).

3.1 Lie Algebras in Field Theory: Purely Infinitesimal Operations

As just seen in the case of the gauge group U(1) of QED, local symmetries are introduced by making the Lie group parameters functions of space and time. In the more general field theory case of a group of unitary transformations in a Hilbert space for internal degrees of freedom, the Lie symmetry group G can be expressed in terms of its infinitesimal generators X^a

$$G = e^{i\theta^a X^a}$$

where θ^a (a =1... N) are the real parameters of the group, the X^a are linearly independent Hilbert space matrices and there is an implied summation. The set of all linear combinations $\theta^a X^a$ form a vector space; the term *generator* refers broadly to an arbitrary element of the vector space or specifically to the basis vectors X^a. Letting t^a=$\{X^a\}$ define a set of such matrices, the Cartan structure condition defines the Lie algebra:

$$\left[t^a, t^b\right] = i f^{abc} t^c$$

where $[t^a, t^b]$ is the commutator (Lie bracket) and f^{abc} are the structure constants of the Lie algebra that define the multiplication properties of the Lie group (at least for elements continuously connected to the group identity), *constants* since the multiplication properties of the group transformations should be independent of any particular representation. As the expression shows, the structure constants themselves generate another representation of the algebra. Local symmetry then requires that the parameters θ^a become functions of x, $\theta^a(x)$. For our purposes, we simply point out that generically one can write the covariant derivative of the field as a linear combination of the ordinary derivative of the field (its infinitesimal displacement in space or spacetime) and a field-dependent infinitesimal gauge transformation

$$D_\mu \psi \equiv \partial_\mu \psi - W_\mu \psi$$

where W_μ is a matrix representing the gauge field whose entries are generated by infinitesimal gauge transformations. Therefore the can be decomposed into its generators t_a,

$$W_\mu = W_\mu^a t_a.$$

This shows that the gauge field takes values in the Lie algebra corresponding to the gauge group G. The characteristic of a gauge field, exemplified in its Lie algebra, is that it carries information regarding group structure from one spacetime point to another, a "purely infinitesimal", hence *evident*, operation.

3.2 *The Gauge Principle Generalized: The SM*

The above gauge invariance of QED is only the simplest example of infinite parameter or Lie group symmetry, and it is not typical as its gauge group U(1) is abelian (commutative). The gauge argument was generalized by Yang and Mills [51] so that Yang-Mills theories of the SM arise in the same logical pattern: Gauging a global symmetry requires (to restore invariance of the field Lagrangian) the introduction of a covariant derivative; the new derivative is required to transform in a manner that introduces a new (gauge) field; the gauge field provides the form of the interaction forces of a matter field. The same mathematical expressions appear with only minor changes, e.g., in place of the phase of the electron field, there are generalized phases associated with the wave functions of multicomponent matter fields. This is the Yang-Mills template for the Standard Model, a spontaneously broken non-abelian gauge theory containing three types of particles: elementary scalars, fermions (spin-1/2) and spin-1 bosons. Spin-1 gauge bosons are the particles of gauge fields. The SM is often represented by its "gauge group", the direct product group SU(3) x SU(2) x U(1) representing the fundamental interactions. Unlike U(1), the special unitary Lie groups SU(2) and SU(3) are non-commutative (non-abelian). SU(2) x U(1) is the symmetry group of the "electro-weak" interactions, where U(1) is the phase symmetry of the weak hypercharge (slightly different from the phase symmetry of the electromagnetic interaction QED) and SU(2) is the isospin symmetry describing weakly interacting particles. The elements of SU(2) are 2×2 matrices; in the weak interaction matter particles (*up* and *down* quarks; electrons and electron neutrinos) are sorted into doublets such that the two particles in a doublet are interchangeable, indistinguishable in that interaction. The group SU(3) plays two roles in the SM: as an exact gauge symmetry associated with color for the strong interaction, and as an approximate global flavor symmetry of the strong interactions (the "eightfold way" of Gell-Mann and Ne'eman). SU(3) elements are represented by 3×3 matrices, so the symmetry operations pertain to a triplet of particles. Quarks are the matter particles of the strong interaction; each with its own mass and fractional charge comes in one of six *flavors* partitioned into three doublets (*up, down*), (*strange, charm*), (*top, bottom*). *Color,* the strong force analogue to electric charge, itself has three manifestations, *red, green, blue.* Within the same flavor, changing e.g., *red* quarks to *green* quarks leaves the interaction energy of the system unchanged. In general, non-abelian Lie groups yield theories of multiple vector particles, whose interactions are strongly constrained by a gauge symmetry.

In sum, a gauge symmetry is a constraint on the Lagrangian L for any quantum field theory; the first step in constructing a quantum field theory is to ask what gauge symmetry it must obey.[25] A gauge symmetry will dramatically reduce the vast number of theoretically possible Lagrangians. Moreover, both massless and massive gauge theories (provided the masses are generated by spontaneous symmetry breaking via the Higgs mechanism[26]) are renormalizable. In fact, the only way to form a relativistic quantum field theory of spin-1 particles (force-carrying *bosons*) is a gauge theory [22].

4 Towards an Elucidation of the Gauge Principle

The gauge principle has been recently described as "the most fundamental cornerstone of modern theoretical physics" [18]. As has been seen, the force fields of the SM are gauge fields, each formally is a variation on the basic template of the gauge principle. Nonetheless, despite its success, few if any theorists believe the SM to be a truly fundamental theory. For one thing, the SM contains approximately 26 free parameters (notably particle masses and Yukawa coupling terms that parameterize the interactions of fermions and the scalar Higgs field). A "free parameter" is one for which there is no theoretical explanation, one whose value has to be put into calculations "by hand", i.e., as determined by experiment rather than predicted by theory. Furthermore, the values of many of these parameters appear suspiciously "fine-tuned" such that if the observed value differed by just a few percent, life as we know it would not be possible. (e.g., Rees [19] In addition, the SM features

[25]The second step is to determine the representations of fermions and scalars under the gauge symmetry; a third step is to postulate the pattern of spontaneous symmetry breaking.

[26]A symmetry of a system is said to be "spontaneously broken" if its lowest energy state is not invariant under the operations of that symmetry. This is an extremely important concept in the weak interaction as the bosons introduced by gauge symmetries are massless, like the photon; their masses arise from the "spontaneous breaking" of the SU(2) x U(1) symmetry through couplings to the scalar Higgs field.

three families of fermions, without any internal account of why this should be so. Moreover, within the SM, there is no clear answer to how many Higgs bosons there are, though only one has been identified (mass 125.1 GeV), discovery announced by CERN on July 4, 2012); this underdetermination increases the artificiality of the Higgs mechanism for SSB. Finally, the SM says nothing about "dark matter" or "dark energy", i.e., nothing at all about approximately 95% of the energy budget of the universe. For these and other reasons, most quantum field theorists and cosmologists view the Standard Model as merely an "effective theory", a provisional stage in the descent to ever smaller distance scales corresponding to an ascent to ever higher energies. Hence it is widely assumed that the SM is a "low energy" consequence of more fundamental physics at the higher energies of the early universe, up to Grand Unification (GUT) scales of 10^{15}–10^{16} GeV (at which the gauge coupling strengths of the three interactions theoretically meet;) or even the most fundamental unification Planck scale 10^{19} GeV which necessarily includes gravitation, the remaining known interaction.

To the inquiring philosopher, the SM presents several challenges. For the default scientific realist, some additional work must be done to explain how the known physical laws with their accompanying ontological posits are not truly fundamental but contingent regularities in the sense that their validity is restricted to certain "low energy" epochs. The *restricted* necessity of these laws, if such there is, may originate only in conditions that are accidental or environmental according to both string theory and most models of inflationary cosmology. More fundamentally, there is the matter of how to understand the empirical success of the gauge principle.

Local symmetry transformations introduce new gauge degrees of freedom that appear to be redundancies or mathematical surplus structure in the physical description of an interaction. The same physical state can be represented by many different solutions of the field equations when these solutions are related by gauge transformations. *Prima facie,* gauge symmetries connect states that are physically the same yet differ in their mathematical description. This prompts an analogy to general relativity, the site of origin of the gauge principle, and a reflection upon Weyl's claim [35] that the principle of gauge invariance has the "character of a 'general' ('*allgemeiner*') relativity".

For just as a specific coordinate system must be chosen to extract physical observations in general relativity, so gauge fixing is necessary to retrieve physical predictions from gauge theories. This would seem to indicate that gauge symmetries are not symmetries of nature but of physical description of nature. The redundancy of gauge descriptions itself raises various puzzles. How it is possible for spontaneous symmetry breaking (SSB) of a gauge symmetry to have physical consequences (e.g., gauge bosons acquire a mass)?[27] Further puzzling is the fact that the gauge redundancy of mathematical description appears to be fortuitous since gauge theories of vector particles (photons, and the massive spin 2 bosons of the electroweak theory) as in the SM are *renormalizable*, which means, roughly, that the generic infinities (non-physical divergent integrals) associated with free parameters (such as particle masses) in quantum field theory can be tamed. Thus, local gauge invariance is required for the Lagrangians of the SM to be mathematically tractable and predictive up to higher and higher energies.

Weyl's "purely infinitesimal" generalization of general relativity issued in the demand that fundamental physical theories, in addition to the requirement of coordinate freedom ("general covariance"), should also satisfy the requirement of gauge (more appropriately in its debut, scale) invariance. Both requirements introduce *arbitrary* mathematical degrees of freedom at each point P of the four-dimensional differential manifold representing spacetime; the arbitrariness can be understood phenomenologically, as each point indifferently can be considered the locus of an experiencing, constructing ego.[28] These degrees of freedom arise from two metaphysical prerequisites: (1) a postulate of transcendental phenomenological idealism, that "Reality [*Wirklichkeit*] is not a being- in- itself [*Sein an sich*] but rather is constituted *for* a consciousness", and (2) the *aspiration,* fortified by the successes of differential calculus in physics and indeed of field physics (*Nahewirkungphysik*) itself, that this "reality", constituted as it is by a *situated* consciousness, "can be understood from its behavior in the infinitesimally small", i.e., mathematically comprehended starting from the *evident* simple linear relations within the tangent space. The arbitrary degrees of freedom represent *at each point* particular magnitudes of physical states, *either* as different mathematical functions (scalar, vector, tensor, spinor) of four independent variables (spacetime coordinates) determined by the field laws *or* in terms of an arbitrary vector function of these spacetime coordinates signifying an internal gauge symmetry. In the latter case, these degrees of freedom serve to represent interactions as proceeding through the exchange of particles of spin 1 such as the photon, or the massive gauge bosons of the electro-weak interaction ($W^{\mp}, Z^0$).

[27] SSB plays two roles in the SM, giving mass to gauge bosons (other than the photon) and giving mass to fermions (leptons and quarks). In the electroweak theory (unifying the weak and electromagnetic interactions), SSB plays a crucial part, breaking the electroweak SU (2) X U (1) symmetry into individual electromagnetic and weak forces while enabling mass (the SU (2) multiplets) to emerge spontaneously in theories where initially, there is no mass. A standard story is that SSB invokes the Higgs mechanism to spontaneously break a local gauge symmetry; of course, this a symmetry connecting states that cannot be physically distinguished. In fact, in SSB it is not the local gauge transformations that are spontaneously broken. Rather it is a global symmetry (a unique vacuum state) that is spontaneously broken while the gauge symmetry is explicitly broken by gauge fixing in the Higgs mechanism in order to extract physical predictions. Elitzur [6] has showed the spontaneous breaking of a local symmetry is logically impossible.

[28] Weyl [25, 72]: "The coordinate system is the unavoidable residue of the ego's annihilation *(das unvermeidliche Residuum der lch- Vernichtung)* in that geometrico-physical world that reason sifts from the given under the norm of 'objectivity' - a final faint token in this objective sphere that existence (*Dasein*) is only given, and *can* only be given as the intentional content of the conscious experience of a pure, sense-giving ego."

At this point we may recall the Weyl-Nozick slogan *objectivity = invariance* [15] but signal our departure from the Nozickian realist (such as Earman [4]) who maintains that *objective* pertains (by definition) to the mind-independent structure of the world. Our conjecture is rather that of Weyl, that general covariance and gauge invariance are but particular demands of objectivity upon any theoretical construction initiated from the standpoint of radical locality: the constructed physical theory must be independent from any particular starting point from which it is constituted. As a constructive requirement of objectivity, the invariance of laws under arbitrary coordinate and local gauge transformations can be understood in the first instance not as symmetries of nature but of a radically local description of nature. In this way, perhaps, we can see how Weyl's central idea of the gauge principle as a "purely infinitesimal" remnant of sense-constitution is preserved even today in the internal (phase) symmetries of the Standard Model.

References

1. M. Atiyah, Hermann Weyl 1885–1955, in *National Academy of Science: Biographical Memoirs*, vol. 82 (National Academies Press, Washington, DC, 2002), 1–14
2. J. Bernard, *L'idéalisme dans l'infinitésimal: Weyl et l'espace à l'époque de la relativité* (Presses universitaires de Paris Nanterre, Nanterre, 2015)
3. O. Darrigol, *Physics and Necessity: Rationalist Pursuits from the Cartesian Past to the Quantum Present* (Oxford University Press, New York, 2014)
4. J. Earman, Laws, symmetry, and symmetry breaking: invariance, conservation principles, and objectivity. Philos. Sci. **71**(December), 1227–1241 (2004)
5. Christophe Eckes, *Les groups de Lie dans l'oeuvre de Hermann Weyl* (Presses Universitaires de Nancy/Éditions Universitaires de Lorraine, Nancy, 2013)
6. S. Elitzur, Impossibility of spontaneously breaking local symmetries. Phys. Rev. D **12**(12) (15 December), 3978-82 (1975)
7. T. Hawkins, *Emergence of the Theory of Lie Groups: An essay in the history of mathematics 1869–1926* (Springer, New York, 2000)
8. D. Hilbert, Neubegründung der Mathematik (Erste Mitteilung). Abhandlungen aus dem mathematischen Seminar der Hamburgischen Universität **1**, 157–177 (1922)
9. D. Hilbert, Über das Unendliche. Math. Ann. **95**, 161–190 (1926)
10. E. Husserl, Zur Auseinandersetzung meiner transzendentalen Phänomenologie mit Kants Transzendentalphilosophie, in *Erste Philosophie (1923–24) Erster Teil*, ed. by Husserliana VII, R. Boehm (Martinus Nijhoff, The Hague, 1956, 1908), 381–95
11. E. Husserl, in *Ideen zu einer reinen Phänomenologie und phänomenologischen Philosophie I*, ed. by Husserliana III, K. Schuhmann (Martinus Nijhoff, The Hague, 1976, 1913), 1–2
12. E. Husserl, in *Formale und Transzendentale Logik*, ed. by Husserliana XVII, P. Janssen (Martinus Nijhoff, The Haag, 1974, 1929)
13. E. Husserl, in *Cartesianische Meditationen und Pariser Vorträge*, ed. by Husserliana I, S. Strasser (Martinus Nijhoff, The Hague, 1963, 1931)
14. E. Husserl, in *Die Krisis der europäischen Wissenschaften und die transzendentale Phänomenologie: Eine Einleitung in die phänomenologische Philosophie*, in Husserliana VI, W. Biemel (Martinus Nijhoff, The Hague, 1954, 1936)
15. R. Nozick, *Invariances*, revised edition edn. (Harvard University Press, Cambridge, MA, 2003)
16. P. Pesic (ed.), *Hermann Weyl: Mind and Nature. Selected Writings on Philosophy, Mathematics, and Physics* (Princeton: Princeton University Press, 2009)

17. P. Pesic (ed.), *Hermann Weyl: Levels of Infinity. Selected Writings on Mathematics and Philosophy* (Dover, Mineola, NY, 2012)
18. M. Redhead, The interpretation of gauge symmetry, in *Ontological Aspects of Quantum Field Theory*, ed. by M. Kuhlmann, H. Lyre, A. Wayne (World Scientific, London, Singapore, Hong Kong, 2002), 281–301
19. M. Rees, *Just Six Numbers. The Deep Forces that Shape the Universe* (Weidenfeld & Nicolson, London, 1999)
20. T. Ryckman, *The Reign of Relativity: Philosophy in Physics 1915–1925* (Oxford University Press (Oxford Studies in Philosophy of Science), Oxford, 2005)
21. E. Scholz, Hermann Weyl's analysis of the "problem of space" and the origin of gauge structures. Sci. Context **17**(1/2), 165–197 (2004)
22. S. Weinberg, *The Quantum Theory of Fields*, vol. 1 (Cambridge University Press, New York, 1995)
23. H. Weyl, *Die Idee der Riemannschen Fläche* (B.G. Teubner, Leipzig, 1913)
24. H. Weyl, *Raum-Zeit-Materie* (Springer, Berlin, 1918)
25. H. Weyl, *Das Kontinuum. Kritische Untersuchungen über die Grundlagen der Analysis* (Veit, Leipzig, 1918)
26. H. Weyl (ed.), *B. Riemann: Über die Hypothesen, welche der Geometrie zu Grunde liegen* (Springer, Berlin, Heidelberg, 1919)
27. H. Weyl (ed.) Das Raumproblem. Jahresbericht der Deutschen Mathematikervereinigung **31**, 205–221 (1922); reprinted in [47] GA II, 328–44
28. H. Weyl, *Raum-Zeit-Materie*, 5 edn. (Springer, Berlin, 1923)
29. H. Weyl, *Mathematische Analyse des Raumproblems: Vorlesungen gehalten in Barcelona und Madrid* (Springer, Berlin, 1923)
30. H. Weyl, Theorie der Darstellung kontinuierlicher halbeinfacher Gruppen durch lineare Transformationen I, I, II, und Nachtrag. Mathematische Zeitschrift 23–24 (1925-6); reprinted in [47] GA II, 543–647
31. H. Weyl, Die Heutige Erkenntnislage in der Mathematik. Sonderdrucke des Symposion, Heft **3**, (1926) 1–32; reprinted in [47] GA II, 511–42
32. H. Weyl, *Philosophie der Mathematik und Naturwissenschaft* (R. Oldenbourg, München und Berlin, 1927)
33. H. Weyl, Diskussionsbemerkungen zu dem zweiten Hilbertschen Vortrag über die Grundlagen der Mathematik. Abhandlungen aus dem mathematischen Seminar der Hamburgischen Universität **6**, 86–88 (1928), reprinted in [47] GA III, 147–9
34. H. Weyl, *Gruppentheorie und Quantenmechanik* (S. Hirzel Verlag, Leipzig, 1928)
35. H. Weyl, Elektron und Gravitation. Zeitschrift für Physik **56**, 330–52 (1929); reprinted in [47] GA III, 245–67
36. H. Weyl, *Levels of Infinity* translation of *Die Stufen des Unendlichen*, (Gustav Fischer, Jena, 2012, 1931a), in [17, 17–31]
37. H. Weyl, *The Theory of Groups and Quantum Mechanic*, 2nd ed., trans. H.P. Robertson (Methuen and Co. Ltd., London, 1931; reprint, Dover, New York, 1950)
38. H. Weyl, Geometrie und Physik. Die Naturwissenschaften **19**, 49–58 (1931); reprinted in [47] GA III, 336–45
39. H. Weyl, Topologie und abstrakte Algebra als zwei Wege mathematischen Verständnisses. Unterrichtsblätter für Mathematik und Naturwissenschaften **38**, 177–88 (1932); reprinted in [47] GA III, 348–58
40. H. Weyl, *The Open World* (Yale University Press, New Haven, 1932b); reprinted in [16, 34–82]
41. H. Weyl, *Mind and Nature* (University of Pennsylvania Press, Philadelphia, 1934); reprinted in [16, 83–150]
42. H. Weyl, *The Structure and Representation of Continuous Groups.* Based on notes by Richard Brauer taken at Weyl's course at The Institute for Advanced Study, reprinted 1955 (1934–5)
43. H. Weyl, Similarity and congruence, in *Epistemology of Science*, Lecture at ETH Zurich. ETH Bibliothek, Hochschularchiv Hs91a (1948–9), 31, 23 pp

44. H. Weyl, *Philosophy of Mathematics and Natural Science* (Princeton University Press, Princeton, 1949); enlarged translation of [32]
45. H. Weyl, Address on the unity of knowledge delivered at the *Bicentennial Conference* of Columbia University (1954); reprinted in [47] GA IV, 623–29
46. H. Weyl, Erkenntnis und Besinnung (Ein Lebensrückblick). Studia Philosophica, Jahrbuch der Schweizerischen Philosophischen Gesellschaft, 153–71 (1954); reprinted in [47] GA IV 631–49
47. H. Weyl, Gesammelte Abhandlungen Bd. I-IV, ed. by K. Chandrasekharan (Springer, Berlin, Heidelberg, New York, 1968)
48. H. Weyl, Constructive versus axiomatic procedures in mathematics. Typescript written after 1953. *The Mathematical Intelligencer* **7**(4) (December), 10–17 (1985)
49. H. Weyl, *Riemanns geometrische Ideen, ihre Auswirkung und ihre Verknüpfung mit der Gruppentheorie* (Springer, Berlin, Heidelberg, New York, 1988). Manuscript written in 1925 for a Russian edition of Lobachevsky's mathematical works
50. C.N. Yang, "Hermann Weyl's contributions to physics", A Hermann Weyl centenary lecture, ETH Zürich, October 24, 1985, in *Hermann Weyl 1885–1985*, ed. by K. Chandrasekharan (Springer, Berlin, Heidelberg, New York, 1988), 7–21
51. C.N. Yang, R.L. Mills, Conservation of isotopic spin and isotopic gauge invariance. Phys. Rev. **90**, 191–195 (1954)

Weyl's *Raum-Zeit-Materie* and the Philosophy of Science

Silvia De Bianchi

Abstract In this contribution I explore the philosophical underpinning of Weyl's interpretation of Relativity as emerging from *Raum-Zeit-Materie.* I emphasize the important distinction between the philosophical and the mathematical methods, as well as the dichotomy and relationship between time and consciousness. Weyl identified the latter as the conceptual engine moving the whole history of Western philosophy. and the revolutionary relevance of relativity for its representation is investigated together with the conceptual underpinning of Weyl's philosophy of science. In identifying the main traits of Weyl's philosophy of science in 1918, I also offer a philosophical analysis of some underlying concepts of unified field theory.

1 Introduction

It is not an easy task to discuss the philosophical underpinning of Weyl's *Raum-Zeit-Materie* and his interpretation of Einstein's relativity theories. After 100 years from its publication, we have a number of excellent studies that dealt with it,[1] underlining the relevance of Weyl's work for the history and development of gauge theory (see [18, 19]. Therefore, what I will try to offer in my contribution is a novel reconstruction of Weyl's philosophical reflection on the foundations of gauge theory. I shall mostly refer to the first and second edition of *Raum-Zeit-Materie* in order to limit the discussion to very concrete and specific topics that however were and still are of capital importance both in the history and in the epistemology of science. Before

[1]We have nowadays various philosophical approaches to the conceptual analysis of Weyl's work, apart from the pivotal work by Thomas Ryckman, *The Reign of Relativity* that takes into account a transcendental approach and confronts Weyl's view with Husserl's Phenomenology [16]. We also find approaches integrating the historical and philosophical perspective, such as Sieroka's [20], or those focusing on Weyl's group-theoretic approach and based on Ontic Structural Realism, such as French's, Ladyman's and Bueno's. A more recent attempt at reading the philosophical underpinning of Weyl's proposal is constituted by studies on four-dimensionalism and eternalism [15].

S. De Bianchi (✉)
Department of Philosophy, Autonomous University of Barcelona, Bellaterra, Spain
e-mail: silvia.debianchi@uab.cat

S. De Bianchi and C. Kiefer (eds.), *One Hundred Years of Gauge Theory*,
Fundamental Theories of Physics 199, https://doi.org/10.1007/978-3-030-51197-5_8

entering the discussion of the text, I would like to draw attention to a passage taken from *Symmetry* (1952) and that became a classic in Weyl's studies:

> Symmetry is a vast subject, significant in art and nature. Mathematics lies at its root, and it would be hard to find a better one on which to demonstrate the working of the mathematical intellect. [26, p. 145]

Now, this passage obviously was commented regarding the definition of symmetry offered therein. This time, however, I would like to emphasize the reference that Weyl makes here to "the working of the mathematical intellect". The mathematical intellect, as we shall see, embodies the metaphor of a specific perspective that through symbolic construction (see [3, 6, 7, 12, 13] leads to the objective understanding of the world. We can say that gauge theory illustrates the power of this mathematical intellect thanks to symmetry that dictates the form of interactions. By means of this epistemological standpoint, major achievements in high-energy physics can be seen as disclosing the connection between art (technology and engineering) and nature. Thus, the interesting aspect emerging from the abovementioned lines is that Weyl does not think of philosophy as a means by which we can build up bridges among sciences or among art and nature. It is mathematics that can do that, symmetry can do that. To say this, however, does not imply the absence of philosophy from any process leading to the understanding of the fundamental link between art and nature, nor its absence from physics.

2 Philosophy, Mathematics and Physics

What is then the exact relationship among philosophy, mathematics and physics that Weyl proposed in 1918? I shall focus my analysis on the *Preface* to *Raum-Zeit-Materie*. In it, we read:

> At the same time it was my wish to present this great subject [Relativity n.d.a] as an illustration of the intermingling of philosophical, mathematical, and physical thought, a study which is dear to my heart. This could be done only by building up the theory systematically from the foundations, and by restricting attention throughout to the principles. But I have not been able to satisfy these self-imposed requirements: the mathematician predominates at the expense of the philosopher. [21 v]

Here Weyl is providing an important hint regarding his view of the philosophical and mathematical approaches. Mathematics and philosophy strongly differ regarding the methodology. Philosophical method is identified with the systematic building up of the theory, and it proceeds only in a deductive way from its foundations. On the contrary, mathematics focuses on the principles of a theory but lacks systematicity, or at least can acquire it a posteriori. However, as we shall see in the next sections, for Weyl this lack of systematic deduction of a theory from its foundations is not something negative. It is just something different. Furthermore, the mathematical approach is what enables us to construct bridges among disciplines and different branches of science.[2] We could interpret this take as due to the separation of the concept of systematicity, reserved to philosophy and physical theories, from that of architectonics. The latter entails ends, scopes. Mathematics is not just a heuristic tool; mathematics, in Weyl's view, is able to disclose the scope of the structure of the whole universe. In *Raum-Zeit-Materie*, Weyl declares to endorse the mathematical approach. The mathematician inside him prevailed, but he did not disregard philosophy at all. Let us consider the following passage:

> Einstein's Theory of Relativity has advanced our ideas of the structure of the cosmos a step further. It is as if a wall which separated us from Truth has collapsed. Wider expanses and greater depths are now exposed to the searching eye of knowledge, regions of which we had not even a presentiment. It has brought us much nearer to grasping the plan that underlies all physical happening (*welche dem physischen Weltgeschehen innewohnt*).[3] [21 v]

At that time, these words could sound prophetic. We now know that relativity disclosed to us objects that we could not even imagine 100 years ago, such as black holes. Weyl explicitly defines relativity as a revolution, as a cataclysm "which has swept away space, time, and matter hitherto regarded as the firmest pillars of natural science" [21p. 2]. There is something mystical in the way in which he presents Einstein and his theory, which for him can "make place for a view of things of wider scope, and entailing a deeper vision" [21, p. 2]. What can these passages mean? What is the 'Truth' that is now in front of us and that we couldn't see before? In order to understand what Weyl had in mind we have to look at the way in which he presents space, time and matter from a philosophical standpoint. Only in this way we can grasp the meaning of the "fall of the wall" that separated us from truth. Indeed, the notions of spacetime and matter as developed by Einstein's theory change our view of "happening".

3 Space-Time-Matter and the Foundations of All Happening

When talking about the "happening" and the revolutionary way of presenting it offered by relativity, is Weyl referring to objective or subjective happening or both of them? To both and in a very literal way, I argue. Consider the following passage:

> Space and time are commonly regarded as the forms of existence of the real world, matter as its substance. A definite portion of matter occupies a definite part of space at a definite moment of time. It is in the composite idea of motion that these three fundamental conceptions enter

[2]For a mature view on this topic, see [24].

[3]The notion of physical happening is of fundamental relevance here. In 1918-1920 Weyl follows very closely the lines of Husserl's *Ideen zu einer reinen Phänomenologie und phänomenologischen Philosophie* [11], and I suggest that he gives a meaning to this expression that is very similar, albeit not identical to Husserl's. For Ryckman's take on this, see "The philosophical roots of the gauge principle: Weyl and transcendental phenomenological idealism" [17]. Further studies on Weyl's affinities and differences with Husserl, see [1, 10, 12].

> into intimate relationship. Descartes defined the objective of the exact sciences as consisting in the description of all happening (*alles Geschehen*) in terms of these three fundamental conceptions, thus referring them to motion. [21, p. 1]

What Einstein's relativity accomplished is the unification of subjective and objective happening. From the standpoint of mathematical physics, Weyl's unified field theory was meant to be the expression of such a revolutionary move in history. The revolutionary character of relativity was given by its capacity of putting together subjective and objective happening by means of a new conception of the relationship of geometry and spacetime and the possibility to unify consciousness and external reality into one action:

> As the doer and endurer of actions I become a single individual with a psychical reality attached to a body which has its place in space among the material things of the external world, and by which I am in communication with other similar individuals. Consciousness, without surrendering its immanence, becomes a piece of reality, becomes this particular person, namely myself, who was born and will die. Moreover, as a result of this, consciousness spreads out its web, in the form of time, over reality. [21, p. 6]

In fact what makes it possible to unify all happening within the new framework of relativity is the revolutionary way of presenting time and it is worth spending the next subsection on it.

3.1 Raum-Zeit-Materie and Philosophy of Time

Already at the beginning of the *Introduction* to *Raum-Zeit-Materie*, we find the following definition:

> Time is the primitive form of the stream of consciousness. It is a fact, however obscure and perplexing to our minds, that the contents of consciousness do not present themselves simply as being (such as conceptions, numbers, etc.), but as being now filling the form of the enduring present with a varying content. So that one does not say this is but this is now, yet now no more. If we project ourselves outside the stream of consciousness and represent its content as an object, it becomes an event happening in time, the separate stages of which stand to one another in the relations of earlier and later. [21, p. 5]

In the previous passage Weyl describes the subjective representation of time as a primitive form of the flow or stream of consciousness. For him, only an enduring present exists and with a varying content. The now is the parameter to which one associates the content of consciousness or the subjective experience. However, we are beings projecting outside both the flow and its content and generate by means of this projection an objective representation of this content as something happening in time and according to an order relationship of 'earlier' and 'later'. In other words, relativity shows that what we call cosmic time is nothing else than the result of the projection of an enduring present with varying content. In this way time from a pure form of the flow of consciousness becomes something pertaining to the objective world.[4]

In *Philosophy of Mathematics and Natural Science,* Weyl further clarified his view:

> The objective world is, it does not happen. Only to the gaze of my consciousness, crawling along the lifeline of my body, does a section of this world come to life as a fleeting image in space which continuously changes in time. [25, p. 116]

It is our specific perception, which includes our bodies that makes us project the flow of time outside onto the external realm. The truth is that we live in an enduring present. About the world, about the universe at its most profound level, we can only say that it is.[5] Weyl pertained to a generation of physicists who thought that relativity describes the world as four-dimensional continuum, but that there was more to be said about other conceptions of the world and that not all happening was absorbed by classical physics.[6] It becomes now clearer the sense in which Weyl thought that his unified field theory could embrace all happening. General Relativity theory changes our conception of space, time, matter and therefore motion, but also sheds new light on the relationship between objectivity and subjectivity. Weyl believes that Einstein's theory can overcome this dualism thereby reshaping our conception of time and consciousness that ultimately is possible by abandoning the dichotomy between form and matter, substance and attributes and by embracing a dynamical view of the interaction of matter and fields.

3.2 **Raum-Zeit-Materie** *and the History of Philosophy*

The topics of time and consciousness are central to the Introduction to *Raum-Zeit-Materie.* This might sound surprising at first, but it is even more surprising the fact that Weyl wants to find the right place of Relativity theories within the history of philosophy, by taking the cue from the discussion of the concepts of time and consciousness and their close relationship. This is clear in the following passage:

> Since the human mind first wakened from slumber, and was allowed to give itself free rein, it has never ceased to feel the profoundly mysterious nature of time-consciousness, of the progression of the world in time, of Becoming. It is one of those ultimate metaphysical problems which philosophy has striven to elucidate and unravel at every stage of its history. The Greeks made Space the subject-matter of a science of supreme simplicity and certainty. Out of it grew, in the mind of classical antiquity, the idea of pure science. Geometry became one of the most powerful expressions of that sovereignty of the intellect that inspired the thought of those times. [21, p. 1]

[4]Eddington [8], for example, concluded that "consciousness, looking out through a private door, can learn by direct insight an underlying character of the world which physical measurements do not betray".

[5]Weyl here supports what in current philosophy is called the Block Universe view.

[6]For the historical roots of such a conception in Riemann and others, see Boi [4].

The problem of Becoming or the dichotomy between Being and Becoming has been at the roots of the history of philosophy and its development throughout the centuries. The solution that the Greek offered was based on the reification of space by means of geometry and the representation of time as a "moving image of eternity", as Plato's *Timaeus* suggested. With relativity not only space but also time constitute the subject matter of science. Spacetime is therefore the object of the theory and General Relativity describes the dynamics and the mutual active and passive interaction of spacetime and matter, by means of curvature and bending. What is certainly striking in some sense is that Weyl believes that for more than 2000 years philosophy tried to solve the mystery of time flow and consciousness. However, one might question whether our modern idea of consciousness—which is not unambiguously defined even today—was really at the roots of the problem of Becoming. I suggest that Weyl is at least correct in his insight regarding the problem of the flow of time as intrinsically related to that of Becoming. This emerged not only in the pre-socratic tradition, but also in Plato's *Timaeus* as one of the first examples in Western thought of a geometrization of the physical world. It is in that dialogue that time (*chronos*) is portrayed as an image, as something generated and not pertaining to the realm of Forms. However, for Plato, time is physical or at least its representation embodied by the trajectories of planetary motion has physical meaning. Furthermore, the category of "coming into being" is used by Plato to grasp the function of time in view of our knowledge and measurement of the universe. The knowledge of the physical world depends on an operational definition of time, but the latter alone is not fundamental from the standpoint of the mathematical intellect. Time results fundamental only together with space in view of the generation of any physical world. For this reason, I claim, Weyl presents the history of philosophy as marked by attempts at resolving the problem of Becoming, as continuous attempts to explain or justify the relationship between time flow and consciousness that ended up in the reification of space as the subject-matter of natural science. Nobody before Einstein, at least in Weyl's view, had the insight of making spacetime and its geometry the fundamental structure from which the physical world could have been represented and in a way that included a new way of portraying the relationship between time and consciousness. In this sense, Relativity introduced a catastrophic revolution in Western thought.

4 The Foundations of Mathematics, Space and Philosophy

We are now in a position to appreciate the relevance attributed by Weyl to philosophy and its history—an approach that he conserved throughout his life thanks to the influence of his wife Helene and the study of Cassirer's works [5], see also [7]. What is left to be clarified is that with the formulation of relativity and its representation of spacetime, the relevance of philosophy is to be ascribed to its role of investigating the foundations of space:

> Now, if on the one hand it is very satisfactory to be able to give a common ground in the theory of knowledge for the many varieties of statements concerning space, spatial configurations, and spatial relations which, taken together, constitute geometry, it must on the other hand be emphasised that this demonstrates very clearly with what little right mathematics may claim to expose the intuitional nature of space. Geometry contains no trace of that which makes the space of intuition what it is in virtue of its own entirely distinctive qualities which are not shared by "states of addition-machines" and "gas-mixtures" and "systems of solutions of linear equations". [21, p. 26]

This passage clarifies that geometry is necessary in order to grasp the fundamental physical reality, but that only outside of geometry, in the realm of philosophy, we can find the answer to the deeper question "what is space?". This position comes as no surprise, considering that Weyl was sympathetic with intuitionism. Geometry, he believes, cannot show the intimate and ultimate nature of space.[7] The following passage clarifies also which branch of philosophy is necessary to make space comprehensible and this is metaphysics:

> It is left to metaphysics to make this "comprehensible" or indeed to show why and in what sense it is incomprehensible. We as mathematicians have reason to be proud of the wonderful insight into the knowledge of space which we gain, but, at the same time, we must recognise with humility that our conceptual theories enable us to grasp only one aspect of the nature of space, that which, moreover, is most formal and superficial. [21, p. 26]

One question that might rise in reading what Weyl suggests here for space is whether this is also valid for time. This is an open question, at least in *Raum-Zeit-Materie*. What we can certainly appreciate in it is that Weyl thought that philosophy is fundamentally different from natural science and mathematics. However, philosophy is fundamental for GR and GR is fundamental for philosophy: there is a mutual and beneficious interaction among the two. On the one hand, one notices the great advancement produced by GR for the philosophy of time and epistemology; on the other hand, it is thanks to philosophy that the great potentialities of GR are understood. The philosophical standpoint, in Weyl's view, can grasp how GR encompasses both subjectivity and objectivity in the world. In order to reach this higher standpoint, Weyl suggests to reinterpret GR by unifying gauge invariance and covariance of laws that are expression of objectivity with the inclusions of the observations of measurements in constructing a theory of gravitation. And it does so through geometry. General relativity as a universal theory of gravitation offers the possibility of unifying different realms through geometry and ultimately this represents 'the sovereignty of the intellect' but only if inserted within the unified field theory extension that Weyl proposed. In Weyl's view the transition from the special to the general theory of relativity is a purely mathematical process. To formulate physical laws so that they remain covariant for arbitrary transformations is a possibility that is purely mathematical in essence and denotes no peculiarity of these laws.[8] However,

---

[7]Note that this position is very close to Eddington's view of space and this is the result, I claim, of neo-Kantian readings and of the influence of Brower's intuitionism.

[8]It would be interesting to compare Weyl's view with Einstein's and Kretschmann's. For a reconstruction of the Kretschmann-Einstein debate on the foundations of relativity and the meaning of general covariance, see Norton [14].

a new gauge factor appears when it is assumed that the metrical structure of the world is not given a priori, but that the quadratic form is related to matter by gauge invariant laws. In other words, Weyl believed that his unified field theory was the most genuine expression of the great revolution introduced by Einstein. This is a consequence of Weyl's assumption according to which to overcome any dualism is the goal of scientific enquiry and of relativity in the highest sense:

> The physical world-picture here described in its first outlines is characterised by the dualism of matter and field, between which there is a reciprocal action. Not till the advent of the theory of relativity was this dualism overcome, and, indeed, in favour of a physics based solely on fields. [21, p. 68]

Therefore, Weyl's attempt to find a unified field theory should be read as a strategy to lead relativity to its more radical consequences by following upon the mathematical method and the underlying fundamental concept of the unity of nature.

5 Weyl's Philosophy of Science?

Whereas in *Philosophy of Mathematics and Natural Science*, as well as in the late period of his life, Weyl was more explicit in drawing the main tenets of his reflections on science and philosophy of science, in *Raum-Zeit-Materie* we have to consider bits of the texts in order to reconstruct his views. Since *Raum-Zeit-Materie* is a book on his interpretation of relativity, we find in it traces of Weyl's view of scientific theories and his take on holism that will become more explicit from 1927 onward (see [9]). However, this position is already present in 1918 and 1920. First, let us consider that there are two ways for Weyl in which we can portray a physical theory:

1. As a system (objectively)
2. Through symbolic construction (including the subjective within it), as the result of a "tower" or sequence of theories.[9]

The two views are not in contrast, only dogmatism counts them as opposite. This assumption is fundamental in order to understand the connotation of holism that Weyl suggests:

> We cannot merely test a single law detached from this theoretical fabric! The connection between direct experience and the objective element behind it, which reason seeks to grasp conceptually in a theory, is not so simple that every single statement of the theory has a meaning which may be verified by direct intuition. We shall see more and more clearly in the sequel that Geometry, Mechanics, and Physics form an inseparable theoretical whole in this way. We must never lose sight of this totality when we enquire whether these sciences interpret rationally the reality which proclaims itself in all subjective experiences of consciousness, and which itself transcends consciousness: that is, truth forms a system. [21, p. 67]

[9]For further details on Weyl's view of levels or tower of levels of reality, see [23].

Both mathematical consistency and truth value principles can be seen as what encompasses consciousness within a theory. It is not possible for Weyl to admit the correspondence with reality of a theory by testing each by each its statements. A theory must be considered in its totality. Another important point emerging from the text is the importance for Weyl to make the passage from SR to GR possible by means of mathematics, and geometry in particular, which however is the result of a deeper level of physical reality yet to be penetrated. This belief is what drove Weyl in his early proposal of unified field theory and even if he dropped his idea out of the theory of relativity, yet he conserved this fundamental assumption in the later editions of *Ram-Zeit-Materie*:

> Only the consciousness that passes on in one portion of this world experiences the detached piece which comes to meet it and passes behind it, as history, that is, as a process that is going forward in time and takes place in space. This four-dimensional space is metrical like Euclidean space, but the quadratic form which determines its metrical structure is not definitely positive, but has one negative dimension. This circumstance is certainly of no mathematical importance, but has a deep significance for reality and the relationship of its action. [21, p. 217]

For Weyl, the great advance in our knowledge consists in recognising that the scene of action of reality is "a four-dimensional world, in which space and time are linked together indissolubly" [21, p. 217]. This assumption allows us to crystallize the physical world and offers an overview that excludes the qualitative and intuitive experience of time and space from objective knowledge (even if knowledge and understanding might need intuition at first). How? This is possible by means of symbolic construction, it is thanks to the deep interaction of the philosophical and mathematical methods that the quadratic form determining the four-dimensional spacetime is related to matter by generally invariant laws. This is the result of a fundamental fact of the theory, namely Weyl notices that according to the form of Pythagoras' Theorem, whereas the potential of the electromagnetic field is built up from the coefficients of an invariant linear differential form of the world-coordinates, the potential of the gravitational field is made up of the coefficients of an invariant quadratic differential form.[10] Now, since the potential of the gravitational field can be given in this form, the necessity of this fact is not empirically based, but, as Weyl underlines, it comes from geometry, from what he calls "the observations of measurements", it does not actually derive from the direct observation of gravitational phenomena. It becomes clearer now why he presents his unified field approach by means of analogy:

> The same fact is indispensable if we wish to solve the problem of the relativity of motion; it also enables us to complete the analogy mentioned above, according to which the metrical field is related to matter in the same way as the electric field to electricity. Only if we accept this fact does the theory briefly quoted at the end of the previous section become possible, according to which gravitation is a mode of expression (*Äusserungsweise*) of the metrical field. [21, p. 226]

[10]For further details on this, see Weyl [22] and the recent collection by Bernard and Lobo [2] on Weyl and the problem of space.

From the philosophical standpoint, Weyl is endorsing an epistemology of science privileging the a priori over the empirical data and indirect proofs over direct proofs. The analogy between the gravitational and the electric potential is only indirect and indeed it will be discarded a few months later by Einstein. However, it was advanced by Weyl in the name of the unity and systematicity of scientific theories, a truth value or at least an aim that also Yang and Mills pursued in 1954. What is certainly interesting in the abovementioned passage is the definition of gravitation as a mode of expression of the metrical field. This term "mode of expression" (*Äusserungsweise*) was highly diffused among the German-speaking philosophers, because it was used by both Husserl and Heidegger. A mode of expression, for instance, can be open or closed, considering that Weyl believed in the systematic view of scientific theories and their progressive inclusion of higher standpoints, he probably meant gravitation to be one mode of the metrical field, susceptible of being modified by another successive theory. Again, one can legitimately ask whether Weyl used philosophy to support his mathematical views or whether at least for the construction of this specific analogy philosophy was one of the ingredients that he needed for the formulation of his theory. To answer this question requires a separate study, but I tried to provide the reader with some hints that will be perhaps useful in pursuing this investigation.

6 Closing Remarks

The aim of this contribution was simply to highlight some philosophical aspects emerging from Weyl's *Raum-Zeit-Materie*. In particular, I wanted to recall his interpretation of relativity theory and underline its importance for current studies in philosophy of time and philosophy of science. I also wanted to point out the presence of Weyl's holism and its characterization before 1927, as well as his view of the structure of scientific theories. Finally, I wanted to stress the importance of considering Weyl's reflection upon the role of geometry and epistemology of science in order to present his unified field theory. One of the major lessons that one can take from Weyl's work is certainly his tendency to overcome dualities in philosophy, such as in the case of elaborating views overcoming the subjective/objective dichotomy or when stating that the process of theory construction is complementary to system analysis. After more than 100 years, Hermann Weyl's work is still stimulating the discussion among physicists, mathematicians and philosophers. A result of which he would certainly be proud.

Acknowledgements I am very thankful to Claus Kiefer, Friedrich Hehl, Erhard Scholz and Gabriel Catren for the comments made on the early draft and presentation of this work. This research has been made possible thanks to the Ramón y Cajal program (RYC-2015-17289) and has received funding from the European Union's Horizon 2020 research and innovation programme under grant agreement No. 758145—PROTEUS "Paradoxes and Metaphors of Time in Early Universe(s)".

References

1. J. Bell, Hermann Weyl's later philosophical views: his divergence from Husserl, in *Husserl and the Sciences*, ed. by R. Feist (University of Ottawa Press, Ottawa, 2004), pp. 173–185
2. J. Bernard, C. Lobo C. (eds.), *Weyl and the Problem of Space. Studies in History and Philosophy of Science*, vol. 49 (Springer, Cham, 2019)
3. F. Biagioli, Intuition and conceptual construction in Weyl's analysis of the problem of space, in *Weyl and the Problem of Space* (Springer, Cham, 2019), pp. 347–368
4. L. Boi, Weyl's Deep insights into the mathematical and physical worlds: his important contribution to the philosophy of space, time and matter, in *Weyl and the Problem of Space* (Springer, Cham, 2019), pp. 231–263
5. E. Cassirer, *Substance and Function and Einstein's Theory of Relativity* (The Open Court Publishing Company, Chicago, 1923)
6. J.J. da Silva, Husserl and Weyl on the constitution of space, in *Weyl and the Problem of Space* (Springer, Cham, 2019), pp. 389–402
7. S. De Bianchi, From the problem of space to the epistemology of science: Hermann Weyl's reflection on the dimensionality of the world, in *Weyl and the Problem of Space* (Springer, Cham, 2019), pp. 189–209
8. Eddington, *The Nature of the Physical World: Gifford Lectures (1927)* (Cambridge University Press, Cambridge, [1927]2012)
9. C. Eckes, Weyl's philosophy of physics: from apriorism to holism (1918–1927). Philos. Sci. **22**(2), 163–184 (2018)
10. R. Feist, Husserl and Weyl: phenomenology, mathematics, and physics, in *Husserl and the Sciences*, ed. by R. Feist (University of Ottawa Press, Ottawa, 2004), pp. 153–172
11. E. Husserl, *Ideas Pertaining to a Pure Phenomenology and to a Phenomenological Philosophy*, First Book, trans. by F. Kersten (Martinus Nijhoff, The Hague, [1913]1983)
12. P. Kerszberg, The scientific implications of epistemology: Weyl and Husserl, in *Weyl and the Problem of Space* (Springer, Cham, 2019), pp. 403–418
13. U. Majer, Knowledge by symbolic constructions, in *Philosophy and the Many Faces of Science*, ed. by D. Anapolitanos, A. Baltas, S. Tsinorema (Rowman & Littlefield, Lanham, 1998), pp. 40–64
14. J.D. Norton, General covariance and the foundations of general relativity: eight decades of dispute. Rep. Prog. Phys. **56**(7), 791 (1993)
15. S. Prosser, *Experiencing Time* (Oxford University Press, Oxford, 2016)
16. T.A. Ryckman, *The Reign of Relativity* (Oxford University Press, Oxford, 2005)
17. T.A. Ryckman, The philosophical roots of the gauge principle: Weyl and transcendental phenomenological idealism, in *Symmetries in Physics: Philosophical Reflections*, ed. by K. Brading, E. Castellani (2003), pp. 61–88
18. E. Scholz, Hermann Weyl's analysis of the "problem of space" and the origin of gauge structures. Sci. Context **17**, 165–197 (2004)
19. E. Scholz, The changing faces of the Problem of Space in the work of Hermann Weyl, in *Weyl and the Problem of Space* (Springer, Cham, 2019), pp. 213–230
20. N. Sieroka, Weyl's 'agens theory' of matter and the Zurich Fichte. Stud. Hist. Philos. Sci. **38**, 84–107 (2007)
21. H. Weyl, *Space-Time-Matter* (Dutton, New York, [1918]1922)
22. H. Weyl, Die Einzigartigkeit der Pythagoreischen Maßbestimmung. Math. Z. **12**, 114–146 (1922)
23. H. Weyl, Weyl levels of infinity, in *Levels of Infinity, Selected Writings on Mathematics And Philosophy*, ed. by P. Pesic (Dover, New York, 1930), pp. 17–32
24. H. Weyl, The mathematical way of thinking, in *Levels of Infinity, Selected Writings on Mathematics and Philosophy*, vol. 2012, ed. by P. Pesic (Dover, New York, 1940), pp. 67–84

25. H. Weyl, *Philosophy of Mathematics and Natural Science* (Princeton University Press, Princeton, 1949)
26. H. Weyl, *Symmetry* (Princeton University Press, Princeton, 1952)

Theoretical and Mathematical Physics of Gauge Theory

Space, Time, Matter in Quantum Gravity

Claus Kiefer

Abstract The concepts of space, time, and matter are of central importance in any theory of the gravitational field. Here I discuss the role that these concepts might play in quantum theories of gravity. To be concrete, I will focus on the most conservative approach, which is quantum geometrodynamics. It turns out that spacetime is absent at the most fundamental level and emerges only in an appropriate limit. It is expected that the dynamics of matter can only be understood from a fundamental quantum theory of all interactions.

1 From Classical to Quantum Gravity

In his famous habilitation colloquium on June 10, 1854, Bernhard Riemann concluded

> The question of the validity of the hypotheses of geometry in the infinitely small is bound up with the question of the ground of the metric relations of space.... Either therefore the reality which underlies space must form a discrete manifoldness, or we must seek the ground of its metric relations outside it, in binding forces which act upon it.... This leads us into the domain of another science, of physic, into which the object of this work does not allow us to go to-day. ([27]; translated by William Kingdon Clifford 1873).[1]

---

[1]The German original reads ([21], p. 43): "Die Frage über die Gültigkeit der Voraussetzungen der Geometrie im Unendlichkleinen hängt zusammen mit der Frage nach dem innern Grunde der Massverhältnisse des Raumes. ...Es muss also entweder das dem Raume zu Grunde liegende Wirkliche eine discrete Mannigfaltigkeit bilden, oder der Grund der Massverhältnisse ausserhalb, in darauf wirkenden bindenden Kräften, gesucht werden. ...Es führt dies hinüber in das Gebiet einer andern Wissenschaft, in das Gebiet der Physik, welches wohl die Natur der heutigen Veranlassung nicht zu betreten erlaubt." The English translation can be found in [22]. For the role of Clifford in the development of these ideas, see e.g. [15].

C. Kiefer (✉)
Institute for Theoretical Physics, University of Cologne, Zülpicher Straße 77a, 50937 Köln, Germany
e-mail: kiefer@thp.uni-koeln.de

S. De Bianchi and C. Kiefer (eds.), *One Hundred Years of Gauge Theory*, Fundamental Theories of Physics 199, https://doi.org/10.1007/978-3-030-51197-5_9

Riemann's pioneering ideas are important for at least two reasons. First, although Riemann did not take into account the time dimension, his ideas led to the mathematical formalism that enabled Albert Einstein to formulate his theory of general relativity (GR) in 1915. In GR, gravity is understood as the manifestation of a dynamical geometry of space and time, which are unified into a four-dimensional spacetime.

Second, as is clear from the sentences quoted above, matter and geometry are no longer imagined as independent from each other; the metric now depends on the "binding forces which act upon it". The metrical field is no longer given rigidly once and for all, but stands in causal dependence on matter. This idea is at the core of GR.

In his commentary on Riemann's text from 1919, Hermann Weyl emphasized the unification of geometry and field theory in physics,

> For geometry, here the same step happened that Faraday and Maxwell performed within physics, in particular electricity theory, which was done by the transition from an action-at-a-distance to a local-action theory: carrying out the principle to understand the world from its behaviour in the infinitely small. See [21], p. 45.[2]

Riemann's approach turned out to be much more powerful than alternative ideas on the foundation of geometry, for example those of Hermann von Helmholtz, see e.g. [22], p. 119. Helmholtz starts from experience[3] and postulates the possibility of free motion of bodies. As he can prove mathematically, for this free motion a space with constant curvature is required. From the later perspective of GR, this turns out to be too narrow. Riemann's idea, that bodies can carry geometry with them, *is* realized in GR, which allows spaces, in fact spacetimes, to have arbitrary curvature, as determined by the Einstein field equations. These equations read

$$R_{\mu\nu} - \frac{1}{2}g_{\mu\nu}R + \Lambda g_{\mu\nu} = \kappa T_{\mu\nu}. \tag{1}$$

Here, $g_{\mu\nu}$ denotes the spacetime metric, $R_{\mu\nu}$ the Ricci tensor, and R the Ricci scalar. Non-gravitational degrees of freedom (for simplicity called 'matter') are described by a symmetric energy–momentum tensor $T_{\mu\nu}$; it obeys the covariant conservation law

$$T_{\mu\nu;}{}^{\nu} = 0. \tag{2}$$

It is important to emphasize that this is not a standard conservation law (with a partial instead of a covariant derivative) from which a conserved current and charge can be derived.[4] If the energy–momentum tensor obeys the dominant energy condition (energy densities dominate over pressures), causality is implemented in the sense that no influence from outside the lightcone can enter its inside.

---

[2] "Für die Geometrie geschah hier der gleiche Schritt, den FARADAY und MAXWELL innerhalb der Physik, speziell der Elektrizitätslehre, vollzogen durch den Übergang von der Fernwirkungs- zur Nahewirkungstheorie: das Prinzip, die Welt aus ihrem Verhalten im Unendlichkleinen zu verstehen, gelangt zur Durchführung."

[3] The title of Helmholtz's article, "Ueber die Thatsachen, die der Geometrie zu Grunde liegen" ("On the Facts which Lie at the Bases of Geometry"), makes a dig at the title of Riemann's work.

[4] This is only possible in the presence of a symmetry, as expressed by a Killing vector.

There are two free parameters in the gravitational sector: κ and Λ. From the Newtonian limit, one can identify

$$\kappa = 8\pi G/c^4, \tag{3}$$

with G the gravitational (Newton) constant and c the speed of light. In 1917, Einstein had recognized that another free parameter is allowed—the cosmological constant Λ which has the physical dimension of an inverse length squared. From observations we find the value $\Lambda \approx 1.2 \times 10^{-52}\text{m}^{-2} \approx 0.12\ (\text{Gpc})^{-2}$.[5] The relation of this value to naive estimates from quantum field theory is an open question.

The Einstein field equations (1) describe a non-linear interaction between geometry and matter. In this sense, $T_{\mu\nu}$ must not be interpreted as the source from which the metric is determined. For the description of matter, the metric is also needed, since it enters the field equations for matter as well as the equation of motion for test bodies[6] given by

$$\ddot{x}^\mu + \Gamma^\mu_{\alpha\beta}\dot{x}^\alpha\dot{x}^\beta = 0. \tag{4}$$

Here, $\Gamma^\mu_{\alpha\beta}$ are the components of the Levi–Civita connection, which is determined by the metric, and the dots denote derivatives with respect to proper time (for timelike geodesics) or with respect to an affine parameter (null geodesics). Equation (4) is the geodesic equation which reflects the universal coupling of gravity to matter. In contrast to its Newtonian analogue, it corresponds to free motion in the geometry described by $g_{\mu\nu}$ ('equivalence principle'). So for the description of matter, the pair $(T_{\mu\nu}, g_{\alpha\beta})$ is needed, and one needs a rather involved initial value formulation to determine the spacetime metric (see next section).

A major feature of GR, and one that is particularly relevant for its quantization, is *background independence*. This must be carefuly distinguished from mere general covariance, which means form invariance of equations under an arbitrary change of coordinates. In contrast, background independence means that there are no absolute (non-dynamical) fields in the theory—this applies to GR, where the metric is a dynamical quantity that acts on matter and is acted upon by it. As Jürgen Ehlers has remarked ([9], p. 91): "Conceptually, the background independence must be seen as the principal achievement of general relativity theory; it is, however, at the same time the main obstacle to overcome if general relativity theory and quantum theory are to be united." In GR, the law of motion (4) cannot be formulated independently from the field equations (1)—in fact, it follows from them by employing (2). This would not be possible in a theory with an absolute background, that is, with an absolute non-dynamical spacetime.

In 1918, Weyl generalized the notion of the Levi-Civita connection that occurs in (4) to a symmetric linear connection, [30], see also the extended discussion in *Raum, Zeit, Materie*, [31]. For this concept, a metrical structure on the manifold is

[5] Recent doubts on this Λ-observation are expressed e.g. in [7].

[6] Test bodies in GR cannot be mass points. The mass-to-radius ratio of objects has an upper bound of $c^2/2G$; the concept of a mass point is replaced by a black hole.

not needed, only the notion of a parallel transport for vectors and tensors, which provides the means to connect different points on the manifold. In contrast to the Levi-Civita connection, his more general connection need not be derivable from a metric. Weyl distinguishes, in fact, three levels of geometry: the first level is the topological manifold (which he calls *situs manifold or empty world*),[7] the second level the affinely connected manifold, and the third level the metric continuum (which he also calls "ether"); see also [28] for a lucid presentation.

The notion of a symmetric linear connection allowed Weyl to construct a generalization of Einstein's theory. In his theory, the magnitude of vectors is not fixed, but the connection allows the comparison of magnitudes in different points. This introduces a new freedom into the theory—the freedom to perform gauge transformations. The metric is here determined only up to a (spacetime-dependent) factor. The exponent of this factor can be connected with a function that behaves as the electromagnetic vector potential (here interpreted as a one-form). Weyl thought that he has constructed in this way a unified theory of gravity and electromagnetism; for details, see [31], p. 121 ff.[8] In the above hierarchy, Weyl's theory can be located between the second and third level: in it, spacetime has a conformal structure, which provides a more general framework than the structure of Riemannian geometry.

Weyl was convinced that fundamental geometric relations should only refer to infinitesimally neighbouring points (*Nahgeometrie* instead of *Ferngeometrie*). This principle plays a key role in both the 1918 and the 1929 versions of gauge theories. In [32], p. 115, he writes (emphasis by Weyl): "*Only in the infinitely small can we expect to encounter everywhere the same elementary laws*, thus the world must be understood from its behaviour in the infinitely small."[9]

In spite of its formal elegance, Weyl's theory is empirically wrong, as was soon realized by Einstein. The reason is that a non-integrable connection leads to path-dependent frequencies for atomic spectra, in contrast to observations. But his theory can nevertheless be seen as the origin of our modern gauge theories. A decade later, in 1929, Weyl came up with a gauge theory of electromagnetism and the Dirac field. Instead of the real conformal factor multiplying the metric, there occurs now a phase factor for which the exponent is a one-dimensional integral over the vector potential. A non-integrable connection is manifested there, for example, in the Aharonov–Bohm effect. In its non-Abelian generalization, gauge invariance is a key ingredient to the Standard Model of particle physics, see e.g. [8] for a review. The Standard Model (extended by massive neutrinos) is experimentally extremely well tested, and no obvious deviation from it is seen so far in experiments at the Large Hadron Collider (LHC) and elsewhere.

---

[7]*Analysis situs* is an older name for topology.

[8]Here, the words 'gauge' (*Eichung*), 'to gauge' (*eichen*), and 'gauge invariance' (*Eich-Invarianz*) enter. Their original meaning arises from providing standards for physical quantities (including distances), which is different from their later abstract use in the description of intrinsic symmetries in gauge theories.

[9]The German original reads: "*Nur im Unendlichkleinen dürfen wir erwarten auf die elementaren, überall gleichen Gesetze zu stoßen*, darum muß die Welt aus ihrem Verhalten im Unendlichkleinen verstanden werden.

But what about gravity and spacetime? In its standard formulation, GR is not a gauge theory. The reason is that the connection $\Gamma^{\mu}_{\alpha\beta}$ is not independent there, but is derived from a metric. On thus has the chain

$$g_{\mu\nu} \longrightarrow \Gamma^{\mu}_{\alpha\beta} \longrightarrow R^{\mu}_{\alpha\beta\gamma},$$

where $R^{\mu}_{\alpha\beta\gamma}$ denotes the Riemann curvature tensor. For gauge theories, the first step in this chain is lacking. Gauge theories of gravity do, however, exist, and they are needed for the consistent implementation of fermions, see [1].[10] Weyl's original theory is a special case of this general class, but it is important to emphasize that the coupling of Weyl's vector potential is not to the electrodynamic current—as its creator believed—but to the dilaton current (because the one-parameter dilation group is gauged), see [19].

One of the striking properties of GR is that it exhibits its own incompleteness. This is expressed in the singularity theorems which state that, under general conditions, singularities in spacetime are unavoidable, see [17]. Singularities are here understood in the sense of geodesic incompleteness—timelike or null geodesics as found from (4) terminate at finite proper time or finite affine parameter value. In most physically relevant cases, the occurrence of singularities is connected with regions of infinite curvature or energy density; notable examples are the singularities characterizing the beginning of the Universe ("big bang") and the interior of black holes. One of the hopes connected with the construction of a quantum theory of gravity is that such a theory will avoid singularities. This hope may be extended to a different type of singularities in our present physical theories – the infinities that arise in almost every local quantum field theory. One has learnt to cope with the latter singularities by employing sophisticated methods of regularization and renormalization. Nevertheless, one would expect that a truly fundamental theory will be finite from the onset. The reason is that the occurrence of singularities is connected with an unsufficient understanding of the microstructure of spacetime. True infinities should not occur in any sensible description of Nature, cf. [11].

One possible solution to the singularity problem is to avoid a continuum for the spacetime structure and to assume instead that spacetime is built up from discrete entities. There are indications for such a discrete structure in some approaches to quantum gravity, but the last work has not yet been spoken. Interestingly, Riemann himself envisaged the possibility of a continuous as well as a discrete manifold; the smallest entities he calls *quanta* ([22], p. 32):

> Definite portions of a manifoldness, distinguished by a mark or by a boundary, are called Quanta. Their comparison with regard to quantity is accomplished in the case of discrete magnitudes by counting, in the case of continuous magnitudes by measuring.
> (Translated by William Kingdon Clifford 1873)[11]

[10] See also the contributions by Hehl and Obukhov and by Scholz to this volume.

[11] The German original reads ([21], p. 31): "Bestimmte, durch ein Merkmal oder eine Grenze unterschiedene Theile einer Mannigfaltigkeit heissen Quanta. Ihre Vergleichung der Quantität nach geschieht bei den discreten Grössen durch Zählung, bei den stetigen durch Messung.".

Weyl, in his commentary to Riemann's text, speculates that the final answer to the problem of space may be found in its discrete nature.[12]

What happens to this when the quantum of action $\hbar$ comes into play? One of the early pioneers of attempts to quantizing gravity, Matvei Bronstein, through the application of thoughts experiments, arrived at the necessity of introducing minimal distances in spacetime, thus abandoning the idea of a metric continuum. He writes[13]

> The elimination of the logical inconsistencies connected with this [his thought experiments] requires a radical reconstruction of the theory, and in particular, the rejection of a Riemannian geometry dealing, as we see here, with values unobservable in principle, and perhaps also the rejection of our ordinary concepts of space and time, modifying them by some much deeper and nonevident concepts. *Wer's nicht glaubt, bezahlt einen Taler.*

In Bronstein's analysis, quantities appear that can be found by combining G, c, and $\hbar$ into units of length, time, and mass (or energy). They were first presented by Max Planck in 1899 (one year before the 'official' introduction of the quantum of action into physics!) and are called *Planck units* in his honour. They read[14]

$$l_{\mathrm{P}} = \sqrt{\frac{\hbar G}{c^3}} \approx 1.62 \times 10^{-35}\ \mathrm{m} \tag{5}$$

$$t_{\mathrm{P}} = \frac{l_{\mathrm{P}}}{c} = \sqrt{\frac{\hbar G}{c^5}} \approx 5.40 \times 10^{-44}\ \mathrm{s} \tag{6}$$

$$m_{\mathrm{P}} = \frac{\hbar}{l_{\mathrm{P}} c} = \sqrt{\frac{\hbar c}{G}} \approx 2.17 \times 10^{-8}\ \mathrm{kg} \approx 1.22 \times 10^{19}\ \mathrm{GeV}/c^2. \tag{7}$$

At the end of his 1899 paper, Planck wrote the following prophetic sentences, see [24], p. 6:

> These quantities retain their natural meaning as long as the laws of gravitation, of light propagation in vacuum, and the two laws of the theory of heat remain valid; they must therefore, if measured in various ways by all kinds of intelligent beings, always turn out to be the same.[15]

One can form a dimensionless number out of these Planck units by bringing the cosmological constant Λ into play. Inserting the present observational value for Λ (see above), this gives

$$l_{\mathrm{P}}^2 \Lambda \equiv \frac{G\hbar\Lambda}{c^3} \approx 3.3 \times 10^{-122}. \tag{8}$$

---

[12]"Sehen wir von der ersten Möglichkeit ab, es könnte 'das dem Raum zugrunde liegende Wirkliche eine diskrete Mannigfaltigkeit bilden' (*obschon in ihr vielleicht einmal die endgültige Antwort auf das Raumproblem enthalten sein wird*, my emphasis) ....

[13]The quotation is from [24], p. 20.

[14]See e.g. [24], p. 5.

[15]The German original reads: "Diese Grössen behalten ihre natürliche Bedeutung so lange bei, als die Gesetze der Gravitation, der Lichtfortpflanzung im Vacuum und die beiden Hauptsätze der Wärmetheorie in Gültigkeit bleiben, sie müssen also, von den verschiedensten Intelligenzen nach den verschiedensten Methoden gemessen, sich immer wieder als die nämlichen ergeben.".

The smallness of this number is one of the biggest open puzzles in fundamental physics. Only a fundamental unified theory of all interactions is expected to provide a satisfactory explanation.

What are the general arguments that speak in favour of a quantum theory of gravity?[16] First, as mentioned above, there is the singularity problem of classical general relativity, which points to the incompleteness of Einstein's theory. Second, the search for a unified theory of all interactions should include quantum gravity: gravity interacts universally to all fields of Nature, and all non-gravitational fields are successfully described by quantum (field) theory so far, so a quantum description should apply to gravity, too. Third, a very general argument was put forward by Richard Feynman in 1957, see [24], p. 18: if we generate a superposition of two masses at different locations, their gravitational fields should also be superposed, unless the superposition principle of quantum theory breaks down. A quantum theory of gravity is needed to describe such superpositions. It is clear that such a state can no longer correspond to a classical spacetime. There are at present interesting suggestions for the possibility to observing the gravitational field generated by a quantum superposition in laboratory experiments, see [4] and references therein.

Several approaches to quantum gravity exist, but there is so far no consensus in the community, see [24]. The ideal case would be to construct a finite quantum theory of all interactions from which present physical theories can be derived as approximations (or "effective field theories") in appropriate limits. The only reasonable candidate so far is string theory. In this theory, the dimension of spacetime assumes the number ten or eleven. Unfortunately, it is so far not clear how to recover the Standard Model from string theory and how to test it by experiments. Connected with this is the difficulty to proceed in a more or less unique way from the ten or eleven spacetime dimensions to the four dimensions of the observed world.

The main alternatives to finding a unified theory are the more modest attempts to construct first a quantum theory of the gravitational field and to relegate unification to a later step. The usual starting point is GR, but quantization methods may be applied to any other gravitational theory. Standard methods are path integral quantization and canonical quantization. We shall focus below on the canonical quantization of GR using metric variables, because conceptual issues dealing with space and time are most transparent in this approach, see [23].

2 The Configuration Space of General Relativity

Besides ordinary three-dimensional space (or four-dimensional spacetime), the concept of *configuration space* plays an eminent role in physics. In mechanics, this is the N-dimensional space generated by all configurations, described by coordinates $\{q^a\}$, $a = 1, \ldots N$, that the system can assume. In field theory, it is the infinite-dimensional space of possible field configurations. In quantum theory, it will enter

[16]See e.g. [24] for a comprehensive discussion.

the argument of the wave function (functional) and lead to the central property of entanglement.

What is the configuration space in general relativity? As John Wheeler writes ([34], p. 245): “A decade and more of work by Dirac, Bergmann, Schild, Pirani, Anderson, Higgs, Arnowitt, Deser, Misner, DeWitt, and others has taught us through many a hard knock that Einstein’s geometrodynamics deals with the dynamics of geometry: of 3-geometry, not 4-geometry.” Most of these developments happened after Weyl’s death in 1955. In fact, upon application of the canonical (or Hamiltonian) formalism, Einstein’s theory can be written as a dynamical system for the *three-metric* h_{ab} and its canonical momentum π^{ab} on a spacelike hypersurface Σ. The ten Einstein equations can be formulated as four constraints, that is, restrictions on initial data h_{ab} and π^{ab} on Σ, and six evolution equations. The four constraints read (per spacepoint)

$$\mathcal{H}_\perp = 2\kappa\, G_{abcd}\pi^{ab}\pi^{cd} - (2\kappa)^{-1}\sqrt{h}({}^{(3)}R - 2\Lambda) + \sqrt{h}\rho \approx 0 \quad (9)$$

$$\mathcal{H}^a = -2\nabla_b\pi^{ab} + \sqrt{h}j^a \approx 0, \quad (10)$$

with the (inverse) DeWitt metric

$$G_{abcd} = \frac{1}{2\sqrt{h}}(h_{ac}h_{bd} + h_{ad}h_{bc} - h_{ab}h_{cd}) \quad (11)$$

and κ given by (3). Here, ${}^{(3)}R$ denotes the three-dimensional Ricci scalar and h the determinant of h_{ab}; ρ and j^a denote matter density and current, respectively. The constraint $\mathcal{H}_\perp \approx 0$ is called “Hamiltonian constraint”, while $\mathcal{H}^a \approx 0$ are called “momentum (diffeomorphism) constraints”. The symbol ≈ 0 is Dirac’s weak equality and means “vanishing as a constraint”. The canonical momentum π^{ab} is related to the extrinsic curvature K_{cd} of Σ by

$$\pi^{ab} = \frac{G^{abcd}K_{cd}}{2\kappa}, \quad (12)$$

where G^{abcd} denotes the DeWitt metric itself (the inverse of the expression in (11)). This quantity plays the role of a metric in the space of all Riemannian three-metrics h_{ab}, a space called Riem Σ.

Is Riem Σ the configuration space of GR? Not yet. The constraints $\mathcal{H}^a \approx 0$ guarantee the invariance of the theory under infinitesimal three-dimensional coordinate transformations. The real configuration space is thus the space of all three-*geometries*, not the space of all three-*metrics*. This is what Wheeler called *superspace*, here denoted by $\mathcal{S}(\Sigma)$, see [34]. It is the arena for classical and quantum geometrodynamics. One can formally write

$$\mathcal{S}(\Sigma) := \text{Riem}\ \Sigma/\text{Diff}\ \Sigma,$$

where Diff Σ denotes the group of three-dimensional diffeomorphisms (“coordinate transformations”). By going to superspace, the momentum constraints are automatically fulfilled. Whereas Riem Σ has a simple topological structure, the topological structure of $\mathcal{S}(\Sigma)$ is very complicated because it inherits (via Diff Σ) some of the topological information contained in Σ; see [14] for details.

The DeWitt metric has pointwise a Lorentzian signature with one negative and five positive directions, that is, it has negative, null, and positive directions. Due to the minus sign, the kinetic term for the gravitational field is indefinite. It is important to note that this minus sign is unrelated to the signature of spacetime; starting with a four-dimensional Euclidean space instead of a four-dimensional spacetime, the same signature for the DeWitt metric is found. The presence of this minus sign is related to the attractive nature of gravity. It is also worth mentioning that the DeWitt metric reveals a surprising analogy with the elasticity tensor in three-dimensional elasticity theory and the local and linear constitutive tensor in four-dimensional electrodynamics, see [18]. This analogy could be of importance for theories of emergent gravity.

Constraints and evolution equations have an intricate relationship; see e.g. [16]. Let me summarize the main features as well as pointing out analogies with electrodynamics. First, there is an important connection with the (covariant) conservation law of energy–momentum. The constraints are preserved in time if and only if the energy–momentum tensor of matter has vanishing covariant divergence. In electrodynamics, the Gauss constraint is preserved in time if and only if electric charge is conserved.

Second, Einstein's equations represent the unique propagation law consistent with the constraints. To be more concrete, if the constraints are valid on an "initial" hypersurface and if the dynamical evolution equations (the pure spatial components of the Einstein equations) hold, the constraints hold on every hypersurface. And if the constraints hold on every hypersurface, the dynamical evolution equations hold. Again, there is an analogy with electrodynamics: Maxwell's equations are the unique propagation law consistent with the Gauss constraint.

It must be emphasized that the picture of a spacetime foliated by a one-parameter family of hypersurfaces only emerges *after* the dynamical equations are solved. Then, spacetime can be interpreted as a "trajectory of spaces". Before this is done, one only has a three-dimensonal manifold Σ with given topology, equipped with the canonical variables satisfying the constraints (9) and (10).

This fact that spacetime is not given from the outset but must be constructed through an initial value formulation, is an expression of the background independence discussed in the previous section. In this sense, the analogy with electrodynamics on a given external spacetime breaks down. Background independence is related with the classical version of what is called the *problem of time*: if we restrict ourselves to compact three-manifolds Σ, the total Hamiltonian of GR is a combination of the constraints (9) and (10).[17] Thus, no external time parameter exists; all physical time parameters are to be constructed from within our system, that is, as functional of the canonical variables. A priori, there is no preferred choice of such an intrinsic time parameter. It is this absence of an external time and the non-preference of an intrinsic

[17]In the asymptotically flat case, additional boundary terms are present.

one that is known as the problem of time in (classical) canonical gravity. Still, after the solution of the dynamical equations, spacetime as a trajectory of spaces exists. This is different in the quantum theory where it leads to the far-reaching quantum version of the problem of time (next section).

The possibility of constructing spacetime in the way just described is also reflected in the closure of the Poisson algebra for the constraints (9) and (10):

$$\{\mathcal{H}_\perp(\mathbf{x}), \mathcal{H}_\perp(\mathbf{y})\} = -\sigma\delta_{,a}(\mathbf{x},\mathbf{y})\left(h^{ab}(\mathbf{x})\mathcal{H}_b(\mathbf{x}) + h^{ab}(\mathbf{y})\mathcal{H}_b(\mathbf{y})\right) \quad (13)$$

$$\{\mathcal{H}_a(\mathbf{x}), \mathcal{H}_\perp(\mathbf{y})\} = \mathcal{H}_\perp(\mathbf{x})\delta_{,a}(\mathbf{x},\mathbf{y}) \quad (14)$$

$$\{\mathcal{H}_a(\mathbf{x}), \mathcal{H}_b(\mathbf{y})\} = \mathcal{H}_b(\mathbf{x})\delta_{,a}(\mathbf{x},\mathbf{y}) + \mathcal{H}_a(\mathbf{y})\delta_{,b}(\mathbf{x},\mathbf{y}) \quad (15)$$

It is not a Lie algebra, though, because the Poisson bracket between two Hamiltonian constraints at different points also contains (the inverse of) the three-metric, h^{ab}. We also remark that the signature of the spacetime metric enters here in the form of the parameter σ: in fact, $\sigma = -1$ corresponds to a four-dimensional spacetime, while $\sigma = 1$ corresponds to a four-dimensional space. It is a fundamental (and open) question whether the closure of this algebra also holds in quantum gravity.

The relation of the transformations generated by the constraints to the spacetime diffeomorphisms is a subtle one and will not be discussed here; see e.g. [24, 29].

3 Quantum Geometrodynamics

In the last section, we have reviewed the canonical (Hamiltonian) formulation of GR. Here, we discuss the quantum version of this, see e.g. [24] for a comprehensive treatment. We follow Dirac's heuristic approach and transform the classical constraints (9) and (10) into conditions on physically allowed wave functionals. These wave functionals are defined on the space of all three-metrics (the above space Riem Σ) and matter fields on Σ. The quantum version of (9) reads

$$\hat{\mathcal{H}}_\perp\Psi \equiv \left(-16\pi G\hbar^2 G_{abcd}\frac{\delta^2}{\delta h_{ab}\delta h_{cd}} - (16\pi G)^{-1}\sqrt{h}\left({}^{(3)}\!R - 2\Lambda\right) + \sqrt{h}\hat{\rho}\right)\Psi = 0 \quad (16)$$

and is called the *Wheeler–DeWitt equation*. We note that the kinetic term in this equation only has formal meaning before the issues of factor ordering and regularization are successfully addressed.[18] The quantum implementation of (10) reads

$$\hat{\mathcal{H}}^a \Psi \equiv -2\nabla_b \frac{\hbar}{\mathrm{i}} \frac{\delta \Psi}{\delta h_{ab}} + \sqrt{h}\, \hat{j}^a \Psi = 0 \tag{17}$$

and is called the momentum (or diffeomorphism) constraints. These latter equations have a simple interpretation. Under a coordinate transformation

$$x^a \mapsto \bar{x}^a = x^a + \delta N^a(\mathbf{x}),$$

the three-metric transforms as

$$h_{ab}(\mathbf{x}) \mapsto \bar{h}_{ab}(\mathbf{x}) = h_{ab}(\mathbf{x}) - D_a \delta N_b(\mathbf{x}) - D_b \delta N_a(\mathbf{x}).$$

The wave functional then transforms according to

$$\Psi[h_{ab}] \mapsto \Psi[h_{ab}] - 2\int \mathrm{d}^3x\, \frac{\delta \Psi}{\delta h_{ab}(\mathbf{x})} D_a \delta N_b(\mathbf{x}).$$

Assuming the invariance of the wave functional under this transformation, one is led to

$$D_a \frac{\delta \Psi}{\delta h_{ab}} = 0.$$

This is exactly (17) (restricted here to the vacuum case).

A simple analogy to (17) is Gauss's law in quantum electrodynamics (or its generalization to the non-Abelian case). The quantized version of the constraint $\nabla \mathbf{E} \approx 0$ reads

$$\frac{\hbar}{\mathrm{i}} \nabla \frac{\delta \Psi[\mathbf{A}]}{\delta \mathbf{A}} = 0,$$

where $\mathbf{A}$ is the vector potential. This equation reflects the invariance of Ψ under spatial gauge transformations of the form $\mathbf{A} \to \mathbf{A} + \nabla\lambda$.

The constraints can only be implemented in the form (16) and (17) if the quantum version of the constraint algebra (13)–(15) holds without extra c-number terms on the right-hand side. Otherwise, only a part of the quantum constraints (or even none) holds in this form. The situation is reminiscent of string theory where the Virasoro algebra displays such extra (central or Schwinger) terms. More general quantum constraints hold there provided the number of spacetime dimensions is restricted to a specific number (ten in the case of superstrings). It is imaginable that a restriction

[18] For a recent attempt into this direction, see [12].

in the number of spacetime dimensions arises also here from a consistent treatment of the quantum constraint algebra. But so far, this is not clear at all.[19]

In the last section, we have seen that we can interpret spacetime as a generalized trajectory of spaces. In its construction, the four constraint equations and the six dynamical equations are inextricably interwoven. What happens in the quantum theory? There, the trajecory of spaces has disappeared, in the same way as the ordinary mechanical trajectory of a particle has disappeared in quantum mechanics. The three-metric h_{ab} and its momentum π^{cd} play the role of the q^i and p_j in mechanics, so it is clear that in quantum gravity h_{ab} and π^{cd} cannot be "determined simultaneously", which means that spacetime is absent at the most fundamental limit, and only the configuration space of all three-metrics respective three-geometries remains. This is clearly displayed in Table 1 on p. 248 in [34].

From this point of view it is clear that in the quantum theory only the constraints survive. The evolution equations lose their meaning in the absence of a spacetime. In a certain sense, this is anticipated in the classical theory by the strong connection between constraints and evolution equations as discussed in the previous section.

The absence of spacetime, and in particular of time, is usually understood as the quantum version of the *problem of time*. It means that the quantum world at the fundamental level is timeless—it just *is*. Weyl has attributed such a static picture already to the classical spacetime of GR. In [32] p. 150, he writes:

> The objective world just *is*, it does not *happen*. Only from the view of the consciousness crawling upwards in the worldline of my life a sector of this world "lives up" and passes by at it as a spatial picture in temporal transformation.[20]

In the quantum theory, there is not even a spacetime and a worldline with a conscious observer, at least not at the most fundamental level. So how can we relate this picture of timelessness, forced upon us by a straightforward extrapolation of established physical theories, with the standard concept of time in physics? There are two points to be discussed here.

First, as already mentioned, the DeWitt metric (12) has an indefinite signature: one *minus* and five *plus*. This means that the Wheeler–DeWitt equation has a local hyperbolic structure through which part of the three-metric is distinguished as an *intrinsic timelike variable*. One can show that this role is played by the "local scale" $\sqrt{h}$. In simple cosmological models of homogeneous and isotropic (Friedmann–Lemaître) universes, this is directly related to the *scale factor*, a. Using units with

---

[19]This problem was already known to Dirac and was the reason why he abandoned working on quantum gravity. In his last contribution to this field, he remarked, [6], p. 543: "The problem of the quantization of the gravitational field is thus left in a rather uncertain state. If one accepts Schwinger's plausible methods, the problem is solved. [Dirac refers to a heuristic regularization proposed by Schwinger in 1962, C.K.] But one cannot be happy with such methods without having a reliable procedure for handling quadratic expressions in the δ-function." Such a reliable procedure is still missing. But see [12].

[20]The German original reads: "Die objektive Welt *ist* schlechthin, sie *geschieht* nicht. Nur vor dem Blick des in der Weltlinie meines Lebens emporkriechenden Bewußtseins "lebt" ein Ausschnitt dieser Welt "auf" und zieht an ihm vorüber als räumliches, in zeitlicher Wandlung begriffenes Bild." [emphasis by Weyl].

$2G/3\pi = 1$, the Wheeler–DeWitt equation for a closed Friedmann–Lemaître universe with a massive scalar field reads

$$\frac{1}{2}\left(\frac{\hbar^2}{a^2}\frac{\partial}{\partial a}\left(a\frac{\partial}{\partial a}\right)-\frac{\hbar^2}{a^3}\frac{\partial^2}{\partial\phi^2}-a+\frac{\Lambda a^3}{3}+m^2a^3\phi^2\right)\psi(a,\phi)=0. \qquad (18)$$

Additional gravitational and matter degrees of freedom[21] come with kinetic terms that differ in sign from the kinetic term with respect to a. For equations such as (18), one can thus formulate an initial value problem with respect to intrinsic time a. The configuration space is here two-dimensional and spanned by the two variables a and ϕ.

Standard quantum theory employs the mathematical structure of a Hilbert space in order to implement the probability interpretation for the quantum state. An important property is the unitary evolution of this state; it guarantees the conservation of the total probability with respect to the external time t. But what happens when there is no external time, as we have seen is the case in quantum gravity? There is no common opinion on this, but it is at least far from clear whether a Hilbert-space structure is needed at all, and if yes, which one. This is also known as the *Hilbert-space problem* and is evidently related to the problem of time.[22]

The second point concerns the recovery of the standard (general relativistic) notion of time from the fundamentally timeless theory of gravity. The standard way proceeds via a Born–Oppenheimer type of approximation scheme, similarly to molecular physics. For this to work, the quantum state, which is a solution of (16) and (17), must be of a special form. For such a state one can recover an approximate notion of semiclassical (WKB) time. One can show that this WKB time (which, in fact, is a "many-fingered time") corresponds to the notion of time in Einstein's theory. Equations (16) and (17) then lead to a functional Schrödinger equation describing the limit of quantum field theory in curved spacetime, the latter given by Einstein's equations. It is in this limit that one can apply the standard Hilbert-space structure and the associated probability interpretation. Higher orders of this approximation allow the derivation of quantum-gravitational corrections terms, which, for example, give corrections to the Cosmic Microwave Background (CMB) anisotropy spectrum proportional to the inverse Planck-mass squared. Such terms follow from a straightforward expansion of (16) and (17) and could in principle give a first observational test of quantum geometrodynamics, [3].

Quantum geometrodynamics, like practically all approaches to quantum gravity, is a linear theory in the quantum states and thus obeys the superposition principle. This means that most states do not correspond to any classical three-geometry. The situation resembles, of course, Schrödinger's cat. Like there, one can employ the process of *decoherence* to understand why such weird superpositons are not observed, see [20]. Decoherence is the irreversible and unavoidable interaction of a quantum

[21]Except *phantom fields*, which play a role in connection with discussions about dark energy, cf. [2, 7].

[22]See e.g. [24, 25] for a detailed discussion of this and the other conceptual issues discussed below.

system with the irrelevant degrees of freedom of its "environment".[23] In quantum cosmology, one can consider, for example, the variables a and ϕ in (18) as describing the (relevant) quantum system, while small density perturbations and tiny gravitational waves can play the role of the environment. The entanglement between system and environment leads to the suppression of interferences between different a and different ϕ (within some limits); in this sense, classical geometry and classical universe emerge. The same holds for the emergence of structure in the universe from primordial quantum fluctuations, see [26].

It is evident from the above that the question about the correct interpretation of quantum theory enters here with its full power. Since by definition the Universe as a whole is a strictly closed quantum system, one cannot invoke any classical measurement agent as acting from the outside. Following [5], the standard interpretation used, at least implicitly, is the Everett interpretation, which states that all components in the linear superposition are real.[24]

It is obvious that at the level of (18) there is no intrinsic difference between big bang and big crunch; both correspond to the region a approaching zero in configuration space. This has important consequences for cosmological models in which classically the universe expands and recollapses, see [36]. In the quantum version, there is no trajectory describing the expansion and the recollapse. The only structure available is an equation of the form (18) in which only the scale factor a (and other variables) enter. The natural way to solve such an equation is to specify initial values on constant-a hypersurfaces in configuration space and to evolve them from smaller a to larger a. In more complicated models, one can evolve also the entanglement entropy between degrees of freedom in this way. If the entropy is low at small a (as is suggested by observations), it will increase all along from small a to large a. There is then a formal reversal of the arrow of time at the classical turning point, although this cannot be noticed by any observer, because the classical evolution comes to an end before the region of the classical turning point is reached.

We have limited the discussion here to quantum geometrodynamics. The main conclusions also hold for the path-integral approach and to loop quantum gravity.[25] In loop quantum gravity, there are analogies with gauge theories, for example with Faradays's lines of forces, see [13]. Still, it is not a gauge theory by itself, and many conceptual issues such as the semiclassical limit are much less clear than in quantum geometrodynamics.

[23] "Environment" is a metaphor here. It stands for other degrees of freedom in configuration space which become entangled with the quantum system, but which cannot be observed themselves.

[24] Alternatives are the de Broglie–Bohm approach and collapse models, which both are more new theories than new interpretations.

[25] The situation in string theory so far is less clear; there are indications that not only the concept of spacetime, but also the concept of space is modified, as is discussed in the context of the AdS/CFT conjecture.

4 The Role of Matter

Very early on, Einstein was concerned with a fundamental duality oberved in the physical description of Nature: the duality between fields and matter. This duality is the prime motivation for introducing the concept of light quanta in his important paper on the photoelectric effect from 1905. At that time, the only known dynamical field was the electromagnetic field; ten years later, with GR, the gravitational field joined in.

In *Raum, Zeit, Materie*, Weyl writes at the end of the main text, see [31], p. 317:

> In the darkness, which still wraps up the problem of matter, perhaps quantum theory is the first dawning light.[26]

Here, the hope is expressed that quantum theory, which in 1918 was still in its infancy, may provide a solution for this duality. This is certainly along the lines of Einstein's 1905 light quanta hypothesis. But, ten years later, the final quantum theory gave a totally different picture: central notions of the theory are wave functions and the probability interpretation. Einstein was repelled by this, especially by the feature of entanglement, which seems to provide a "spooky" action at a distance. This is why he focused on a unified theory of gravity and electrodynamics. He hoped to understand "particles" as solitonic solutions of field equations. His project did not succeed.

A somewhat different direction to understand 'matter from space' was pursued by John Wheeler in the 1950s, see [33]. The idea is that mass, charge, and other particle properties originate from a non-trivial topological structure of space, the most famous example being Wheeler's wormhole. This is most interesting, but has not led to anything close to a fundamental theory.[27]

Weyl's 1929 idea of understanding the interaction of electrons with the electromagnetic field by the gauge principle turned out to be more promising. The Standard Model of particle physics is an extremely successful gauge theory, and virtually all of its extensions make use of this principle, too. Gauge fields can also be described in a geometric way by adopting the mathematical structure of fibre bundles. Still, this is relatively far from the geometric concepts of GR, which deal with spacetime and not with the internal degrees of freedom of gauge theories. Perhaps gauge theories of gravity may help in finding a unified field theory, see [1].

Our physical theories all employ a metric to represent matter fields and their interactions, so GR is always relevant, even in situations where its effects are small. As Jürgen Ehlers writes in [9], p. 91: "Since inertial mass is inseparable from active, gravity-producing mass, an ultimate understanding of mass can be expected only from a theory comprising inertia and gravity." This should also apply for the origin of the masses in the Standard Model. The Higgs mechanism provides only a partial answer; the masses of elementary (non-composite) particles are given by the coupling to the Higgs, but the masses of composite particles such as proton and neutron cannot

[26]The German original reads: "In dem Dunkel, welches das Problem der Materie annoch umhüllt, ist vielleicht die Quantentheorie das erste anbrechende Licht.".

[27]For a recent account of matter from (the topology of) space, see e.g. [15].

be explained. In fact, it seems that the mass of the proton mostly arises from the binding energy of its constituents—quarks and gluons—and not from their masses, which to first order are negligible. Invoking the inverse of Einstein's famous formula, $m = E/c^2$, one can speculate that mass ultimately originates from energy, see [35]. It is hard to imagine that this origin can be understood without gravity. Perhaps a unified theory at the fundamental level is conformally invariant, similar to Weyl's 1918 theory, expressing the irrelevance of masses at high energies (small scales); masses would then only emerge as an effective, low-energy concept.

Unfortunately, despite many attempts, the duality of matter and fields remains unresolved, even in present approaches to quantum gravity. An exception may be string theory, but this approach has its own problems and it is far from clear whether it can be tested empirically. Perhaps the solution to the problem of matter may arrive from a completely unexpected direction. Space, time, and matter continue to be central concepts for research in the 21st century. The question posed in the title of [10], "Do gravitational fields play an essential role in the constitution of material elementary particles?" will most likely have to be answered by a definite *yes*.

Acknowledgements I am grateful to Silvia De Bianchi and Friedrich Hehl for their comments on my manuscript.

References

1. M. Blagojević, F.W. Hehl, *Gauge Theories of Gravitation*. A Reader with Commentaries (Imperial College, London, 2013)
2. M. Bouhmadi-López, C. Kiefer, P. Martín-Moruno, Phantom singularities and their quantum fate: general relativity and beyond. Gen. Relativ. Grav. **51**, article number 135 (2019)
3. D. Brizuela, C. Kiefer, M. Krämer, Quantum-gravitational effects on gauge-invariant scalar and tensor perturbations during inflation: the slow-roll approximation. Phys. Rev. D **94**, article number 123527 (2016)
4. M. Carlesso, A. Bassi, M. Paternostro, H. Ulricht, Testing the gravitational field generated by a quantum superposition. New. J. Phys. **21**, article number 093052 (2019)
5. B.S. DeWitt, Quantum theory of gravity I. The canonical theory. Phys. Rev. **160**, 1113–1148 (1967)
6. P.A.M. Dirac, The quantization of the gravitational field, in *Contemporary physics: Trieste Symposium* ed. by D. W. Duke and J. F. Owens (AIP Conference Proceedings, New York, 1968), pp. 539–543
7. E. Di Valentino, A. Melchiorri, J. Silk, Cosmic Discordance: Planck and luminosity distance data exclude LCDM (2020). arXiv:2003.04935v1 [astro-ph.CO]
8. H.G. Dosch, The Standard Model of Particle Physics, in *Approaches to Fundamental Physics*, ed. by E. Seiler, I.-O. Stamatescu (Springer, Berlin, 2007), pp. 21–50
9. J. Ehlers, General Relativity, in *Approaches to Fundamental Physics*, ed. by E. Seiler, I.-O. Stamatescu (Springer, Berlin, 2007), pp. 91–104
10. A. Einstein, Spielen Gravitationsfelder im Aufbau der materiellen Elementarteilchen eine wesentliche Rolle? Sitzber. Preuss. Akad. Wiss. XX, pp. 349–356 (1919)
11. G.F.R. Ellis, K. Meissner, H. Nicolai, The physics of infinity. Nat. Phys. **14**, 770–772 (2018)
12. J.C. Feng, Volume average regularization for the Wheeler-DeWitt equation. Phys. Rev. D **98**, article number 026024 (2018)

13. S. Frittelli, S. Koshti, E.T. Newman, C. Rovelli, Classical and quantum dynamics of the Faraday lines of force. Phys. Rev. D **49**, 6883–6891 (1994)
14. D. Giulini, The superspace of geometrodynamics. Gen. Relativ. Gravit. **41**, 785–815 (2009)
15. D. Giulini, Matter from Space, in *Beyond Einstein*, ed. by D. Rowe, T. Sauer, S. Walter (New York, NY, Birkhäuser, 2018), pp. 363–399
16. D. Giulini, C. Kiefer, The Canonical Approach to Quantum Gravity: General Ideas and Geometrodynamics, in *Approaches to Fundamental Physics*, ed. by E. Seiler, I.-O. Stamatescu (Springer, Berlin, 2007), pp. 131–150
17. S. Hawking, R. Penrose, *The Nature of Space and Time* (Princeton University Press, Princeton, New Jersey, 1996)
18. F.W. Hehl, C. Kiefer, Comparison of the DeWitt metric in general relativity with the fourth-rank constitutive tensors in electrodynamics and in elasticity theory. Gen. Relativ. Gravit. **50**, article number 8 (2018)
19. F.W. Hehl, J.D. McCrea, E.W. Mielke, Weyl spacetimes, the dilation current, and creation of gravitating mass by symmetry breaking, in *Exact Sciences and their Philosophical Foundations*, ed. by W. Deppert, et al. (Verlag Peter Lang, Frankfurt a. M., 1988), pp. 241–310
20. E. Joos, H.D. Zeh, C. Kiefer, D. Giulini, J. Kupsch, I.-O. Stamatescu, *Decoherence and the Appearance of a Classical World in Quantum Theory*, 2nd edn. (Springer, Berlin, 2003)
21. J. Jost (ed.), *Bernhard Riemann: Über die Hypothesen, welche der Geometrie zu Grunde liegen* (Springer, Berlin, 2013)
22. J. Jost (ed.), *Bernhard Riemann: On the Hypotheses Which Lie at the Bases of Geometry* (Birkhäuser, Basel, 2016)
23. C. Kiefer, Quantum geometrodynamics: whence, whither? Gen. Relativ. Gravit. **41**, 877–901 (2009)
24. C. Kiefer, *Quantum Gravity*, 3rd edn. (Oxford University Press, Oxford, 2012)
25. C. Kiefer, Conceptual problems in quantum gravity and quantum cosmology. ISRN Math. Phys., article ID 509316, 17 (2013)
26. C. Kiefer, D. Polarski, Why do cosmological perturbations look classical to us? Adv. Sci. Lett. **2**, 164–173 (2009)
27. B. Riemann, Über die Hypothesen, welche der Geometrie zu Grunde liegen. (Aus dem Nachlaß des Verfassers mitgetheilt durch R. Dedekind). Abh. Ges. Gött., Math. Kl. **13**, 133–152 (1868). Reprinted and annotated in Jost (2013) [German version] and Jost (2016) [English version]
28. E. Schrödinger, *Space-Time Structure* (Cambridge University Press, Cambridge, 1954)
29. K. Sundermeyer, *Symmetries in Fundamental Physics*, 2nd edn. (Springer, Cham, 2014)
30. H. Weyl, Reine Infinitesimalgeometrie. Math. Z. **2**, 384–411 (1918)
31. H. Weyl, Raum, Zeit, Materie. Vorlesungen über allgemeine Relativitätstheorie, 7th edn. (edited and complemented by J. Ehlers) (Springer, Berlin, 1993)
32. H. Weyl, *Philosophie der Mathematik und Naturwissenschaft*, 7th edn. (Oldenbourg Verlag, München, 2000)
33. J.A. Wheeler, *Geometrodynamics* (Academic Press, New York and London, 1962)
34. J.A. Wheeler, Superspace and the nature of quantum geometrodynamics, in *Battelle rencontres*, ed. by C.M. DeWitt, J.A. Wheeler (Benjamin, New York, 1968), pp. 242–307
35. F. Wilczek, Mass without mass I: most of matter. Phys. Today, 11–13 (November 1999); F. Wilczek, Mass without mass II: the medium is the mass-age. Phys. Today, 13–14 (January 2000)
36. H.D. Zeh, *The Physical Basis of the Direction of Time*, 5th edn. (Springer, Berlin, 2007)

Conservation of Energy-Momentum of Matter as the Basis for the Gauge Theory of Gravitation

Friedrich W. Hehl and Yuri N. Obukhov

Abstract According to Yang and Mills (1954), a *conserved* current and a related rigid ('global') symmetry lie at the foundations of gauge theory. When the rigid symmetry is extended to a *local* one, a so-called gauge symmetry, a new interaction emerges as gauge potential A; its field strength is $F \sim \mathrm{curl} A$. In gravity, the conservation of the energy-momentum current of matter and the rigid translation symmetry in the Minkowski space of special relativity lie at the foundations of a gravitational gauge theory. If the translation invariance is made local, a gravitational potential ϑ arises together with its field strength $T \sim \mathrm{curl}\,\vartheta$. Thereby the Minkowski space deforms into a Weitzenböck space with nonvanishing torsion T but vanishing curvature. The corresponding theory is reviewed and its equivalence to general relativity pointed out. Since translations form a subgroup of the Poincaré group, the group of motion of special relativity, one ought to straightforwardly extend the gauging of the translations to the gauging of full Poincaré group thereby also including the conservation law of the *angular momentum* current. The emerging Poincaré gauge (theory of) gravity, starting from the viable Einstein-Cartan theory of 1961, will be shortly reviewed and its prospects for further developments assessed.

1 Yang-Mills Theory, Gauge Theory

In the 1920s and 1930s it became clear that the atomic nuclei consist of protons (p) and neutrons (n) which interact with each other via a *strong* nuclear force. The masses of proton and neutron are nearly equal. The proton carries a positive elementary electric charge whereas the neutron is electrically neutral (but still caries a magnetic moment). Otherwise, in particular with respect to their nuclear interaction, they

F. W. Hehl (✉)
Institute of Theoretical Physics, University of Cologne, 50923 Köln, Germany
e-mail: hehl@thp.uni-koeln.de

Y. N. Obukhov
Nuclear Safety Institute, Russian Academy of Sciences, 115191 Moscow, Russia
e-mail: obukhov@ibrae.ac.ru

S. De Bianchi and C. Kiefer (eds.), *One Hundred Years of Gauge Theory*, Fundamental Theories of Physics 199, https://doi.org/10.1007/978-3-030-51197-5_10

behave very similar. This charge independence of the nuclear interaction of p-p, n-p, and n-n was an important *experimental* result.

Heisenberg [1] was led to the hypothesis that there exists a new particle called *nucleon* that has two different *states*, a positively charged one, the proton, and a neutral one, the neutron. These two different states were put in analogy to an electron which can have a state with spin up and one with spin down. Accordingly, Heisenberg attributed to the nucleon the new quantum number $\boldsymbol{I}$ of isospin, which is *conserved* in nuclear interactions. And the isospin up, $I_3 = +\frac{1}{2}$, represents the proton and the one down $I_3 = -\frac{1}{2}$, the neutron.

After Yukawa [2] had introduced the pion π as mediator of the strong nuclear force, it eventually turned out that the pion exists in three differently charged states, namely as π^+, π^-, and as π^0. Thus, one had to attribute to it the isospin $I = 1$. With the help of this insight, one got a consistent and experimentally verified framework for the nuclear force. At the same time, the new quantum number isospin found its way from nuclear physics into the systematics of elementary particle physics, as proposed by Kemmer [3].

Considering the nucleon together with the pion, it became clear that the invariance group of the strong nuclear interaction at the level of the nucleon is the unitary Lie group $SU(2)$ and the charge independence of the nuclear interaction translates into the requirement that no direction in the isospin space is distinguished. In other words, the corresponding action is invariant under rigid $SU(2)$ transformations and we have an associated *conservation* of the isospin $\boldsymbol{I}$.

Here Yang and Mills [4] set in, proposing "Conservation of Isotopic Spin and Isotopic Gauge Invariance" as the foundation for establishing a hypothetical $SU(2)$ gauge theory of *strong interaction* [4]. The conserved isospin current, via the reciprocal of the Noether theorem [5], yields a rigid ('global') $SU(2)$-invariance. Insisting, as Yang and Mills did, that a rigid symmetry is inconsistent with field-theoretical ideas, the $SU(2)$-invariance is postulated to be valid *locally.* This enforces to introduce a compensating (or gauge) field A, the gauge potential,[1] which upholds the $SU(2)$-invariance even under these generalized local transformations. Then the curl of A turns out to be the field strength of the emerging gauge field. The prototypical procedure for the conserved electric current of the Dirac Lagrangian and its $U(1)$ gauge invariance had already been executed by Weyl [6] and Fock [7] in 1929, see also [8]. Accordingly, we can define a gauge theory as follows:

A gauge theory is a heuristic scheme within the Lagrange formalism in the Minkowski space of special relativity for the purpose of deriving a new interaction from a conserved current and the attached rigid symmetry group. This new 'gauge' interaction is induced by demanding that the rigid symmetry should be extended to a locally valid symmetry.

Explicitly, the Yang-Mills type gauging works as follows, see, e.g., O'Raifertaigh [9], Mack [10], or Chaichian and Nelipa [11]. Let $I = \int d^4x\, \mathfrak{L}$ be an action for the matter field ψ^A with the Lagrangian density $\mathfrak{L} = \mathfrak{L}(\psi^A, \partial_i \psi^A)$. We transform the matter field under the rigid action of an N-parameter internal symmetry group G

[1]Yang and Mills denoted it with B in their original paper [4].

$$\psi^A \longrightarrow \psi'^A = \psi^A + \delta\psi^A, \qquad \delta\psi^A = \varepsilon^I (t_I)^A_B \, \psi^B, \tag{1}$$

with the generators t_I, $I = 1, \dots, N$, and $\partial_i \varepsilon^I = 0$. We suppose that the action does not change under the transformation of the matter field: $\delta I = 0$. We assume G to be a Lie group, and the generators $t_I \in \mathcal{G}$ form the basis of the corresponding Lie algebra with the commutator

$$[t_I \, , t_J] = f^K{}_{IJ} \, t_K. \tag{2}$$

The structure constants $f^K{}_{IJ} = -\, f^K{}_{JI}$ satisfy the Jacobi identity

$$f^N{}_{IL} f^L{}_{JK} + f^N{}_{JL} f^L{}_{KI} + f^N{}_{KL} f^L{}_{IJ} \equiv 0. \tag{3}$$

The *Noether theorem* tells us that, provided the matter variables satisfy the field equations, the invariance of the action under (1) yields a conservation law

$$\delta I = 0 \qquad \Longrightarrow \qquad \partial_i J^i{}_I = 0 \tag{4}$$

of the canonical *Noether current*

$$J^i{}_I := (t_I)^A_B \, \psi^B \, \frac{\partial \mathfrak{L}}{\partial \partial_i \psi^A}. \tag{5}$$

As a result, for an N-parameter symmetry group there exist N conserved charges

$$Q_I = \int d^3x \, J^0{}_I, \qquad I = 1, \dots, N, \tag{6}$$

where integral is taken over the spatial 3-surface $t =$const.

When the symmetry is made local, $\partial_i \varepsilon^I \neq 0$, the action with the matter Lagrangian $\mathfrak{L}(\psi^A, \partial_i \psi^A)$ is no longer invariant. One needs a gauge (compensating) field $A_i{}^I$ to be introduced via the *minimal coupling* recipe

$$\mathfrak{L}(\psi^A, \partial_i \psi^A) \longrightarrow \mathfrak{L}(\psi^A, D_i \psi^A), \tag{7}$$

with the partial derivative replaced $\partial_i \to D_i$ by the *covariant* one:

$$D_i \psi^A = \partial_i \psi^A + A_i{}^I \, (t_I)^A_B \, \psi^B. \tag{8}$$

Then the invariance of the modified action $I = \int d^4x \, \mathfrak{L}(\psi^A, D_i \psi^A)$ is recovered because the crucial covariance property

$$\delta(D_i \psi^A) = \varepsilon^I (x) \, (t_I)^A_B \, D_i \psi^B \tag{9}$$

is guaranteed by the inhomogeneous transformation law of the gauge field:

$$\delta A_i{}^I = -D_i\varepsilon^I = -(\partial_i\varepsilon^I + A_i{}^K f^I{}_{KJ}\varepsilon^J). \tag{10}$$

This completes the kinematics of the gauge theory. The gauge field $A_i{}^I$ becomes a true dynamical variable by adding a suitable kinetic term, $\mathfrak{V}$, to the minimally coupled matter Lagrangian: $\mathfrak{L} \to \mathfrak{L} + \mathfrak{V}$. This supplementary term has to be gauge invariant, such that the gauge invariance of the total action is kept. The gauge invariance of $\mathfrak{V}$ is obtained by constructing it in terms of the *gauge field strength*:

$$F_{ij}{}^I = \partial_i A_j{}^I - \partial_j A_i{}^I + f^I{}_{JK} A_i{}^J A_j{}^K. \tag{11}$$

Using (10) and (3) we straightforwardly verify the transformation law $\delta F_{ij}{}^I = \varepsilon^K(x)\, f^I{}_{KJ} F_{ij}{}^J$. The important property of the gauge field strength is the *Bianchi identity*

$$D_{[k}F_{ij]}{}^I = 0, \tag{12}$$

which can be naturally interpreted as the homogeneous field equation.

Since the gauge field Lagrangian $\mathfrak{V}$ should be also invariant under the local symmetry group, it should be a function of $F_{ij}{}^I$. The (inhomogeneous) Yang–Mills field equation is derived from the total action

$$I_{\text{tot}} = \int d^4x \left\{\mathfrak{L}(\psi^A, D_i\psi^A) + \mathfrak{V}(F_{ij}{}^I)\right\}. \tag{13}$$

Variation with respect to the gauge field potential yields explicitly

$$D_j H^{ij}{}_I = J^i{}_I, \qquad \text{with} \qquad H^{ij}{}_I := -2\frac{\partial \mathfrak{V}}{\partial F_{ij}{}^I}. \tag{14}$$

Quite remarkably, the matter source of the gauge field turns out to be a covariant Noether current (5). However, in the locally gauge invariant theory, the original conservation law (4) is replaced by the covariant one

$$D_i J^i{}_I = 0. \tag{15}$$

By recasting (14) into

$$\partial_j H^{ij}{}_I = \overset{A}{J}{}^i{}_I, \qquad \overset{A}{J}{}^i{}_I = J^i{}_I + A_j{}^K f^J{}_{KI} H^{ij}{}_J, \tag{16}$$

we can derive the modified conservation law

$$\partial_i \overset{A}{J}{}^i{}_I = 0, \tag{17}$$

which reflects the fact that the gauge field couples not only to matter, but also to itself. In other words, the *gauge field carries its own charge*.

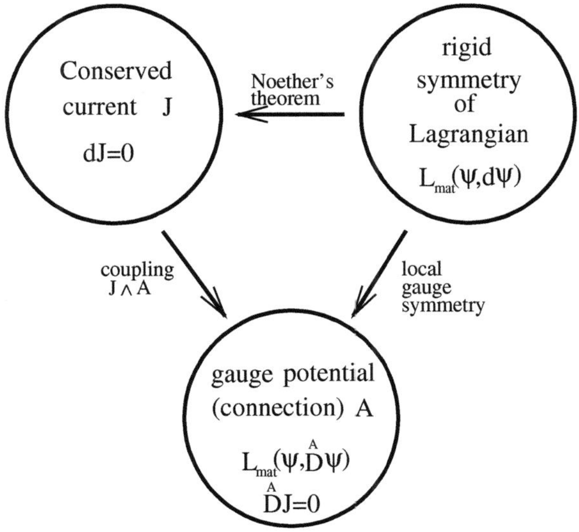

Fig. 1 The structure of a gauge theory à la Yang–Mills is depicted in this diagram, which is adapted from Mills [13]

As we see, the formal structure of a general gauge field theory (15), (12), and (14) appears as a generalization of the Maxwell theory [12]. The final "building block" of this generalization is the explicit form of the *constitutive relation* $H = H(F)$ between the gauge field strength and the excitation, see above in (14):

$$H^{ij}{}_I = -2\frac{\partial \mathfrak{V}}{\partial F_{ij}{}^I}. \tag{18}$$

In the original Yang-Mills theory [4], the Lagrangian was constructed as a quadratic Maxwell type invariant of the gauge field strength, and the resulting constitutive law is linear: $H^{ij}{}_I = F^{ij}{}_I$. Later, Mills [14] also discussed a *non*linear, Born–Infeld type "constitutive" relation between H and F. But this didn't prove to be useful.

Schematically, we represented the gauge procedure in Fig. 1. Let us stress our main points: A gauge theory is based on a *conserved current* and the *symmetry* connected with it. The symmetry is first rigid—and there is no interaction—then, subsequently, made local, and the gauge potential A and the gauge field strength $F \sim \overset{A}{D} A$ emerge in this procedure.

Incidentally, here we confine our attention only to classical field theory and we do *not* investigate quantum field theoretical consequences [15]. However, a consistent particle picture does arise in the quasiclassical approximation. A point particle with the 4-velocity u^i carries internal degrees of freedom in the form of a "color charge" q_I. It couples with the gauge field via the interaction Lagrangian $q_I u^i A_i{}^I$, see Wong [16]. Accordingly, particle's motion is affected by the generalized Lorentz force

$$f_i = q_I F_{ij}{}^I u^j. \tag{19}$$

2 Newton-Einstein Gravity

Turning now to gravity, the fundamental question is: What is the conserved current and what is the gauge group of gravity? Our starting point will be, of course, Newtonian gravity. There mass is the source of gravity or rather the mass density $\rho(\boldsymbol{r}, t)$ in its quasi field-theoretical formulation as Poisson equation for the gravitational potential $\phi(\boldsymbol{r}, t)$:

$$\Delta\phi(\boldsymbol{r}, t) = 4\pi G\rho(\boldsymbol{r}, t)\,. \tag{20}$$

Here Δ is the Laplace operator and G Newton's gravitational constant. In Newtonian mechanics, the motion of a material continuum with the mass density ρ is described by the velocity vector field $\boldsymbol{v}$. The mass inside the volume $\Omega(t)$ is given by the integral

$$m(t) = \int\limits_{\Omega(t)} \rho(\boldsymbol{r}, t)\, d^3x\,. \tag{21}$$

The change is straightforwardly evaluated

$$\frac{dm(t)}{dt} = \int\limits_{\Omega(t)} \left\{ \frac{\partial\rho}{\partial t} + \boldsymbol{div}(\rho\boldsymbol{v}) \right\} d^3x\,. \tag{22}$$

Mass is a conserved quantity, $\frac{dm(t)}{dt} = 0$, that is, we have a continuity equation for ρ:

$$\frac{\partial\rho}{\partial t} + \boldsymbol{div}(\rho\boldsymbol{v}) = 0\,. \tag{23}$$

Lavoisier (1789) checked this conservation law of mass successfully in experiments.

At the beginning of the 20th century, Newtonian mechanics was supplanted by the special relativity theory (SR) with its four-dimensional (4d) Minkowski space. Accordingly, also a gravitational gauge theory has to take the special-relativistic framework as a starting point. This procedure is reminiscent of Einstein's heuristic derivation of general relativity theory (GR). He started in flat Minkowski space, went over to accelerated frames, and applied subsequently the equivalence principle. Hereby he had to relax the rigidity of the Minkowski space ending up with the Riemannian space of GR. Einstein [17] worked out this procedure in considerable detail in his Princeton lectures of 1921.

Let us stress a point that is often misunderstood. In heuristically deriving a gauge theory of gravity, the physical system under consideration is embedded in a Minkowski space with its rigid 4-parameter translation group $T(4)$ and its rigid 6-parameter Lorentz group $SO(1, 3)$. The relaxation of the rigidity of the Minkowski space is achieved by the postulate of *local* (instead of rigid) translational and, at a later stage, of Lorentz invariance. The appropriate geometrical framework of spacetime is induced by the gauge principle alone.

We should recall that the Minkowski geometry of SR and the corresponding group of motion, the semidirect product of the translation and the Lorentz groups, the Poincaré group $P(1,3) = T(4) \rtimes SO(1,3)$, is supported by all high-energy experiments with great accuracy. Moreover, Wigner (1939) has shown that all elementary quantum mechanical objects obey a *mass-spin classification* [18]; massless particles are classified according to mass-helicity. The particle attributes mass m and spin s correspond in field theory to the energy-momentum current $\mathfrak{T}$ and the spin (angular-momentum) current $\mathfrak{S}$. Accordingly, the mass density of Newton's theory translates field-theoretically into the energy-momentum current of matter: $m \to \mathfrak{T}$. Thus, the energy-momentum current of matter must be the source of gravity. Later, in Sect. 4 we will see that additionally also the spin current may play a role: $s \to \mathfrak{S}$. Jointly with the substitution $m \to \mathfrak{T}$, the mass conservation theorem (23) is dissolved and the conservation of the energy-momentum current replaces it:

$$\partial_j \mathfrak{T}_i{}^j = 0 . \tag{24}$$

The nuclear explosion of Alamogordo (1945) is an unmistakable proof of the violation of the mass conservation law.

By Noether's theorem, energy-momentum conservation is induced by translational invariance of the Lagrangian of an isolated system. Hence without further ado, we can now rephrase the title of the Yang-Mills paper for gravity as follows: *Conservation of the energy-momentum current and translational gauge invariance*. In this way we recognize that the translation group $T(4)$ is the gauge group of ordinary gravity. This point was already made in the beginning of 1960s by many well-known physicists:

- Sakurai (1960): "...there exists a deep connection between energy conservation and the very existence of the gravitational coupling. The gravitational field, being the dynamical manifestation of energy, is to be coupled to energy-momentum density....hence the gravitational field can interact with itself in the same way as the $T = 1$ Yang-Mills $B_\mu^{(T)}$ field (which is the dynamical manifestation of isospin) can interact with itself." [19] (T denotes the isospin.)
- Glashow and Gell-Mann (1961): "...if we set up the Einstein theory by gauge methods then the conclusions are slightly different. Instead of an isotopic rotation, we perform a 4-dimensional translation at each point of space..." [20]
- Feynman (1962): "The equations of physics are invariant when we make coordinate displacements [by] any constant amount a^μ...it is possible to investigate how we might make the equations of physics invariant when we allow space dependent *variable* displacements..." [21]

Isn't this clear enough? We propose to consider those three statements as support for our point of view.

The energy-momentum current in exterior and in tensor calculus
The canonical energy-momentum tensor density $\mathfrak{T}_\alpha{}^\beta$ of tensor calculus and the canonical energy-momentum current 3-form of exterior calculus Σ_α are equivalent. We have

$$\Sigma_\alpha = \mathfrak{T}_\alpha{}^\beta \epsilon_\beta \,, \qquad \mathfrak{T}_\alpha{}^\beta = {}^\diamond(\vartheta^\beta \wedge \Sigma_\alpha)\,, \tag{25}$$

with $\epsilon_\beta = \frac{1}{3!}\epsilon_{\beta\mu\nu\rho}\vartheta^\mu \wedge \vartheta^\nu \wedge \vartheta^\rho$ and $\epsilon_{\beta\mu\nu\rho}$ as the totally antisymmetric Levi-Civita symbol with values $(0, \pm 1)$. The 'diamond dual' $^\diamond$ is defined by means of the metric-free Levi-Civita symbol [12], and ϑ^β denotes the coframe, see Appendix 1. The different components of $\mathfrak{T}_\alpha{}^\beta$ carry the following physical interpretations:

$$\mathfrak{T}_\alpha{}^\beta = \begin{pmatrix} \mathfrak{T}_0{}^0 \sim \text{energy density} & \mathfrak{T}_0{}^b \sim \text{energy flux density} \\ \mathfrak{T}_a{}^0 \sim \text{momentum density} & \mathfrak{T}_a{}^b \sim \text{momentum flux density} = \text{stress} \end{pmatrix}, \tag{26}$$

see, e.g., Rezzolla and Zanotti [22] or [12].

In the hydrodynamic approximation of a relativistic continuum, the non-interacting (dust) matter elements carry momentum density p_α, and the energy-momentum tensor reads $\mathfrak{T}_\alpha{}^\beta = p_\alpha u^\beta$, with the 4-velocity vector field u^β. We immediately recognize a direct analogy between the energy-momentum $\mathfrak{T}_\alpha{}^\beta = p_\alpha u^\beta$ and the Yang-Mills current $J^i{}_I = u^i q_I$:

$$\text{(electric charge)} \longleftrightarrow \text{(color charge)} \longleftrightarrow \text{(momentum)}. \tag{27}$$

Sciama [23] developed "The analogy between charge and spin in general relativity" (the title of his paper) even further within a gauge approach to gravity. There the natural material source of the gravitational field is a continuum of non-interacting elements with momentum p_α and internal angular momentum (spin) $s_{\alpha\beta} = -s_{\beta\alpha}$. The corresponding Poincaré matter current encompasses the energy-momentum and the spin tensor densities

$$\left\{\mathfrak{T}_\alpha{}^\beta = p_\alpha u^\beta, \;\; \mathfrak{S}_{\alpha\beta}{}^\gamma = s_{\alpha\beta}\, u^\gamma\right\}. \tag{28}$$

In exterior language, the motion of a relativistic continuum is described by the flow 3-form $u = u^\alpha \epsilon_\alpha$, and the Poincaré currents for the spinning dust matter read

$$\left\{\Sigma_\alpha = p_\alpha\, u, \quad \tau_{\alpha\beta} = s_{\alpha\beta}\, u\right\}. \tag{29}$$

A priori, Σ_α and $\mathfrak{T}_\alpha{}^\beta$ have 16 independent components. If a *metric* $g_{\alpha\beta}$ is available—and this is always the case for a Minkowski space we started with—we can lower the second index and define a tensor of type (0,2):

$$\mathfrak{T}_{\alpha\beta} := g_{\beta\gamma}\,\mathfrak{T}_\alpha{}^\gamma\,. \tag{30}$$

It can be decomposed into symmetric and antisymmetric pieces according to $16 = 10 \oplus 6$: $\mathfrak{T}_{\alpha\beta} = \mathfrak{T}_{(\alpha\beta)} + \mathfrak{T}_{[\alpha\beta]}$. Furthermore, the trace can be extracted from the symmetric piece $\check{\mathfrak{T}}_{\alpha\beta} := \mathfrak{T}_{(\alpha\beta)} - \frac{1}{4}g_{\alpha\beta}\mathfrak{T}_\gamma{}^\gamma$, with $\check{\mathfrak{T}}_{[\alpha\beta]} = 0$ and $\check{\mathfrak{T}}_\gamma{}^\gamma = 0$. Thus, we arrive at the following decomposition of the canonical energy-momentum tensor:

$$\mathfrak{T}_{\alpha\beta} = \check{\mathfrak{T}}_{\alpha\beta} + \mathfrak{T}_{[\alpha\beta]} + \frac{1}{4} g_{\alpha\beta} \mathfrak{T}_{\gamma}{}^{\gamma}\,, \qquad 16 = 9 \oplus 6 \oplus 1\,. \tag{31}$$

In his deduction of GR, Einstein considered as a model for 'matter' the classical Euler fluid and the electromagnetic field in vacuum; in the former case $\mathfrak{T}_{(\alpha\beta)}$ is sufficient, in the latter one $\check{\mathfrak{T}}_{\alpha\beta}$. Accordingly, in GR matter is described by the symmetric energy-momentum tensor $\mathfrak{T}_{(\alpha\beta)}$. If matter with spin is involved, the canonical tensor $\mathfrak{T}_{\alpha}{}^{\beta}$ is indispensable. We will come back to the material currents in Sect. 4.3.

3 Translational Gauge Theory (TG)

As we saw in the last section, at the beginning of the 1960s it was already clear to Sakurai, Glashow and Gell-Mann, and to Feynman that a gravitational gauge theory should be based on translation (or displacement) invariance. Accordingly, the task was to investigate the conservation of the material energy-momentum current and the related invariance under rigid and, subsequently, under local translations. The localization of the translational invariance should create the gravitational field!

Soon thereafter, in the 1970s, a translational gauge theory (TG) was set up. It turned out to be a teleparallelism theory [24, 25]. We delineated the historical development and an up-to-date formalism already in Blagojević and Hehl [26].[2]

The paper of Cho [27], see also [28], may be taken as a concise description of a translational gauge theory of gravity. Its structure is revisited from a modern geometrical point of view in the more recent papers of Obukhov and Pereira [29, 30], see also [31]. We abstain from publishing once more this well-known formalism of TG, but refer to the literature [32] instead.

Let us recall that rigid translational invariance is made *local* at the price of introducing 4 translational gauge potentials—the coframe $\vartheta^{\alpha} = e_i{}^{\alpha} dx^i$—which compensate the violation of the rigid invariance:

$$\text{rigid transl. inv.} \overset{\text{heur. princ.}}{\longrightarrow} \text{local transl. inv.} \longrightarrow \text{coframe } \vartheta^{\alpha} \text{ compensates.} \tag{32}$$

Thus ϑ^{α} is the analog of $A_i{}^I$ above. The curl of ϑ^{α}, the torsion, T^{α} —the analog of $F_{ij}{}^I$ above—, arises as the gravitational field strength,

$$T^{\alpha} := D\,\vartheta^{\alpha} = d\vartheta^{\alpha} + \Gamma_{\beta}{}^{\alpha} \wedge \vartheta^{\beta}\,, \tag{33}$$

with $\Gamma_{\alpha}{}^{\beta}$ as the Lorentz connection. The corresponding curvature vanishes:

$$R_{\alpha}{}^{\beta} := d\Gamma_{\alpha}{}^{\beta} - \Gamma_{\alpha}{}^{\gamma} \wedge \Gamma_{\gamma}{}^{\beta} = 0\,. \tag{34}$$

[2]See in particular the pages 195, 236, and 241 to 249.

This signifies that a vector, for instance, can be parallelly transported around in an integrable way. We have a distant parallelism, a teleparallelism. It takes place in a so-called Weitzenböck geometry [33, 34], see also [35–37].

Analogously to the Yang-Mills case, $T^\alpha \neq 0$ is the criterion for the emerging of a new non-trivial gravitational/translational gauge field. It can be shown that the teleparallelism theory, for a suitable Lagrangian quadratic in the torsion, is equivalent to general relativity of 1916, provided a *symmetric* energy-momentum tensor is chosen, see [38]. This is, in our opinion, a major achievement which demonstrates that translational gauging leads, via a Weitzenböck spacetime, directly to general relativity with its Riemannian spacetime. In the subsequent Sect. 4, it will turn out that TG is a special case of a Poincaré gauge theory of gravity (PG) that we will discuss in quite some detail.

In order to get a bird's eye of view on TG, we would like to display the structure of the Lagrange-Noether formalism of TG in a Tonti-diagram [40], see Fig. 2. The left column is built up from the *configuration variables:* The coordinates x^i (four 0-forms), the coframe ϑ^α (four 1-forms), the torsion T^β (four 2-forms, in the figure called F^β), and eventually $DT^\beta \stackrel{*}{=} dT^\beta$ (four 3-forms). In the right column which depicts *source variables,* we start with the 4-forms, the volume force f_α (four 4-forms), continue with the material energy-momentum ${}^{\mathrm{m}}\Sigma_\alpha$ (four 3-forms) and end up with the translational excitation H_α (four 2-forms). The corresponding four 1-forms should be the potential of the excitation, but we don't know much about such a quantity; compare, however, with Rund [41]. The *constitutive relation* links the configuration variable 2-form in a linear way to the source variable 2-form, much like in electrodynamics $F = (\boldsymbol{E}, \boldsymbol{B})$ is related linearly to $H = (\mathcal{D}, \mathcal{H})$.

It is remarkable that the *configuration* variables are exclusively *premetric* concepts, that is, whereas the existence of a linear connection $\Gamma_\alpha{}^\beta$ is necessary, a metric does not enter anywhere.[3] The analogous is true for the *source* variables. In the constitutive laws, however, in which the field strength 2-forms as configuration variables are related to the excitation 2-forms as source variables, a metric tensor is indispensable. This is a lesson which one can take over from the premetric version of electrodynamics, see Post [42].

Note that the motion of a point particle in TG is described by the teleparallel analog of the Lorentz force in the Yang-Mills theory (19)

$$f_i = p_\alpha F_{ij}{}^\alpha u^j \,. \tag{35}$$

One can prove that the corresponding equation of motion turns out to be the usual geodesic curve [43].

We know that the Minkowski space of SR is an *affine space,* see Kopczyński and Trautman [44], that is, "a vector space which has lost its origin." A translation is an affine concept unrelated to a metric. Consequently, the gauging of translations

---

[3] Enzo Tonti disagrees on this point since in his way of setting up the basic definitions of his configuration and source variables, the existence of a Euclidean metric is assumed a priori. We, however, take the view that a suitable premetric generalization is possible.

Tonti diagram of the premetric teleparallel theory of gravity* (TG)

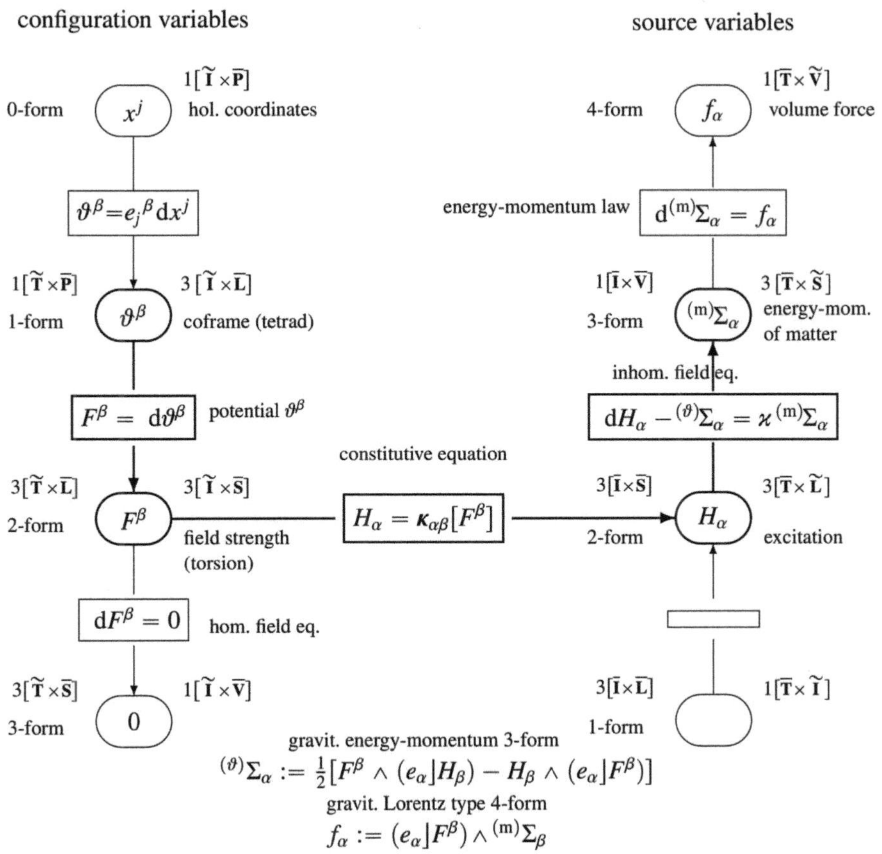

Fig. 2 Patterned after Tonti [39], pages 402 and 315. We denoted here the torsion 2-form with F^β in order to underline its function as a translational gauge field strength; for full notation details see [25, 39]. *Also known as translation gauge theory of gravity

happens in an affine space with the canonical energy-momentum 3-form ${}^{(m)}\Sigma_\alpha$ (16 independent components) as source. No metric is involved at all in this. However, this teleparallelism scheme cannot be directly compared with nature.

In gravity, as we have discussed above, we start with a Minkowski space and apply the gauge procedure with this background. Minkowski space is indispensable as a starting point for treating gravity, as Einstein [17] has taught us. For defining a symmetric energy-momentum tensor we need a metric, as we saw already in (30). Hence the premetric teleparallelism scheme does not qualify as a bona fide physical theory. However, since we started from SR, we have a metric available and we can use it for formulating the constitutive law of a teleparallelism theory. Then TG becomes the *teleparallel equivalent of general relativity* $GR_{||}$, as we discussed above.

Why did Einstein arrive in 1915/16 at a Riemann and not at a Weitzenböck space? *An afterthought*

- Einstein (1916) gauged the *direction* of a vector.
- Weyl (1918) gauged the *magnitude* (modulus) of a vector; however, instead of coupling it to the dilation current (as we know today), he coupled it to the unrelated electric current [45].
- E.Cartan (1923) recognized that Einstein took the flat Minkowski space as a vector space instead of an affine space. A Minkowski space has no preferred point. But Einstein took in his construction a preferred point in order to gauge the direction. In order to get rid of this preferred point, Cartan rolled without gliding a Minkowski space along a contour of the contorted and curved space under consideration (Cartan circuit, see Kröner [46], Sharpe [47], and Sternberg [48]). This is the meaning of this procedure which provides more insight, in our opinion, than all those theories using fiber bundles. Fiber bundles were successfully applied for internal symmetries, like $U(1)$, $SU(2)$, and $SU(3)$, but for external, i.e. spacetime symmetries, they did not provide any further insight. Unfortunately, no bundle theorist has essentially contributed to the understanding of torsion and/or constructively developed teleparallelism (with the possible exception of Schücking and Surowitz [49]), as far as we can see.

4 Poincaré Gauge Gravity (PG)

In the next step we will discuss now the gauging of the Poincaré group $P(1,3) = T(4) \rtimes SO(1,3)$. Before we do so, we would like to look at a prototypical experiment by Colella, Overhauser, and Werner (COW) on the 'behavior' of a neutron beam in a gravitational field. In our understanding of gravity, Newton's apple should nowadays be substituted by a neutron beam as it is used in the COW experiment as quantum system with mass m and spin $s = \frac{\hbar}{2}$. As J. L. Synge formulated it so beautifully: "Newton successfully wrote apple = moon, but you cannot write apple = neutron."

4.1 Colella-Overhauser-Werner (COW) Experiment Heralds a New Era in Gravitational Physics: The Kibble Laboratory

The quantum mechanical properties of a neutron wave/particle in interaction with the ordinary Newtonian gravitational field were first observed in the Colella-Overhauser-

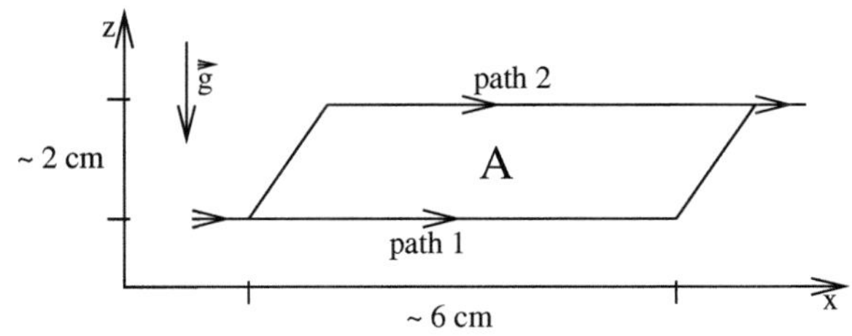

Fig. 3 COW experiment schematically: A neutron beam is split into two beams which travel in different gravitational potentials. Eventually the two beams are reunited and an interferometric picture is observed due to their relative phase shift

Werner neutron interferometer[4] in 1975 [50], see also Rauch and Werner [51] for a corresponding textbook presentation. The scheme of the experiment is sketched in Fig. 3. Later more accurate measurements with atomic beams were performed by Kasevich and Chu [52], see also the most recent developments by Asenbaum et al. [53] and Overstreet et al. [54]. Unfortunately, in most textbooks on GR, these important experiments are not even mentioned and the old Einstein procedure of using Newton's apple for a heuristic derivation of GR is simply copied.

To predict the result of the COW experiment was not complicated: provided the neutron spin is not polarized, one couples the Schrödinger equation for a neutron to the Newtonian gravitational potential. The quantum phase shift predicted, and experimentally confirmed by COW and Kasevich and Chu, was mass dependent. Then it was soon argued that this would violate the equivalence principle. However, this interpretation turned out to be incorrect, see Audretsch et al. [55]. It was necessary to take the wave function of the neutron—in generalization of the Newtonian point particle—as a new basic ingredient for the discussion of the equivalence principle: the neutron wave/particle of the COW-experiment supplanted the Newtonian point particle moving in a gravitational field.[5] Here quantum mechanical results for the test matter are needed for the foundation of the theory of gravity.

Einstein [17] heuristically derived GR from SR by considering mass points and electromagnetic fields in an accelerating and thus in a noninertial frame of reference. In the *Einstein laboratory*, in which he executed his thought experiments, the acceleration was described by using curvilinear coordinates x^i, with $i = 0, 1, 2, 3$. The inertial forces, which emerge in an accelerated reference system, were locally equivalent to the corresponding gravitational forces; for a detailed discussion see, for instance, Audretsch et al. [55] or Blagojević and Hehl [26], in particular Fig. 4.1 therein, and our Table 1.

[4]The interferometer, built from a silicon monocrystal, had a linear size of about 10 cm. In the energy range covered by the COW experiment, the neutron can be considered to be elementary, that is, its quark structure can be neglected.

[5]For a very down-to-earth and highly interesting discussion of the interaction between a quantum system and a classical gravitational field, one should compare Nesvizhevsky and Voronin [56].

Table 1 Einstein's approach to GR versus the gauge approach to gravity. Used are a mass point m or a Dirac matter field Ψ (referred to a local frame), respectively. IF means inertial frame, NIF non-inertial frame. The table refers to special relativity up to the second horizontal double line. Below, gravity will be switched on. Note that for the Dirac spinor already the force-free motion in an *inertial* frame *does depend* on the mass parameter m. The tilde $\tilde{}$ denotes the quantities in a Riemannian spacetime

	Einstein's approach: Einstein laboratory	**Gauge approach:** Kibble laboratory
Elementary object in Special Relativity (SR)	mass point m with velocity u^i	Dirac spinor $\Psi(x)$ of mass m (with four components)
Inertial frame (IF)	Cartesian coord. system x^i $ds^2 \stackrel{*}{=} o_{ij}\, dx^i dx^j$	holonomic orth. frame $e_\alpha = \delta^i_\alpha\, \partial_i, \quad e_\alpha \cdot e_\beta = o_{\alpha\beta}$
Force-free motion in IF	$\dot{u}^i \stackrel{*}{=} 0$	$(i\gamma^i \partial_i - m)\Psi \stackrel{*}{=} 0$
Non-inertial frame (NIF)	arbitrary curvilinear coord. system $x^{i'}$	anholonomic orth. frame $e_\alpha = e^i{}_\alpha \partial_i$ or coframe $\vartheta^\alpha = e_i{}^\alpha dx^i$
Force-free motion in NIF	$\dot{u}^i + u^j u^k \tilde{\Gamma}_{jk}{}^i = 0$	$\left[i\gamma^\alpha e^i{}_\alpha(\partial_i + \Gamma_i) - m\right]\Psi = 0$ $\Gamma_i := \frac{1}{2}\Gamma_i{}^{\beta\gamma}\rho_{\beta\gamma}$ Lorentz
Non-inertial geometrical objects	$\tilde{\Gamma}_{jk}{}^i$ 40	$\vartheta^\alpha, \quad \Gamma^{\alpha\beta} = -\Gamma^{\beta\alpha}$ 16 + 24
Constraints in SR	$\tilde{R}^{\alpha\beta}(\partial\tilde{\Gamma}, \tilde{\Gamma}) = 0$ 20	$T^\alpha(\partial e, e, \Gamma) = 0, \; R^{\alpha\beta}(\partial\Gamma, \Gamma) = 0$ 24 + 36
Global IF	$g_{ij} \stackrel{*}{=} o_{ij}, \quad \tilde{\Gamma}_{jk}{}^i \stackrel{*}{=} 0$	$(e_i{}^\alpha, \Gamma_i{}^{\alpha\beta}) \stackrel{*}{=} (\delta_i^\alpha, 0)$
Archetypal experiment	Apple in grav. field (Newton)	Neutron in grav. field (COW)
Switch on gravity	$\tilde{R}^{\alpha\beta} \neq 0$ Riemann spacetime	$T^\alpha \neq 0, \quad R^{\alpha\beta} \neq 0$ Riemann-Cartan spacetime
Local IF ('Einstein elevator')	$g_{ij}\vert_P \stackrel{*}{=} o_{ij}, \quad \tilde{\Gamma}_{jk}{}^i\vert_P \stackrel{*}{=} 0$	$(e_i{}^\alpha, \Gamma_i{}^{\alpha\beta})\vert_P \stackrel{*}{=} (\delta_i^\alpha, 0)$
Gravitational field equations	$\widetilde{\text{Ric}} - \frac{1}{2}\text{tr}(\widetilde{\text{Ric}}) \sim$ mass **GR**	$\text{Ric} - \frac{1}{2}\text{tr Ric} \sim$ mass $\text{Tor} + 2\,\text{tr Tor} \sim$ spin **EC**

In short, we can characterize the Einstein lab(oratory) as follows: Specified are

{E1} a neutral point particle with mass m;
{E2} an inertial frame K span by Cartesian coordinates;
{E3} an accelerated (i.e., non-inertial) frame K' span by curvilinear coordinates;
{E4} a homogeneous gravitational field described by g_{ij} referring to the curvilinear coordinates; and
{E5} light rays.

Einstein's procedure, by means of which he deduced GR, could have a broader domain of application than that encompassed by the tools and constructs employed in setting up the theory. Still, it is hard for us to believe that Einstein's discussion would also cover the COW experiment. After all, the neutron has quantum mechanical properties, it is a fermion, and it has spin $s = \frac{\hbar}{2}$.

Accordingly, we consider the *Kibble laboratory*, in which a fermion is described by a Dirac wave function with respect to a local reference (co)frame (vierbein) $e_i{}^\alpha$, where $\alpha = 0, 1, 2, 3$ numbers the frame vectors, see Kibble [57]. As soon as the reference frames are accelerated, they are no longer aligned and pick up a non-vanishing curl of these frames, the so-called object of anholonomity. In a non-relativistic approximation and neglecting its spin, the neutron obeys the stationary Schrödinger equation in the external homogeneous Newtonian gravitational field. If one solves this equation, the experimentally observed gravitational phase shift is described successfully.

The rationale of all of this is that we should simply study, in Minkowski space, a Dirac field in a Kibble lab, that is, in an accelerated frame of reference $e_i{}^\alpha(x)$ with $\partial_{[i}e_{j]}{}^\alpha \neq 0$, and read off its inertial forces. This is what we did, see [58] and the literature quoted there. Subsequently, we executed a Foldy-Wouthuysen transformation in order to determine the non-relativistic limit of the Dirac equation. If one neglects the spin, one recovers the COW term thereby certifying the correctness of this procedure. If the spin is kept, as one should do since the fundamental building blocks of our universe are fermions, one recovers additionally a spin-rotation coupling predicted earlier by Mashhoon [59, 60] and experimentally confirmed in the meantime, see Danner et al. [61].

Thus, the Kibble lab(oratory) can be described as follows: it contains

{K1} an unquantized Dirac spinor[6] (fermionic field with mass m and spin $s = \frac{\hbar}{2}$);
{K2} an inertial frame, $\vartheta^\alpha = \delta_i^\alpha dx^i$ spanned by Cartesian coordinates;
{K3} a translationally and rotationally accelerated frame, $\vartheta^{\alpha'}$ spanned by an arbitrary orthonormal frame;
{K4} homogeneous gravitational fields described by $(e_i{}^\alpha, \Gamma_i{}^{\alpha\beta})$; and

[6] It is demonstrated in the textbooks of Ivanenko and Sokolov [62] and Hund [63] that it is useful at relatively low energies to consider, in an approximate way, the Dirac spinor as a *classical* field—in the so-called first quantized version. Thus, quantum field theory is excluded from our considerations. In this context, one should also compare the discussion of Cao [64] in his Sec.11.3: One is searching for a background independent quantum theory of gravity. For that purpose a Riemann-Cartan background à la Kibble et al. seems to be a reasonable starting point.

{K5} light rays.

If we compare the different labs, we recognize that (i) the objects considered are different, (ii) the notions of inertial systems are different, and (iii), see Mashhoon [65], the rotational acceleration plays an additional role in the Kibble lab. The rest, in particular the application of the equivalence principle, is similar. Because we consider spinors in the Kibble lab, we get, by using Einstein's original ideas, a modified outcome.

Conventionally, the equivalence principle is only discussed in the Einstein laboratory. The Kibble laboratory, which is really based on Élie Cartan's moving (co)frames (repères mobiles), is, in our opinion, a necessity if Dirac particles are considered. This amounts to a generalization of the equivalence principle to a 'more local' neighborhood, see, in particular von der Heyde [66]—the curvilinear coordinates x^i are generalized to arbitrary orthonormal coframes $\vartheta^\alpha = e_i{}^\alpha dx^i$.

According to the equivalence principle, the Riemann-Cartan (RC) spacetime looks Minkowskian from a local point of view. In a RC-spacetime, at any fixed point with coordinates $\overset{\circ}{x}{}^k$ it is possible to trivialize the gravitational gauge potentials [66]:

$$\left.\begin{array}{l} e_i{}^\alpha \Big|_{x^k = \overset{\circ}{x}{}^k} \stackrel{*}{=} \delta_i^\alpha \\ \Gamma_i{}^{\alpha\beta} \Big|_{x^k = \overset{\circ}{x}{}^k} \stackrel{*}{=} 0 \end{array}\right\} . \tag{36}$$

This is important to recognize: In spite of the presence of torsion in a RC-spacetime, at any fixed point, the local connection $\Gamma_i{}^{\alpha\beta}$ can be transformed to zero [67–69]. What in an Einstein lab is the geodesic coordinate system becomes the trivialized reference frame (36) in a Kibble lab; this is a widely underestimated point. The frame (36) in RC-space supersedes the geodesic coordinate system in a Riemannian space.

4.2 *Poincaré Gauge Gravity Kinematics*

The standard model of particle physics is based on gauge theories for the internal symmetries $U(1)$, $SU(2)$, $SU(3)$, see O'Raifeartaigh [9]. Accordingly, apart from GR, the gauge idea seems to underlie all physical theories. However, already fairly early also gravity was understood as a gauge theory. It was Utiyama [70] who paved the way in this direction by using the Lorentz group $SO(1,3)$ as a gauge group for gravity. It turned out to be unsuccessful, though, since the current coupling to the Lorentz group is the angular momentum current. However, as we know from Newton's theory of gravity, it is the mass density or—according to SR—the energy-momentum current that gravity has as its source. And energy-momentum couples to the translation group $T(4)$.

Flat gravity-free Minkowski space has the Poincaré group $T(4) \rtimes SO(1,3)$, the semi-direct product of the four-parameter translation group[7] $T(4)$ and the six-parameter Lorentz group $SO(1,3)$, as its group of motions. Accordingly, Minkowski space is invariant under rigid ('global') Poincaré transformations. Consequently, as found by Wigner [18], a quantum mechanical system in a Minkowski space can be classified according to *mass and spin.* The corresponding field-theoretical currents are the material energy-momentum and spin angular momentum currents Σ_α and $\tau_{\alpha\beta} = -\tau_{\beta\alpha}$, respectively.

Thus, if we want to apply Einstein's recipe for setting up a gravitational theory based on the equivalence principle, we have to introduce accelerated frames in Minkowski space. Due to the involvement of a quantum mechanical system, see the COW neutrons or the Kasevich and Chu atoms mentioned above, we have to turn to Kibble's laboratory and to introduce coframes as reference systems. This yields, as was shown by Sciama [23] and Kibble [57], via *local* Poincaré invariance a Riemann-Cartan spacetime with torsion $T_{\alpha\beta}{}^{\gamma}$ and curvature $R_{\alpha\beta}{}^{\gamma\delta}$. The rigid Lie algebra of the Poincaré group is extended to a so-called deformed, soft, or local "Lie algebra" (D_α and $\rho_{\alpha\beta} = -\rho_{\beta\alpha}$ generate translations and Lorentz transformations, respectively):

$$\left.\begin{aligned} [D_\alpha, D_\beta] &= -T_{\alpha\beta}{}^{\gamma} D_\gamma + R_{\alpha\beta}{}^{\gamma\delta} \rho_{\delta\gamma} \\ [\rho_{\alpha\beta}, D_\gamma] &= -g_{\gamma\alpha} D_\beta + g_{\gamma\beta} D_\alpha \\ [\rho_{\alpha\beta}, \rho_{\mu\nu}] &= -g_{\alpha\mu}\rho_{\beta\nu} + g_{\alpha\nu}\rho_{\beta\mu} + g_{\beta\mu}\rho_{\alpha\nu} - g_{\beta\nu}\rho_{\alpha\mu} \end{aligned}\right\} . \tag{37}$$

The rigid Lie algebra of Minkowski space is recovered for $T_{\alpha\beta}{}^{\gamma} = 0$ and $R_{\alpha\beta}{}^{\gamma\delta} = 0$; then, in Cartesian coordinates, $D_\alpha \to \partial_a$, for details see [38].[8] Thus the Riemannian spacetime of GR is generalized to the Riemann-Cartan spacetime of the Poincaré gauge theory. And the underlying reason for this generalization is evident: It is the application of the Einstein procedure to a quantum mechanical system instead of to a classical point particle. The method remains the same, the objects to which it is applied to were generalized. Instead of an Einstein laboratory with curvilinear coordinates, we use a Kibble lab with frames in order to encompass also fermionic fields.

Up-to-date reviews of the Poincaré gauge theory of gravity can be found in [73, 74], and for more historic and technical details readers may refer to [26, 38, 75–77]. Here we briefly outline the most essential notions and constructions.

Following the general Yang-Mills-Utiyama-Sciama-Kibble gauge-theoretic scheme, the 10-parameter Poincaré group $T_4 \rtimes SO(1,3)$ gives rise to the 10-plet of the gauge potentials which are consistently identified with the coefficients $e_i{}^\alpha$ of the orthonormal coframe $\vartheta^\alpha = e_i{}^\alpha dx^i$ (4 potentials corresponding to the translation subgroup T_4) and the components $\Gamma_i{}^{\alpha\beta} = -\Gamma_i{}^{\beta\alpha}$ of the local connection $\Gamma^{\alpha\beta} = \Gamma_i{}^{\alpha\beta} dx^i$

[7] A highly original contribution to the understanding of translation gauge invariance was provided by Tresguerres [71].

[8] This local Lie algebra structure has also been found in the gauge theory of the de Sitter group $SO(2,3)$, see Stelle and West [72]. Note that the cosmological model favored by present observations seems to favor an underlying anti-de Sitter universe.

(6 potentials for the Lorentz subgroup $SO(1,3)$). The corresponding covariant curls, the field strengths of translations and Lorentz rotations,

$$T_{ij}{}^{\alpha} = \partial_i e_j{}^{\alpha} - \partial_j e_i{}^{\alpha} + \Gamma_{i\beta}{}^{\alpha} e_j{}^{\beta} - \Gamma_{j\beta}{}^{\alpha} e_i{}^{\beta}, \tag{38}$$

$$R_{ij}{}^{\alpha\beta} = \partial_i \Gamma_j{}^{\alpha\beta} - \partial_j \Gamma_i{}^{\alpha\beta} + \Gamma_{i\gamma}{}^{\beta} \Gamma_j{}^{\alpha\gamma} - \Gamma_{j\gamma}{}^{\beta} \Gamma_i{}^{\alpha\gamma}, \tag{39}$$

are the torsion $T_{ij}{}^{\alpha}$ and the curvature $R_{ij}{}^{\alpha\beta}$, both antisymmetric in i and j. This naturally introduces the Riemann-Cartan geometry [78–81] on the spacetime manifold.

Obviously, both gravitational field potentials transform covariantly as covectors

$$e_{i'}{}^{\alpha} = \frac{\partial x^j}{\partial x^{i'}} e_j{}^{\alpha}, \qquad \Gamma_{i'}{}^{\alpha\beta} = \frac{\partial x^j}{\partial x^{i'}} \Gamma_j{}^{\alpha\beta}, \tag{40}$$

under arbitrary local coordinate transformations (diffeomorphisms)

$$x^i \longrightarrow x^{i'} = x^{i'}(x^j). \tag{41}$$

As a result, the gravitational gauge field strengths (38) and (39) transform covariantly

$$T_{i'j'}{}^{\alpha} = \frac{\partial x^k}{\partial x^{i'}} \frac{\partial x^l}{\partial x^{j'}} T_{kl}{}^{\alpha}, \qquad R_{i'j'}{}^{\alpha\beta} = \frac{\partial x^k}{\partial x^{i'}} \frac{\partial x^l}{\partial x^{j'}} R_{kl}{}^{\alpha\beta}, \tag{42}$$

as second rank skew-symmetric tensors under the change of coordinates (41).

The action of the local Lorentz group is nontrivial. The Lorentz transformation, by definition, leaves the metric invariant:

$$\Lambda(x)_{\alpha}{}^{\mu} \Lambda(x)_{\beta}{}^{\nu} g_{\mu\nu} = g_{\alpha\beta}, \tag{43}$$

which means that the Lorentz transformed ("rotated") translational potential

$$e'_i{}^{\mu} = \Lambda(x)_{\alpha}{}^{\mu} e_i{}^{\alpha}, \tag{44}$$

remains orthonormal under (43). The corresponding transformation law for the local Lorentz connection is inhomogeneous:

$$\Gamma'_{i\alpha}{}^{\beta} = \Lambda(x)_{\nu}{}^{\beta} \Gamma_{i\mu}{}^{\nu} \Lambda^{-1}(x)_{\alpha}{}^{\mu} + \Lambda(x)_{\mu}{}^{\beta} \partial_i \Lambda^{-1}(x)_{\alpha}{}^{\mu}. \tag{45}$$

Accordingly, for the torsion and the curvature we find

$$T'_{ij}{}^{\mu} = \Lambda(x)_{\alpha}{}^{\mu} T_{ij}{}^{\alpha}, \qquad R'_{ij}{}^{\alpha\beta} = \Lambda(x)_{\mu}{}^{\alpha} \Lambda(x)_{\nu}{}^{\beta} R_{ij}{}^{\mu\nu}. \tag{46}$$

An infinitesimal Lorentz transformation

$$\Lambda(x)_{\alpha}{}^{\mu} = \delta_{\alpha}^{\mu} + \varepsilon(x)_{\alpha}{}^{\mu} \tag{47}$$

is described by the parameters $\varepsilon(x)_\alpha{}^\mu$ which, in view of (43), satisfy

$$\varepsilon_{\alpha\beta} + \varepsilon_{\beta\alpha} = 0. \tag{48}$$

Noting that $\Lambda^{-1}(x)_\alpha{}^\mu = \delta^\mu_\alpha - \varepsilon(x)_\alpha{}^\mu$, we derive the infinitesimal form of the transformation laws for the coframe (44) and the connection as (45):

$$\delta e_i{}^\alpha = \varepsilon_\beta{}^\alpha e_i{}^\beta, \tag{49}$$

$$\delta \Gamma_i{}^{\alpha\beta} = -\,\partial_i \varepsilon^{\alpha\beta} - \Gamma_{i\gamma}{}^\alpha \varepsilon^{\gamma\beta} - \Gamma_{i\gamma}{}^\beta \varepsilon^{\alpha\gamma}. \tag{50}$$

On account of the explicit generators $(\rho^\mu{}_\nu)_{\alpha\beta} = \delta^\mu_\alpha g_{\nu\beta} - \delta^\mu_\beta g_{\nu\alpha}$ for the vector representation of the Lorentz group, we can recast (49) into

$$\delta e_i{}^\mu = -\frac{1}{2}\varepsilon^{\alpha\beta}(\rho^\mu{}_\nu)_{\alpha\beta}\, e_i{}^\nu. \tag{51}$$

This can directly be extended to the infinitesimal transformation of any field ψ^A which belongs to an arbitrary representation of the Lorentz group with the generators $(\rho^A{}_B)_{\alpha\beta}$:

$$\delta \psi^A = -\frac{1}{2}\varepsilon^{\alpha\beta}(\rho^A{}_B)_{\alpha\beta}\, \psi^B. \tag{52}$$

In accordance with the transformation laws (52) and (50), the covariant derivative of an arbitrary field is defined by

$$D_i \psi^A = \partial_i \psi^A - \frac{1}{2}\Gamma_i{}^{\alpha\beta}(\rho^A{}_B)_{\alpha\beta}\, \psi^B. \tag{53}$$

We conclude this section by noticing that the gravitational Poincaré gauge field strengths satisfy the two Bianchi identities:

$$D_{[i} T_{jk]}{}^\alpha = R_{[ijk]}{}^\alpha, \tag{54}$$

$$D_{[i} R_{jk]}{}^{\alpha\beta} = 0. \tag{55}$$

Here, $R_{ijk}{}^\alpha = R_{ij\beta}{}^\alpha e_k{}^\beta$, and the covariant derivative D_i is defined by (53).

Although the tensor language is, so to say, a mother-tongue for relativists, the use of the modern coordinate free formalism of exterior forms proves to be extremely convenient in the gauge gravity theory. Accordingly, *we switch here to exterior forms*. Then the first (54) and the second (55) Bianchi identities read, respectively, $DT^\alpha = R_\gamma{}^\alpha \wedge \vartheta^\gamma$ and $DR^{\alpha\beta} = 0$.

4.3 Matter Lagrangian: Currents and Conservation Laws

Before we discuss the dynamics, we need to revisit the cornerstones of a gauge theory, i.e. the conservation laws.

For the sake of generality, we will assume a matter field ψ^A to be a p-form that transforms according to an arbitrary representation of the Lorentz group, cf. (52). Let us consider a general matter Lagrangian 4-form

$$L = L(\vartheta^\alpha\,, d\vartheta^\alpha\,, \Gamma^{\alpha\beta}\,, d\Gamma^{\alpha\beta}\,, \psi^A, d\psi^A) = L(\psi^A, D\psi^A, \vartheta^\alpha, T^\alpha, R^{\alpha\beta})\,. \tag{56}$$

Note that we take into account the possibility of a nonminimal coupling between matter and the gravitational field by allowing the dependence of L on the Poincaré gauge field strengths. In the minimal coupling scheme, $L = L(\psi^A, D\psi^A, \vartheta^\alpha)$, so that the matter interacts with the gravity only via the Poincaré gauge field potentials $(\vartheta^\alpha\,, \Gamma^{\alpha\beta})$ which contribute to the Lagrangian either directly or via the covariant derivatives $D\psi^A = d\psi^A - \frac{1}{2}\Gamma^{\alpha\beta}\wedge(\rho^A{}_B)_{\alpha\beta}\psi^B$.

Independent variations of the matter and gravitational arguments $\psi^A, \vartheta^\alpha, \Gamma^{\alpha\beta}$ yield for the matter Lagrangian

$$\begin{aligned} \delta L = &- \delta\vartheta^\alpha \wedge \Sigma_\alpha - \frac{1}{2}\delta\Gamma^{\alpha\beta}\wedge\tau_{\alpha\beta} + \delta\psi^A \wedge \frac{\delta L}{\delta\psi^A} \qquad (57)\\ &+ d\left[\delta\vartheta^\alpha \wedge \frac{\partial L}{\partial T^\alpha} + \delta\Gamma^{\alpha\beta}\wedge\frac{\partial L}{\partial R^{\alpha\beta}} + \delta\psi^A\wedge\frac{\partial L}{\partial D\psi^A}\right]. \end{aligned}$$

Here, for a gauge–invariant Lagrangian L, the expression

$$\frac{\delta L}{\delta\psi^A} = \frac{\partial L}{\partial\psi^A} - (-1)^p D\,\frac{\partial L}{\partial(D\psi^A)} \tag{58}$$

is the covariant *variational derivative* of L with respect to the matter p–form ψ^A. The *matter currents* in (57) are introduced by

$$\Sigma_\alpha := -\,\frac{\delta L}{\delta\vartheta^\alpha} = -\,\frac{\partial L}{\partial\vartheta^\alpha} - D\,\frac{\partial L}{\partial T^\alpha}\,, \tag{59}$$

$$\begin{aligned} \tau_{\alpha\beta} := &-2\frac{\delta L}{\delta\Gamma^{\alpha\beta}} = (\rho^A{}_B)_{\alpha\beta}\psi^B\wedge\frac{\partial L}{\partial(D\psi^A)}\\ &-\vartheta_\alpha\wedge\frac{\partial L}{\partial T^\beta} + \vartheta_\beta\wedge\frac{\partial L}{\partial T^\alpha} - 2D\frac{\partial L}{\partial R_\alpha{}^\beta}\,. \qquad (60) \end{aligned}$$

These are the energy-momentum current and the spin angular momentum current of matter, respectively. Using the master formula (57), we can derive the conservation laws for them.

The local *translations*, or general coordinate transformations (diffeomorphisms), are generated by the Lie derivatives along arbitrary vector fields $\xi = \xi^\alpha e_\alpha$ on the spacetime manifold. When the Lagrangian L is invariant under the local diffeomorphisms, the master formula (57) gives rise to two identities.[9] One consequence of the diffeomorphism invariance is the *first Noether identity*

$$D\Sigma_\alpha \equiv (e_\alpha \rfloor T^\beta) \wedge \Sigma_\beta + \frac{1}{2}(e_\alpha \rfloor R^{\beta\gamma}) \wedge \tau_{\beta\gamma} + W_\alpha, \tag{61}$$

where the generalized force is $W_\alpha := -(e_\alpha \rfloor D\psi^A) \wedge \frac{\delta L}{\delta\psi^A} - (-1)^p (e_\alpha \rfloor \psi^A) \wedge D\frac{\delta L}{\delta\psi^A}$. As another consequence of the translational invariance one finds the explicit form of the *canonical energy–momentum current*:

$$\begin{aligned}\Sigma_\alpha = &(e_\alpha \rfloor D\psi^A) \wedge \frac{\partial L}{\partial D\psi^A} + (e_\alpha \rfloor \psi^A) \wedge \frac{\partial L}{\partial \psi^A} - e_\alpha \rfloor L \\ &- D\frac{\partial L}{\partial T^\alpha} + (e_\alpha \rfloor T^\beta) \wedge \frac{\partial L}{\partial T^\beta} + (e_\alpha \rfloor R^{\beta\gamma}) \wedge \frac{\partial L}{\partial R^{\beta\gamma}}.\end{aligned} \tag{62}$$

Note that the second lines in (60) and (62) describe the nonminimal coupling contributions to the spin angular momentum current and the canonical energy-momentum current of matter, respectively.

The identity (61) is given in the *strong* form, without using the field equations. The generalized force $W_\alpha = 0$ vanishes when the ψ^A satisfy the Euler-Lagrange equations $\frac{\delta L}{\delta\psi^A} = 0$. Then the Noether identity (61) reduces to the conservation law of energy-momentum in the framework of the Poincaré gauge theory:

$$D\Sigma_\alpha \cong (e_\alpha \rfloor T^\beta) \wedge \Sigma_\beta + \frac{1}{2}(e_\alpha \rfloor R^{\beta\gamma}) \wedge \tau_{\beta\gamma}. \tag{63}$$

The sign $\cong$ denotes 'on shell,' that is, the field equations are assumed to be fulfilled. On the right-hand side of the identity (61) and of the conservation law (63) for the canonical energy-momentum current, we find the typical *Lorentz-type force* terms. They have the general structure *field strength* $\times$ *current.*

We assume that the Lagrangian L is invariant under a local Lorentz transformations $\delta\vartheta^\alpha = \varepsilon_\beta{}^\alpha\, \vartheta^\beta$, $\delta\Gamma^{\alpha\beta} = -D\varepsilon^{\alpha\beta}$, $\delta\psi^A = -\frac{1}{2}\varepsilon^{\alpha\beta}\,(\rho^A{}_B)_{\alpha\beta}\,\psi^B$ —recall (49), (50) and (52). Then, from the master formula (57), we find the *second Noether identity*

$$D\tau_{\alpha\beta} + \vartheta_\alpha \wedge \Sigma_\beta - \vartheta_\beta \wedge \Sigma_\alpha \equiv W_{\alpha\beta}. \tag{64}$$

The generalized torque is defined here as $W_{\alpha\beta} := -(\rho^A{}_B)_{\alpha\beta}\psi^B \wedge \frac{\delta L}{\delta\psi^A}$. The latter vanishes when the matter field equation $\frac{\delta L}{\delta\psi^A} = 0$ is satisfied, and then (64) reduces to the weak conservation law of the total angular momentum

$$D\tau_{\alpha\beta} + \vartheta_\alpha \wedge \Sigma_\beta - \vartheta_\beta \wedge \Sigma_\alpha \cong 0. \tag{65}$$

[9]Hint: one obtains two identities because *both,* ξ^α *and* $d\xi^\alpha$, are *point-wise arbitrary.*

4.4 Gravitational Lagrangian: Noether Identities

We assume that the gravitational Lagrangian 4-form

$$V = V(\vartheta^\alpha, T^\alpha, R_\alpha{}^\beta) \tag{66}$$

is an arbitrary function of the gravitational field variables. Its variation can be computed with the help of the master formula (57), and we recast the result into

$$\delta V = \delta\vartheta^\alpha \wedge \mathcal{E}_\alpha + \delta\Gamma^{\alpha\beta} \wedge \mathcal{C}_{\alpha\beta} + d\left[\delta\vartheta^\alpha \wedge H_\alpha + \delta\Gamma^{\alpha\beta} \wedge H_{\alpha\beta}\right] \tag{67}$$

by writing the variational derivatives with respect to the Poincaré gauge gravitational potentials as

$$\mathcal{E}_\alpha := \frac{\delta V}{\delta\vartheta^\alpha} = DH_\alpha - E_\alpha, \tag{68}$$

$$\mathcal{C}_{\alpha\beta} := \frac{\delta V}{\delta\Gamma^{\alpha\beta}} = DH_{\alpha\beta} - E_{\alpha\beta}\,. \tag{69}$$

Here we introduced the *gauge field momenta* 2-forms

$$H_\alpha := \frac{\partial V}{\partial T^\alpha}, \qquad H_{\alpha\beta} := \frac{\partial V}{\partial R^{\alpha\beta}}, \tag{70}$$

and defined the 3–forms of the *canonical* energy–momentum and the canonical spin for the gravitational gauge fields:

$$E_\alpha := -\frac{\partial V}{\partial\vartheta^\alpha}, \qquad E_{\alpha\beta} := -\frac{\partial V}{\partial\Gamma^{\alpha\beta}} = -\vartheta_{[\alpha} \wedge H_{\beta]}\,. \tag{71}$$

Diffeomorphism invariance of V yields the *Noether identities*

$$E_\alpha \equiv -e_\alpha\rfloor V + (e_\alpha\rfloor T^\beta) \wedge H_\beta + (e_\alpha\rfloor R^{\beta\gamma}) \wedge H_{\beta\gamma}, \tag{72}$$

$$D\,\mathcal{E}_\alpha \equiv (e_\alpha\rfloor T^\beta) \wedge \mathcal{E}_\beta + (e_\alpha\rfloor R^{\beta\gamma}) \wedge \mathcal{C}_{\beta\gamma}, \tag{73}$$

whereas the local Lorentz invariance results in the *Noether identity*

$$2D\mathcal{C}_{\alpha\beta} + \vartheta_\alpha \wedge \mathcal{E}_\beta - \vartheta_\beta \wedge \mathcal{E}_\alpha \equiv 0\,. \tag{74}$$

These relations are easily derived from (62), (61), and (64) by replacing L with V and dropping the dependence on ψ^A. All the relations (72)-(74) are *strong identities*; they are always valid independently of the field equations.

4.5 The General Field Equations of Poincaré Gauge Gravity

The field equations for the system of interacting matter and gravitational fields are derived from the total Lagrangian

$$V(\vartheta^\alpha, T^\alpha, R^{\alpha\beta}) + L(\psi^A, D\psi^A, \vartheta^\alpha, T^\alpha, R^{\alpha\beta}). \tag{75}$$

Independent variations of the total Lagrangian with respect to the coframe ϑ^α, the local Lorentz connection $\Gamma^{\alpha\beta}$, and the matter field ψ^A yield the system [77, 82–84]

$$DH_\alpha - E_\alpha = \Sigma_\alpha\,, \tag{76}$$

$$DH_{\alpha\beta} - E_{\alpha\beta} = \frac{1}{2}\,\tau_{\alpha\beta}\,, \tag{77}$$

$$\frac{\partial L}{\partial \psi^A} - (-1)^p D\,\frac{\partial L}{\partial(D\psi^A)} = 0\,. \tag{78}$$

From the dimension of the action we conclude that the Lagrangian 4-form has the same dimension $[L] = [\hbar]$, and one can easily find the dimension of the currents. Taking into account that $[\vartheta^\alpha] = [\ell]$ and $[\Gamma^{\alpha\beta}] = [1]$ (dimensionless) and using the definitions (59) and (60), we derive the dimensions: $[\Sigma_\alpha] = [\frac{\hbar}{\ell}]$ = [momentum] and $[\tau_{\alpha\beta}] = [\hbar]$ = [spin].

By expanding the matter currents with respect to the basis of the 3-forms, we find the energy-momentum and the spin tensors, respectively, as

$$\Sigma_\alpha = \mathcal{T}_\alpha{}^\mu\,\eta_\mu, \qquad \tau_{\alpha\beta} = \mathcal{S}_{\alpha\beta}{}^\mu\,\eta_\mu, \tag{79}$$

with $\eta_\mu = \sqrt{-g}\,\epsilon_\mu$. Since $[\eta_\mu] = [\ell^3]$, we have the dimensions: $[\mathcal{T}_\alpha{}^\mu] = [\frac{\hbar}{\ell^4}] =$ [momentum/volume] and $[\mathcal{S}_{\alpha\beta}{}^\mu] = [\frac{\hbar}{\ell^3}]$ = [spin/volume].

4.6 Einstein-Cartan Theory

The simplest Poincaré gauge theory is the Einstein-Cartan theory of gravity (EC), which results from choosing the *curvature scalar* of the Riemann-Cartan space as gravitational Hilbert-Einstein Lagrangian ($\eta_{\alpha\beta} = {}^*(\vartheta_\alpha \wedge \vartheta_\beta)$, see Appendix 1)

$$V_{\rm HE} = \frac{1}{2\kappa}\eta_{\alpha\beta} \wedge R^{\alpha\beta}. \tag{80}$$

Here, $\kappa = \frac{8\pi G}{c^3}$ is Einstein's gravitational constant with the dimension of $[\kappa] = \mathrm{s\,kg^{-1}}$. $G = 6.67 \times 10^{-11}\,\mathrm{m^3\,kg^{-1}\,s^{-2}}$ is Newton's gravitational constant. The speed of light $c = 2.9 \times 10^8$ m/s. Consistency check for the dimension: $[\frac{1}{\kappa}] = \mathrm{kg\,s^{-1}} = [\frac{m}{t}]$. Since $[R^{\alpha\beta}] = 1$ and $[\eta_{\alpha\beta}] = \ell^2$, we have $[\frac{1}{2\kappa}\eta_{\alpha\beta} \wedge R^{\alpha\beta}] = [\frac{m\ell^2}{t}] = [\hbar]$.

For the Lagrangian (80) we find from (70), (71) and (72):

$$H_\alpha = 0, \quad H_{\alpha\beta} = \frac{1}{2\kappa}\eta_{\alpha\beta}, \quad E_\alpha = -\frac{1}{2\kappa}\eta_{\alpha\beta\gamma}\wedge R^{\beta\gamma}, \quad E_{\alpha\beta} = 0. \tag{81}$$

As a result,

$$\mathcal{E}_\alpha = \frac{1}{2\kappa}\eta_{\alpha\beta\gamma}\wedge R^{\beta\gamma}, \qquad \mathcal{C}_{\alpha\beta} = \frac{1}{2\kappa}\eta_{\alpha\beta\gamma}\wedge T^{\gamma}, \tag{82}$$

and hence the Einstein-Cartan gravitational field equations read

$$\frac{1}{2}\eta_{\alpha\beta\gamma}\wedge R^{\beta\gamma} = \kappa\,\Sigma_\alpha, \qquad \eta_{\alpha\beta\gamma}\wedge T^{\gamma} = \kappa\,\tau_{\alpha\beta}. \tag{83}$$

Substituting $R^{\alpha\beta} = \frac{1}{2}R_{\mu\nu}{}^{\alpha\beta}\,\vartheta^\mu\wedge\vartheta^\nu$ and $T^\alpha = \frac{1}{2}T_{\mu\nu}{}^{\alpha}\,\vartheta^\mu\wedge\vartheta^\nu$ into the left-hand side of (83) and using (79), we find the Einstein-Cartan field equations in components,

$$\mathrm{Ric}_\alpha{}^\beta - \frac{1}{2}\delta_\alpha^\beta\,\mathrm{Ric}_\gamma{}^\gamma = \kappa\,\mathcal{T}_\alpha{}^\beta, \tag{84}$$

$$T_{\alpha\beta}{}^\gamma - \delta_\alpha^\gamma T_{\mu\beta}{}^\mu + \delta_\beta^\gamma T_{\mu\alpha}{}^\mu = \kappa\,\mathcal{S}_{\alpha\beta}{}^\gamma. \tag{85}$$

From the curvature 2-form we derive the Ricci 1-form $\mathrm{Ric}_\alpha = e_\beta\rfloor R_\alpha{}^\beta = \mathrm{Ric}_{i\alpha}dx^i$, components of which constitute the Ricci tensor $\mathrm{Ric}_{i\alpha} = e^j_\beta R_{ji\alpha}{}^\beta$. The curvature scalar is defined as usual by $R = e^\alpha\rfloor\mathrm{Ric}_\alpha = e_\alpha\rfloor e_\beta\rfloor R^{\alpha\beta} = e^i_\beta e^j_\alpha R_{ij}{}^{\alpha\beta}$.

One can "minimally" extend the EC theory by modifying the Hilbert-Einstein Lagrangian (80) with the simplest possible parity-odd term. Such a generalization was proposed by Hojman, Mukku and Sayed [85] (and is known as the Holst Lagrangian [86] in some literature) as

$$V_{\mathrm{HMS}} = \frac{1}{2\kappa}\left(\eta_{\alpha\beta} + \overline{a}_0\vartheta_\alpha\wedge\vartheta_\beta\right)\wedge R^{\alpha\beta}. \tag{86}$$

Here $\overline{a}_0 = \frac{1}{\xi}$, and the dimensionless constant ξ is often called the Barbero-Immirzi parameter.

For the Lagrangian (86)—due to (70), (71), and (72)—we find $H_\alpha = 0$ and $E_{\alpha\beta} = 0$ as in (81); however, now

$$H_{\alpha\beta} = \frac{1}{2\kappa}\left(\eta_{\alpha\beta} + \overline{a}_0\vartheta_\alpha\wedge\vartheta_\beta\right), \quad E_\alpha = -\frac{1}{2\kappa}\left(\eta_{\alpha\beta\gamma}\wedge R^{\beta\gamma} + 2\overline{a}_0 R_{\alpha\beta}\wedge\vartheta^\beta\right). \tag{87}$$

As a result, $\mathcal{E}_\alpha = -E_\alpha$, and

$$\mathcal{C}_{\alpha\beta} = DH_{\alpha\beta} = \frac{1}{2\kappa}\left[\eta_{\alpha\beta\gamma}\wedge T^\gamma + \overline{a}_0\left(T_\alpha\wedge\vartheta_\beta - T_\beta\wedge\vartheta_\alpha\right)\right]. \tag{88}$$

The Einstein-Cartan field equations (83) are then replaced by

$$\frac{1}{2}\eta_{\alpha\beta\gamma} \wedge R^{\beta\gamma} + \overline{a}_0 R_{\alpha\beta} \wedge \vartheta^\beta = \kappa\, \Sigma_\alpha, \tag{89}$$

$$\eta_{\alpha\beta\gamma} \wedge T^\gamma + \overline{a}_0 \left(T_\alpha \wedge \vartheta_\beta - T_\beta \wedge \vartheta_\alpha\right) = \kappa\, \tau_{\alpha\beta}. \tag{90}$$

The Einstein-Cartan theory is a viable gravitational theory that deviates from GR only at very high matter density, see [87] and the references given there. For matter without spin, $\mathcal{S}_{\alpha\beta}{}^\gamma = 0$, the torsion vanishes because of the second field equation (85) ,and the Einstein-Cartan theory coincides with GR. The same applies to the parity-odd HMS model (86).

4.7 *Quadratic Poincaré Gauge Gravity Models*

It is a unique feature of the gauge approach to gravity that the field equations (84) and (85) are *algebraic* in the field strength rather than first order partial differential equations as in any other gauge theory of an internal group $U(1)$, $SU(2)$, $SU(3)$,... . This is due to the fact that, because of the existence of the metric $g_{\alpha\beta}$ of spacetime and of the frame $e_i{}^\alpha$, with $e_i{}^\alpha\, e^i{}_\beta = \delta^\alpha_\beta$, the Lorentz field strength $R_{\alpha\beta}{}^{\gamma\delta} = e^i{}_\alpha e^j{}_\beta\, R_{ij}{}^{\gamma\delta}$ can be contracted to a scalar, namely to the curvature scalar.

This is impossible in Yang-Mills theories since the internal group indices cannot be related to the spacetime indices. Thus, the simplest Lagrangian in Yang-Mills theory is *quadratic* in the field strength yielding first-order differential equations as field equations. Accordingly, the gauge doctrine would suggest for the gauge theory of the Poincaré group a general Lagrangian 4-form that contains all possible quadratic invariants of the torsion and the curvature:

$$\begin{aligned} V = &\frac{1}{2\kappa}\Big\{\Big(a_0\eta_{\alpha\beta} + \overline{a}_0\vartheta_\alpha \wedge \vartheta_\beta\Big) \wedge R^{\alpha\beta} - 2\lambda_0\eta \\ &- T^\alpha \wedge \sum_{I=1}^{3}\left[a_I\, {}^*({}^{(I)}T_\alpha) + \overline{a}_I\, {}^{(I)}T_\alpha\right]\Big\} \\ &- \frac{1}{2\rho} R^{\alpha\beta} \wedge \sum_{I=1}^{6}\left[b_I\, {}^*({}^{(I)}R_{\alpha\beta}) + \overline{b}_I\, {}^{(I)}R_{\alpha\beta}\right]. \end{aligned} \tag{91}$$

The irreducible parts of the torsion and the curvature are defined in Appendix 2. The Lagrangian (91) has a clear structure: the first line is *linear in the curvature,* the second line collects the terms *quadratic in the torsion*, while the third line contains the invariants *quadratic in the curvature*. Furthermore, each line is composed of a parity even piece (first term on each line), and a parity odd part (last term on each line). A special case $a_0 = 0$ and $\overline{a}_0 = 0$ describes the purely quadratic model without the Hilbert-Einstein linear term in the Lagrangian. To recover the Einstein-Cartan model, one puts $a_0 = 1$ and $\overline{a}_0 = 0$.

Besides that, the general PG model contains a set of the coupling constants which determine the structure of quadratic part of the Lagrangian: ρ, a_1, a_2, a_3 and $\overline{a}_1, \overline{a}_2, \overline{a}_3, b_1, \cdots, b_6$ and $\overline{b}_1, \cdots, \overline{b}_6$. The overbar denotes the constants responsible for the parity odd interaction. We have the dimension $[\frac{1}{\rho}] = [\hbar]$, whereas $a_I, \overline{a}_I, b_I$ and $\overline{b}_I$ are dimensionless. Moreover, not all of these constants are independent: we take $\overline{a}_2 = \overline{a}_3$, $\overline{b}_2 = \overline{b}_4$ and $\overline{b}_3 = \overline{b}_6$ because some of terms in (91) are the same,

$$T^\alpha \wedge {}^{(2)}T_\alpha = T^\alpha \wedge {}^{(3)}T_\alpha = {}^{(2)}T^\alpha \wedge {}^{(3)}T_\alpha, \tag{92}$$

$$R^{\alpha\beta} \wedge {}^{(2)}R_{\alpha\beta} = R^{\alpha\beta} \wedge {}^{(4)}R_{\alpha\beta} = {}^{(2)}R^{\alpha\beta} \wedge {}^{(4)}R_{\alpha\beta}, \tag{93}$$

$$R^{\alpha\beta} \wedge {}^{(3)}R_{\alpha\beta} = R^{\alpha\beta} \wedge {}^{(6)}R_{\alpha\beta} = {}^{(3)}R^{\alpha\beta} \wedge {}^{(6)}R_{\alpha\beta}. \tag{94}$$

For the Lagrangian (91), by means of (70)-(71), we can derive the gravitational field momenta

$$H_\alpha = -\frac{1}{\kappa}\, h_\alpha\,, \qquad H_{\alpha\beta} = \frac{1}{2\kappa}\left(a_0\, \eta_{\alpha\beta} + \overline{a}_0 \vartheta_\alpha \wedge \vartheta_\beta\right) - \frac{1}{\rho}\, h_{\alpha\beta}, \tag{95}$$

and, furthermore, the canonical energy-momentum and spin currents of the gravitational field as

$$E_\alpha = -\frac{1}{\kappa}\Big(\frac{a_0}{2}\, \eta_{\alpha\beta\gamma} \wedge R^{\beta\gamma} + \overline{a}_0\, R_{\alpha\beta} \wedge \vartheta^\beta - \lambda_0 \eta_\alpha + q^{(T)}_\alpha\Big) - \frac{1}{\rho}\, q^{(R)}_\alpha, \tag{96}$$

$$E_{\alpha\beta} = \frac{1}{2}\left(H_\alpha \wedge \vartheta_\beta - H_\beta \wedge \vartheta_\alpha\right). \tag{97}$$

For convenience, we introduced here the 2-forms which are linear functions of the torsion and the curvature, respectively, by

$$h_\alpha = \sum_{I=1}^{3} \left[a_I\, {}^*({}^{(I)}T_\alpha) + \overline{a}_I\, {}^{(I)}T_\alpha\right], \quad h_{\alpha\beta} = \sum_{I=1}^{6} \left[b_I\, {}^*({}^{(I)}R_{\alpha\beta}) + \overline{b}_I\, {}^{(I)}R_{\alpha\beta}\right], \tag{98}$$

and 3-forms quadratic in the torsion and in the curvature, respectively:

$$q^{(T)}_\alpha = \frac{1}{2}\left[(e_\alpha \rfloor T^\beta) \wedge h_\beta - T^\beta \wedge e_\alpha \rfloor h_\beta\right], \tag{99}$$

$$q^{(R)}_\alpha = \frac{1}{2}\left[(e_\alpha \rfloor R^{\beta\gamma}) \wedge h_{\beta\gamma} - R^{\beta\gamma} \wedge e_\alpha \rfloor h_{\beta\gamma}\right]. \tag{100}$$

By construction, the first 2-form in (98) has the dimension of a length, $[h_\alpha] = [\ell]$, whereas the second one is obviously dimensionless, $[h_{\alpha\beta}] = 1$. Similarly, we find for (99) the dimension of length $[q^{(T)}_\alpha] = [\ell]$, and the dimension of the inverse length, $[q^{(R)}_\alpha] = [1/\ell]$ for (100).

The resulting Poincaré gravity field equations (68) and (69) then read:

$$\frac{a_0}{2}\eta_{\alpha\beta\gamma} \wedge R^{\beta\gamma} + \overline{a}_0 R_{\alpha\beta} \wedge \vartheta^\beta - \lambda_0 \eta_\alpha$$
$$+ q_\alpha^{(T)} + \ell_\rho^2 \, q_\alpha^{(R)} - Dh_\alpha = \kappa \, \Sigma_\alpha, \tag{101}$$
$$a_0 \, \eta_{\alpha\beta\gamma} \wedge T^\gamma + \overline{a}_0 \left(T_\alpha \wedge \vartheta_\beta - T_\beta \wedge \vartheta_\alpha\right)$$
$$+ h_\alpha \wedge \vartheta_\beta - h_\beta \wedge \vartheta_\alpha - 2\ell_\rho^2 \, Dh_{\alpha\beta} = \kappa \, \tau_{\alpha\beta}. \tag{102}$$

The contribution of the curvature square terms in the Lagrangian (91) to the gravitational field dynamics in the equations (101) and (102) is characterized by the new coupling parameter with the dimension of the area (recall that $[\frac{1}{\rho}] = [\hbar]$):

$$\ell_\rho^2 = \frac{\kappa}{\rho}. \tag{103}$$

The parity-odd sector in PG gravity has been recently analyzed in [88–94], with a particular attention to the cosmological issues. A major progress was made in this domain with the computation of the particle spectrum of general quadratic PG models by Karananas [95] and Blagojević and Cvetković [96].

4.8 *Tonti-Diagram of Quadratic Poincaré Gauge Gravity*

The structure of Poincaré gauge gravity can be visualized by means of the Tonti diagram in Fig. 4. It is constructed as a direct generalization of the Tonti diagram for TG in Fig. 2. In TG we started with 4 translation potentials ϑ^α, in PG we start instead with 4 translation and 6 Lorentz 1-form potentials as *configuration* variables: $\{\vartheta^\alpha, \Gamma^{\alpha\beta} = -\Gamma^{\beta\alpha}\}$. Besides the translation group $T(4)$, now the Lorentz group $SO(1,3)$ is gauged, too. On the side of the *source* variables, besides the energy-momentum current Σ_α, we have the spin current $\tau_{\alpha\beta} = -\tau_{\beta\alpha}$.

Since the Lorentz group depends on the metric, the present Tonti diagram does not contain premetric structures. However, if we generalize the Lorentz connection $\Gamma^{\alpha\beta}$ to a linear one, $\Gamma_\alpha{}^\beta$, that is, if we substitute the $SO(1,3)$ by the linear group $GL(4, R)$, then the configuration and source variables are all premetric again, similar as in the Tonti diagram for TG. Note that $\Gamma^{\alpha\beta} := g^{\alpha\gamma}\,\Gamma_\gamma{}^\beta$. The affine generalization of PG has been discussed as metric-affine gravity in detail in [77], for example.

As we recognize from the Tonti diagram for PG, the *linear* constitutive equations correspond to quadratic Lagrangians, and the Einstein-Cartan theory and Einstein's general relativity are simple subcases of this scheme.

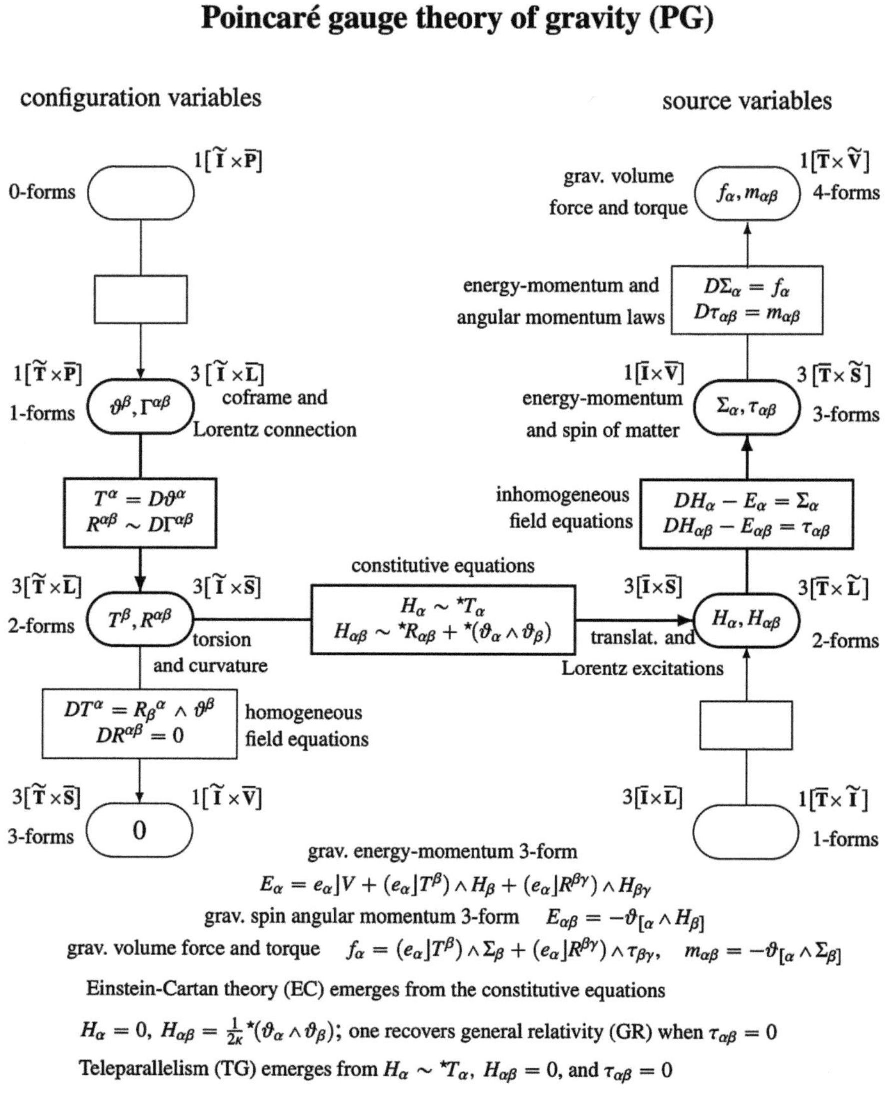

Fig. 4 Tonti diagram for Poincaré gauge theory of gravity

5 Discussion and Outlook

A gauge theory is a heuristic scheme within the Lagrange formalism in the Minkowski space of special relativity for the purpose of deriving a new interaction from a conserved current and the attached rigid symmetry group. This new 'gauge' interaction is induced by demanding that the rigid symmetry should be extended to a locally valid symmetry. We demonstrate that, when applied to gravity theory, this heuristic approach should be based on the conserved energy-momentum current and the associated translational symmetry, consistently supplemented by the local Lorentz symmetry.

We postpone a detailed review of the physical contents of the Poincaré gauge gravity theory to a different publication. Earlier, this subject was intensively studied and the relevant results can be found in [82–84, 97].

At present time, we are again witnessing a considerable growth of interest in gravitational gauge issues. Among other directions of research, the search and analysis of *exact solutions* of the field equations is essential for improvement of understanding of the nature of gravitational interaction; see [98–101] for relevant discussions. Also the topological invariants related to torsion have been reinvestigated [102, 103].

It is important to recall Einstein's view [104] that "...the question whether this continuum has a Euclidean, Riemannian, or any other structure is a question of physics proper which must be answered by experience, and not a question of a convention to be chosen on grounds of mere expediency." Accordingly, when aiming at an experimental probing of the geometrical structure of spacetime (in particular, searching for possible deviations beyond the Riemannian geometry), one should study how do test particles move under the influence of the gravitational field in the framework of the gauge gravity theory. Remarkably, the *propagation equations* should not be postulated in an ad hoc way; they are the consequence of the conservation laws. Importantly, the analysis of the equations of motion reveals that the torsion, in the context of PG, is related only to the *elementary particle spin* and under no circumstances to the orbital angular momentum of test particles [105–108].

Recent advances in *cosmology* have drawn renewed attention to the framework of Poincaré gauge gravity. In this sense, the early results of Trautman [109], and Minkevich [110, 111], which predict a possible aversion of a singularity in the early universe and possible modifications of the late stage of the cosmological evolution [112–115] are currently revisited and extended in the most recent works of Barrow et al. [116, 117] and of Nikiforova, Randjbar-Daemi, Rubakov and Damour [118–121].

Last but not least we should mention the progress in the study of the Hamiltonian approach to PG by Struckmeier et al. [122–124] and, most remarkably, by Blagojević and Cvetković [125, 126]. Toller [127] has proposed a highly interesting generalization of the PG framework by taking the symplectic group $Sp(4, R)$ as gauge group, which is locally isomorphic to the anti-de Sitter group $SO(2, 3)$.

Finally, let us have a look back at the Tonti diagram for TG, see Fig. 2. The constitutive equation is conventionally assumed to be a linear relation. Only then we can recover GR by taking a suitable ansatz. Mashhoon, already since the early 1990s, followed up the idea that the locality principle in special relativity—the clock hypothesis is one particular example—is washed out at extremely high translational and rotational accelerations. This is expected to happen long before quantum effects set in. A detailed discussion can be found in Mashhoon's book [128]. This nonlocality should also have consequences for gravitational theory. The linear constitutive equation in Fig. 2 should then become a *nonlocal* relation, as was proposed by Mashhoon and one of us [129, 130] and has been discussed in detail in [128]. A certain simplification of the nonlocal ansatz has been proposed by Puetzfeld et al. [131]. The nonlocal idea, as applied to TG, has even be generalized to the complete quadratic PG framework, see [132, Appendix]. But a detailed discussion is still missing.

Acknowledgements This article is based on a lecture given at the Weyl Conference in Bad Honnef in 2018. FWH is very grateful to Claus Kiefer, Silvia De Bianchi, and to the Wilhelm and Else Heraeus Foundation for the invitation. The material of the paper is partly overlapping with lectures presented in Tartu in 2017 at the conference "Geometric Foundations of Gravity" organized by Manuel Hohmann, Christian Pfeifer, Laur Järv and Martin Krššák. We thank Yakov Itin (Jerusalem) and Jens Boos (Edmonton, Alberta) for many fruitful discussions. We are specially thankful to Prof. Enzo Tonti (Trieste) for his help in setting up the two Tonti diagrams. We are most grateful to Claus Kiefer for numerous suggestions for improving our article.

Appendix 1

Our basic notation and conventions are consistent with [12, 77]. In particular, Greek indices $\alpha, \beta, \dots = 0, \dots, 3$, denote the anholonomic components (for example, of a coframe ϑ^α), while the Latin indices $i, j, \dots = 0, \dots, 3$, label the holonomic components (e.g., dx^i). Spatial components are numbered by Latin indices from the beginning of the alphabet $a, b, \dots = 1, 2, 3$. The anholonomic vector frame basis e_α is dual to the coframe basis in the sense that $e_\alpha \rfloor \vartheta^\beta = \delta^\beta_\alpha$, where $\rfloor$ denotes the interior product. The volume 4-form is denoted by η, and the η-basis in the space of exterior forms is constructed with the help of the interior products as $\eta_{\alpha_1 \dots \alpha_p} := e_{\alpha_p} \rfloor \dots e_{\alpha_1} \rfloor \eta$, $p = 1, \dots, 4$. They are related to the ϑ-basis via the Hodge dual operator *, for example, $\eta_{\alpha\beta} = {}^*(\vartheta_\alpha \wedge \vartheta_\beta)$. The Minkowski metric is $g_{\alpha\beta} = \mathrm{diag}(c^2, -1, -1, -1)$. All the objects related to the parity-odd sector (coupling constants, irreducible pieces of the curvature, etc.) are marked by an overline to distinguish them from the corresponding parity-even objects.

Appendix 2

The torsion 2-form can be decomposed into the three irreducible pieces, $T^\alpha = {}^{(1)}T^\alpha + {}^{(2)}T^\alpha + {}^{(3)}T^\alpha$, where

$$ {}^{(2)}T^\alpha = \frac{1}{3}\vartheta^\alpha \wedge T, \qquad {}^{(3)}T^\alpha = \frac{1}{3} e^\alpha \rfloor {}^*\overline{T}, \tag{104} $$

$$ {}^{(1)}T^\alpha = T^\alpha - {}^{(2)}T^\alpha - {}^{(3)}T^\alpha. \tag{105} $$

Here the 1-forms of the torsion trace and axial trace are introduced:

$$ T := e_\nu \rfloor T^\nu, \qquad \overline{T} := {}^*(T^\nu \wedge \vartheta_\nu). \tag{106} $$

For the irreducible pieces of the dual torsion ${}^*T^\alpha = {}^{(1)}({}^*T^\alpha) + {}^{(2)}({}^*T^\alpha) + {}^{(3)}({}^*T^\alpha)$, we have the properties

$$ {}^{(1)}({}^*T^\alpha) = {}^*({}^{(1)}T^\alpha), \quad {}^{(2)}({}^*T^\alpha) = {}^*({}^{(3)}T^\alpha), \quad {}^{(3)}({}^*T^\alpha) = {}^*({}^{(2)}T^\alpha). \tag{107} $$

The Riemann-Cartan curvature 2-form is decomposed $R^{\alpha\beta} = \sum_{I=1}^{6} {}^{(I)}R^{\alpha\beta}$ into the 6 irreducible parts

$$ {}^{(2)}R^{\alpha\beta} = -{}^*(\vartheta^{[\alpha} \wedge \overline{\Psi}^{\beta]}), \qquad {}^{(4)}R^{\alpha\beta} = -\vartheta^{[\alpha} \wedge \Psi^{\beta]}, \tag{108} $$

$$ {}^{(3)}R^{\alpha\beta} = -\tfrac{1}{12}\,\overline{X}\,{}^*(\vartheta^\alpha \wedge \vartheta^\beta), \qquad {}^{(6)}R^{\alpha\beta} = -\tfrac{1}{12}\,X\,\vartheta^\alpha \wedge \vartheta^\beta, \tag{109} $$

$$ {}^{(5)}R^{\alpha\beta} = -\tfrac{1}{2}\vartheta^{[\alpha} \wedge e^{\beta]}\rfloor(\vartheta^\gamma \wedge X_\gamma), \tag{110} $$

$$ {}^{(1)}R^{\alpha\beta} = R^{\alpha\beta} - \sum_{I=2}^{6} {}^{(I)}R^{\alpha\beta}, \tag{111} $$

where

$$ X^\alpha := e_\beta \rfloor R^{\alpha\beta}, \quad X := e_\alpha \rfloor X^\alpha, \quad \overline{X}^\alpha := {}^*(R^{\beta\alpha} \wedge \vartheta_\beta), \quad \overline{X} := e_\alpha \rfloor \overline{X}^\alpha, \tag{112} $$

and

$$ \Psi_\alpha := X_\alpha - \tfrac{1}{4}\,\vartheta_\alpha\,X - \tfrac{1}{2}\,e_\alpha\rfloor(\vartheta^\beta \wedge X_\beta), \tag{113} $$

$$ \overline{\Psi}_\alpha := \overline{X}_\alpha - \tfrac{1}{4}\,\vartheta_\alpha\,\overline{X} - \tfrac{1}{2}\,e_\alpha\rfloor(\vartheta^\beta \wedge \overline{X}_\beta). \tag{114} $$

The 1-forms X^α and $\overline{X}^\alpha$ are not completely independent: $\vartheta_\alpha \wedge X^\alpha = {}^*(\vartheta_\alpha \wedge \overline{X}^\alpha)$.

References

1. W. Heisenberg, Über den Bau der Atomkerne. I. Z. Phys. **77**, 1–11 (1932)
2. H. Yukawa, On the interaction of elementary particles. I. Proc. Phys.-Math. Soc. Japan **17**, 48–57 (1935)
3. N. Kemmer, The particle aspect of meson theory. Proc. Roy. Soc. London. Ser. A. Math. Phys. Sci. **173**, 91–116 (1939)
4. C.N. Yang, R.L. Mills, Conservation of isotopic spin and isotopic gauge invariance. Phys. Rev. **96**, 191–195 (1954)
5. Y. Kosmann-Schwarzbach, *The Noether Theorems* (Springer, New York, 2011)
6. H. Weyl, Elektron und Gravitation. Zeitschr. Phys. **56**, 330–352 (1929)
7. V. Fock, Geometrisierung der Diracschen Theorie des Elektrons. Zeitschr. Phys. **57**, 261–277 (1929)
8. H. Weyl, A remark on the coupling of gravitation and electron. Phys. Rev. **77**, 699–701 (1950)
9. L. O'Raifeartaigh, *Group Structure of Gauge Theories* (Cambridge University Press, Cambridge, UK, 1986)
10. G. Mack, Physical principles, geometrical aspects, and locality properties of gauge field theories. Fortsch. Phys. **29**, 135–185 (1981)
11. M. Chaichian, N.F. Nelipa, *Introduction to Gauge Field Theories* (Springer, Berlin, 1984)

12. F.W. Hehl, Yu.N. Obukhov, *Foundations of Classical Electrodynamics: Charge, Flux, and Metric* (Birkhäuser, Boston, MA, 2003)
13. R. Mills, Gauge fields. Am. J. Phys. **57**, 493–507 (1989)
14. R. Mills, Model of confinement for gauge theories. Phys. Rev. Lett. **43**, 549–551 (1979)
15. C. Kiefer, *Quantum Gravity*, 3rd edn. (Oxford University Press, Oxford, UK, 2012)
16. S.K. Wong, Field and particle equations for the classical Yang-Mills field and particles with Isotopic Spin. Nuovo Cim. A **65**, 689–694 (1970)
17. A. Einstein, *The Meaning of Relativity, Princeton Lectures of May 1921*, 5th edn. (Princeton University Press, Princeton, NJ, 1955)
18. E.P. Wigner, On unitary representations of the inhomogeneous Lorentz group. Ann. Math. **40**, 149–204 (1939)
19. J.J. Sakurai, Theory of strong interactions. Ann. Phys. (N.Y.) **11**, 1–48 (1960)
20. S.L. Glashow, M. Gell-Mann, Gauge theories of vector particles. Ann. Phys. (USA) **15**, 296–297 (1961)
21. R. Feynman, F.B. Morinigo, W.G. Wagner, *Feynman Lectures on Gravitation*, Lectures given 1962/63, ed. by B. Hatfield (Addison-Wesley, Reading, MA, 1995)
22. L. Rezzolla, O. Zanotti, *Relativistic Hydrodynamics* (Oxford University Press, Oxford, 2013), p. 140
23. D.W. Sciama, The analogy between charge and spin in general relativity, in *Recent Developments in General Relativity, Festschrift for Infeld* (Pergamon Press, Oxford; PWN, Warsaw, 1962), pp. 415–439
24. Y. Itin, Energy momentum current for coframe gravity. Class. Quantum Grav. **19**, 173 (2002). arXiv:gr-qc/0111036
25. F. W. Hehl, Y. Itin, Yu. N. Obukhov, On Kottler's path: origin and evolution of the premetric program in gravity and in electrodynamics. Int. J. Mod. Phys. D **25**, 1640016 (2016). arXiv:1607.06159
26. M. Blagojević, F.W. Hehl (eds.), *Gauge Theories of Gravitation: A Reader with Commentaries* (Imperial College Press, London, 2013)
27. Y.M. Cho, Einstein Lagrangian as the translational Yang-Mills Lagrangian. Phys. Rev. D **14**, 2521–2525 (1976)
28. J. Nitsch, F.W. Hehl, Translational gauge theory of gravity: postnewtonian approximation and spin precession. Phys. Lett. B **90**, 98 (1980)
29. Yu.N. Obukhov, J.G. Pereira, Metric affine approach to teleparallel gravity. Phys. Rev. D **67**, 044016 (2003)
30. J.G. Pereira, Yu.N. Obukhov, Gauge structure of teleparallel gravity. Universe **5**(6), 139 (2019)
31. T. Koivisto, M. Hohmann, T. Złośnik, The general linear Cartan khronon. Universe **5**(6), 168 (2019)
32. R. Aldrovandi, J.G. Pereira, *Teleparallel Gravity: An Introduction* (Springer, Dordrecht, The Netherlands, 2013)
33. R. Weitzenböck, *Invariantentheorie* (Noordhoff, Groningen, 1923)
34. R. Weitzenböck, Differentialinvarianten in der Einsteinschen Theorie des Fernparallelismus. Sitzungsber. Preuss. Akad. Wiss. Berlin Phys.-Math. Klasse 466–474 (1928)
35. A. Einstein, Riemann-Geometrie mit Aufrechterhaltung des Begriffes des Fernparallelismus. Sitzungsber. Preuss. Akad. Wiss. Berlin, Phys.-Math. Klasse 217–221 (1928)
36. C. Møller, Further remarks on the localization of the energy in the general theory of relativity. Ann. Phys. (N.Y.) **12**, 118–133 (1961)
37. C. Møller, Conservation laws and absolute parallelism in general relativity. Mat.-Fys. Skr. Dan. Vidensk. Selsk. **1**(10) (1961)
38. F.W. Hehl, Four lectures on Poincaré gauge field theory, in *Proceedings of the 6th Course of the School of Cosmology and Gravitation on Spin, Torsion, Rotation, and Supergravity*, Erice, Italy, May 1979, ed. by P.G. Bergmann, V. de Sabbata (Plenum, New York 1980) pp. 5–61. See the author's homepage http://www.thp.uni-koeln.de/gravitation/mitarbeiter/hehl.html
39. E. Tonti, *The Mathematical Structure of Classical and Relativistic Physics, A General Classification Diagram* (Birkhäuser-Springer, New York, 2013)

40. Y. Itin, Yu.N. Obukhov, J. Boos, F.W. Hehl, Premetric teleparallel theory of gravity and its local and linear constitutive law. Eur. Phys. J. C **78**, 907 (2018). arXiv:1808.08048
41. H. Rund, Representations of the duals of gauge field tensors. J. Math. Phys. **20**, 1392 (1979)
42. E.J. Post, *Formal Structure of Electromagnetics—General Covariance and Electromagnetics (North Holland, Amsterdam, 1962, and Dover,* Mineola, NY, 1997)
43. Y. Itin, F.W. Hehl, Yu.N. Obukhov, Premetric equivalent of general relativity: Teleparallelism. Phys. Rev. D **95**, 084020 (2017). arXiv:1611.05759
44. W. Kopczyński, A. Trautman, *Space-Time and Gravitation* (Wiley, Chichester, UK, 1992)
45. F.W. Hehl, J.D. McCrea, E.W. Mielke, Weyl spacetimes, the dilation current, and creation of gravitating mass by symmetry breaking, in *Exact Sciences and their Philosophical Foundations, Hermann Weyl Congress 1985*, ed. by W. Deppert et al. (Lang, Frankfurt am Main, 1988), pp. 241–310
46. E. Kröner, Continuum theory of defects, in "*Physique des Defaults*" *Les Houches, 1980, Session 35*, ed. by R. Balian, et al. (North-Holland, Amsterdam, 1981), pp. 215–315
47. R.W. Sharpe, *Differential Geometry: Cartan's generalization of Klein's Erlangen program* (Springer, New York, 1997)
48. S. Sternberg, *Curvature in Mathematics and Physics* (Dover Publications, Minneola, New York, 2012)
49. E.L. Schücking, E.J. Surowitz, *Einstein's Apple: Homogeneous Einstein Fields* (World Scientific, Singapore, 2015)
50. R. Colella, A.W. Overhauser, S.A. Werner, Observation of gravitationally induced quantum interference. Phys. Rev. Lett. **34**, 1472–1474 (1975)
51. H. Rauch, S.A. Werner, *Neutron Interferometry: Lessons in Experimental Quantum Mechanics, Wave-Particle Duality, and Entanglement*, 2nd edn. (Oxford University Press, Oxford, UK, 2015)
52. M. Kasevich, S. Chu, Atomic interferometry using stimulated Raman transitions. Phys. Rev. Lett. **67**, 181–184 (1991)
53. P. Asenbaum, C. Overstreet, T. Kovachy, D.D. Brown, J.M. Hogan, M.A. Kasevich, Phase shift in an atom interferometer due to spacetime curvature across its wave function. Phys. Rev. Lett. **118**, 183602 (2017). arXiv:1610.03832
54. C. Overstreet, P. Asenbaum, T. Kovachy, R. Notermans, J.M. Hogan, M.A. Kasevich, Effective inertial frame in an atom interferometric test of the equivalence principle. Phys. Rev. Lett. **120**, 183604 (2018). arXiv:1711.09986
55. J. Audretsch, F.W. Hehl, C. Lämmerzahl, Matter wave interferometry and why quantum objects are fundamental for establishing a gravitational theory, in *Relativistic Gravity Research: With Emphasis on Experiments and Observation: Proceedings*, ed. by J. Ehlers, G. Schaefer, Lecture Notes in Physics, vol. 410 (Springer, 1992), pp. 368–407
56. V. Nesvizhevsky, A. Voronin, *Surprising Quantum Bounces* (Imperial College Press, London, 2015)
57. T.W.B. Kibble, Lorentz invariance and the gravitational field. J. Math. Phys. **2**, 212–221 (1961)
58. F.W. Hehl, W.-T. Ni, Inertial effects of a Dirac particle. Phys. Rev. D **42**, 2045–2048 (1990)
59. B. Mashhoon, Neutron interferometry in a rotating frame of reference. Phys. Rev. Lett. **61**, 2639–2542 (1988)
60. B. Mashhoon, On the spin-rotation-gravity coupling. Gen. Relat. Gravit. **31**, 681–691 (1999). arXiv:gr-qc/9803017
61. A. Danner, B. Demirel, W. Kersten, H. Lemmel, R. Wagner, S. Sponar, Y. Hasegawa, Spin-rotation coupling observed in neutron interferometry. NPJ Quantum Inf. **6**, 23 (2020)
62. D. Iwanenko, A. Sokolow, *Klassische Feldtheorie* (Akademie-Verlag, Berlin, 1953)
63. F. Hund, *Materie als Feld* (Springer, Berlin, 1954)
64. Tian Yu Cao, *Conceptual Developments of 20th Century Field Theories*, 2nd edn. (Cambridge University Press, Cambridge, UK, 2019)
65. B. Mashhoon, Gravitoelectromagnetism: a brief review, in *The Measurement of Gravitomagnetism: A Challenging Enterprise*, ed. by L. Iorio (Nova, Hauppauge, NY, 2007), pp. 29–39. arXiv:gr-qc/0311030

66. P. von der Heyde, The equivalence principle in the U_4 theory of gravitation. Nuovo Cim. Lett. **14**, 250–252 (1975)
67. D. Hartley, Normal frames for non-Riemannian connections. Class. Quantum Grav. **12**, L103–L105 (1995)
68. B.Z. Iliev, Normal frames and the validity of the equivalence principle: I. Cases in a neighborhood and at a point. J. Phys. A **29**, 6895–6902 (1996)
69. J.M. Nester, Normal frames for general connections. Ann. Phys. (Berlin) **19**, 45–52 (2010)
70. R. Utiyama, Invariant theoretical interpretation of interaction. Phys. Rev. **101**, 1597–1607 (1956)
71. R. Tresguerres, Translations and dynamics. Int. J. Geom. Meth. Mod. Phys. **5**, 905–945 (2008). arXiv:0707.0296
72. K.S. Stelle, P.C. West, Spontaneously broken de Sitter symmetry and the gravitational holonomy group. Phys. Rev. D **21**, 1466–1488 (1980)
73. Yu.N. Obukhov, Poincaré gauge gravity: selected topics. Int. J. Geom. Meth. Mod. Phys. **3**, 95 (2006). arXiv:gr-qc/0601090
74. Yu.N. Obukhov, Poincaré gauge gravity: an overview. Int. J. Geom. Meth. Mod. Phys. **15**, Supp. 1, 1840005 (2018). arXiv:1805.07385
75. V.N. Ponomarev, A.O. Barvinsky, Yu.N. Obukhov, *Gauge Approach and Quantization Methods in Gravity Theory* (Nauka, Moscow, 2017)
76. E.W. Mielke, *Geometrodynamics of Gauge Fields: On the Geometry of Yang-Mills and Gravitational Gauge Theories*, 2nd edn. (Springer, Cham, Switzerland, 2017)
77. F.W. Hehl, J.D. McCrea, E.W. Mielke, Y. Ne'eman, Metric affine-gauge theory of gravity: field equations, Noether identities, world spinors, and breaking of dilation invariance. Phys. Rept. **258**, 1–171 (1995). arXiv:gr-qc/9402012
78. E. Schrödinger, *Space-Time Structure, Reprinted with Corrections* (Cambridge University Press, London, UK, 1960)
79. J.A. Schouten, *Ricci Calculus*, 2nd edn. (Springer, Berlin, 1954)
80. J.A. Schouten, *Tensor Analysis for Physicists*, 2nd edn. (Dover, Mineola, NY, 1989)
81. F.W. Hehl, Yu.N. Obukhov, Élie Cartan's torsion in geometry and in field theory, an essay. Ann de la Fond. Louis de Broglie **32**, 157–194 (2007). arXiv:0711.1535
82. P. Von der Heyde, The field equations of the Poincaré gauge theory of gravitation. Phys. Lett. A **58**, 141–143 (1976)
83. F.W. Hehl, P. von der Heyde, G.D. Kerlick, J.M. Nester, General relativity with spin and torsion: foundations and prospects. Rev. Mod. Phys. **48**, 393–416 (1976)
84. F.W. Hehl, J. Nitsch, P. von der Heyde, Gravitation and Poincaré gauge field theory with quadratic Lagrangian, in *General relativity and Gravitation—One Hundred Years after the Birth of Albert Einstein*, ed. by A. Held, vol. 1 (Plenum Press, New York, 1980), pp. 329–355
85. R. Hojman, C. Mukku, W.A. Sayed, Parity violation in metric torsion theories of gravitation. Phys. Rev. D **22**, 1915–1921 (1980)
86. S. Holst, Barbero's Hamiltonian derived from a generalized Hilbert-Palatini action. Phys. Rev. D **53**, 5966–5969 (1996)
87. J. Boos, F.W. Hehl, Gravity-induced four-fermion contact interaction implies gravitational intermediate W and Z type gauge bosons. Int. J. Theor. Phys. **56**, 751 (2017). arXiv:1606.09273
88. D. Diakonov, A.G. Tumanov, A.A. Vladimirov, Low-energy general relativity with torsion: a systematic derivative expansion. Phys. Rev. D **84**, 124042 (2011)
89. P. Baekler, F.W. Hehl, Beyond Einstein-Cartan gravity: quadratic torsion and curvature invariants with even and odd parity including all boundary terms. Class. Quantum Grav. **28**, 215017 (2011). arXiv:1105.3504
90. P. Baekler, F.W. Hehl, J.M. Nester, Poincaré gauge theory of gravity: Friedman cosmology with even and odd parity modes: analytic part. Phys. Rev. D **83**, 024001 (2011)
91. H. Chen, F.-H. Ho, J.M. Nester, C.-H. Wang, H.-J. Yo, Cosmological dynamics with propagating Lorentz connection modes of spin zero. JCAP **10**, 027 (2009)

92. F.H. Ho, J.M. Nester, Poincaré gauge theory with coupled even and odd parity spin-0 modes: cosmological normal modes. Ann. d. Physik (Berlin) **524**, 97–106 (2012)
93. F.H. Ho, J.M. Nester, Poincaré gauge theory with coupled even and odd parity dynamic spin-0 modes: dynamical equations for isotropic Bianchi cosmologies. Int. J. Mod. Phys. D **20**, 2125 (2011)
94. F.H. Ho, H. Chen, J.M. Nester, H.J. Yo, General Poincaré gauge theory cosmology. Chin. J. Phys. **53**, 110109 (2015)
95. G.K. Karananas, The particle spectrum of parity-violating Poincaré gravitational theory. Class. Quantum Grav. **32**, 055012 (2015). arXiv:1411.5613. Corrigendum: Class. Quantum Grav. **32**, 089501 (2015)
96. M. Blagojević, B. Cvetković, General Poincaré gauge theory: Hamiltonian structure and particle spectrum. Phys. Rev. D **98**, 104018 (2018). arXiv:1804.05556
97. I.L. Shapiro, Physical aspects of the space-time torsion. Phys. Repts. **357**, 113–213 (2002)
98. J.A.R. Cembranos, J.G. Valcarcel, New torsion black hole solutions in Poincaré gauge theory. JCAP **01**, 014 (2017)
99. J.A.R. Cembranos, J.G. Valcarcel, Extended Reissner-Nordström solutions sourced by dynamical torsion. Phys. Lett. B **779**, 143–150 (2018)
100. C. Heinicke, F.W. Hehl, Schwarzschild and Kerr solutions of Einstein's field equation—an introduction. Int. J. Mod. Phys. D **24**, 1530006 (2014). arXiv:1503.02172
101. Yu.N. Obukhov, Exact solutions in Poincaré gauge gravity theory. Universe **5(5)**, 127 (2019). arXiv:1905.11906
102. H.T. Nieh, Torsion in gauge theory. Phys. Rev. D **97**, 044027 (2018). arXiv:1712.00901
103. H.T. Nieh, Torsional topological invariants. Phys. Rev. D **98**, 104045 (2018). arXiv:1804.07440
104. A. Einstein, Geometrie und Erfahrung. Sitzungsber. Preuss. Akad. Wiss. Phys.-Math. Klasse **1**, 123–130 (1921)
105. P. von der Heyde, Is gravitation mediated by the torsion of spacetime? Z. Naturf. **31a**, 1725–1726 (1976)
106. P.B. Yasskin, W.R. Stoeger, Propagating equations for test bodies with spin and rotation in theories of gravity with torsion. Phys. Rev. D **21**, 2081–2094 (1980)
107. F.W. Hehl, Y.N. Obukhov, D. Puetzfeld, On Poincaré gauge theory of gravity, its equations of motion, and Gravity Probe B. Phys. Lett. A **377**, 1775–1781 (2013). arXiv:1304.2769
108. Y.N. Obukhov, D. Puetzfeld, Multipolar test body equations of motion in generalized gravity theories. *Fundamental Theories of Physics*, vol. 179 (Springer, Cham, 2015), pp. 67–119. arXiv:1505.01680
109. A. Trautman, Spin and torsion may avert gravitational singularity. Nature **242**, 7–8 (1973)
110. A.V. Minkevich, Generalized cosmological Friedmann equations without gravitational singularity. Phys. Lett. A **80**, 232–234 (1980)
111. A.V. Minkevich, Towards the theory of regular accelerating Universe in Riemann-Cartan space-time. Int. J. Mod. Phys. A **31**, 1641011 (2016)
112. J. Magueijo, T.G. Złośnik, T.W.B. Kibble, Cosmology with a spin. Phys. Rev. D **87**, 063504 (2013)
113. N. Popławski, Big bounce from spin and torsion. Gen. Relat. Gravit. **44**, 1007–1014 (2012)
114. D. Puetzfeld, Status of non-Riemannian cosmology. New Astron. Rev. **49**, 59–64 (2005)
115. H. Zhang, L. Xu, Late-time acceleration and inflation in a Poincaré gauge cosmological model. JCAP **09**, 050 (2019)
116. D. Kranas, C.G. Tsagas, J.D. Barrow, D. Iosifidis, Friedmann-like universes with torsion. Eur. Phys. J. C **79**, 341 (2019). arXiv:1809.10064
117. J.D. Barrow, C.G. Tsagas, G. Fanaras, Friedmann-like universes with torsion: a dynamical system approach Eur. Phys. J. C **79**, 764 (2019)
118. V. Nikiforova, S. Randjbar-Daemi, V. Rubakov, Infrared modified gravity with dynamical torsion. Phys. Rev. D **80**, 124050 (2009)
119. V. Nikiforova, S. Randjbar-Daemi, V. Rubakov, Self-accelerating universe in modified gravity with dynamical torsion. Phys. Rev. D **95**, 024013 (2017)

120. V. Nikiforova, T. Damour, Infrared modified gravity with propagating torsion: instability of torsionfull de Sitter-like solutions. Phys. Rev. D **97**, 124014 (2018)
121. T. Damour, V. Nikiforova, Spherically symmetric solutions in torsion bigravity. Phys. Rev. D **100**, 024065 (2019)
122. J. Struckmeier, D. Vasak, H. Stoecker, Extended canonical field theory of matter and spacetime. Astron. Nachr. **336**, 731 (2015)
123. J. Struckmeier, J. Muench, D. Vasak, J. Kirsch, M. Hanauske, H. Stoecker, Canonical transformation path to gauge theories of gravity. Phys. Rev. D **95**, 124048 (2017)
124. J. Struckmeier, D. Vasak, J. Kirsch, Generic theory of geometrodynamics from Noether's theorem for the Diff(M) symmetry group. In: *Discoveries at the Frontiers of Science*, ed. by J. Kirsch, S. Schramm, J. Steinheimer-Froschauer, H. Stöcker. FIAS Interdisciplinary Science Series (Springer, Cham, Switzerland, 2020), pp. 143–181. arXiv:1807.03000
125. M. Blagojević, B. Cvetković, Entropy in Poincaré gauge theory: Hamiltonian approach. Phys. Rev. D **99**, 104058 (2019)
126. M. Blagojević, B. Cvetković, Hamiltonian approach to black hole entropy: Kerr-like spacetimes. Phys. Rev. D **100**, 044029 (2019)
127. M. Toller, A theory of gravitation covariant under $Sp(4, R)$. arXiv:1706.07470
128. B. Mashhoon, *Nonlocal Gravity* (Oxford University Press, Oxford, UK, 2017)
129. F.W. Hehl, B. Mashhoon, Nonlocal gravity simulates dark matter. Phys. Lett. B **673**, 279–282 (2009). arXiv:0812.1059
130. F.W. Hehl, B. Mashhoon, A formal framework for a nonlocal generalization of Einstein's theory of gravitation. Phys. Rev. D **79**, 064028 (2009). arXiv:0902.0560
131. D. Puetzfeld, Yu.N. Obukhov, F.W. Hehl, Constitutive law of nonlocal gravity. Phys. Rev. D **99**, 104013 (2019). arXiv:1903.04023
132. H.J. Blome, C. Chicone, F.W. Hehl, B. Mashhoon, Nonlocal modification of Newtonian gravity. Phys. Rev. D **81**, 065020 (2010). arXiv:1002.1425

Higgs Field in Cosmology

Christian F. Steinwachs

Abstract The accelerated expansion of the early universe is an integral part of modern cosmology and dynamically realized by the mechanism of inflation. The simplest theoretical description of the inflationary paradigm is based on the assumption of an additional propagating scalar degree of freedom which drives inflation—the inflaton. In most models of inflation, the fundamental nature of the inflaton remains unexplained. In the model of Higgs inflation, the inflaton is identified with the Standard Model Higgs boson and connects cosmology with elementary particle physics. A characteristic feature of this model is a non-minimal coupling of the Higgs boson to gravity. I review and discuss several phenomenological and fundamental aspects of this model, including the impact of quantum corrections and the renormalization group, the derivation of initial conditions for Higgs inflation in a quantum cosmological framework and the classical and quantum equivalence of different field parametrizations.

1 Introduction

Hermann Weyl's unique style, weaving together physical ideas with mathematical rigour in an encompassing philosophical framework, has led to many important contributions that still play a major role in theoretical physics today. Several aspects discussed in this contribution are strongly influenced by his ideas on scale invariance, Weyl transformations and on the "geometrization of physics". Back in 1918 when the first edition of *Raum-Zeit-Materie* appeared only three years after Einstein has presented the final form of his theory of General Relativity, Weyl's natural intention was a geometrical unification of gravity and electromagnetism. In this contribution, I discuss a somewhat different kind of unification: the combination of elementary particle physics with cosmology in one unified model. On the one hand, fundamental

C. F. Steinwachs (✉)
Department of Physics, University of Freiburg, Hermann-Herder-Str. 3, 79104 Freiburg, Germany
e-mail: christian.steinwachs@physik.uni-freiburg.de

S. De Bianchi and C. Kiefer (eds.), *One Hundred Years of Gauge Theory*, Fundamental Theories of Physics 199, https://doi.org/10.1007/978-3-030-51197-5_11

physics at the electroweak (EW) scale is well described by the Standard Model (SM) of particle physics. On the other hand, the cosmic history, including a quasi de Sitter stage of the early Universe, is well described by the standard model of cosmology, which incorporates the mechanism of inflation. In both models a scalar field plays a central role: the SM Higgs boson, which provides the masses to the elementary particles, as well as the cosmological inflaton, which drives the accelerated expansion of the early universe. In this contribution, I discuss the theoretical and observational consequences of the idea that the Higgs boson and the inflaton are one and the same particle.

In Sect. 2, I review the formalism of slow-roll inflation. In Sect. 3, I provide the required details about the SM which are relevant for Higgs inflation. In Sect. 4, I discuss the model of Higgs inflation. First, I show why a simple identification of the SM Higgs boson with the inflaton fails. Then, I demonstrate how a non-minimal coupling of the Higgs boson to gravity can improve the situation. I emphasize the importance of quantum corrections, both from a fundamental as well as from a phenomenological perspective and discuss the aspects related to the renormalization group (RG) improvement of the model. Finally, I show how the initial conditions for the model of non-minimal Higgs inflation can be derived from quantum cosmological considerations in a self-consistent way. In Sect. 5, I discuss the classical and quantum equivalence of different field parametrizations in cosmological models and a field covariant geometric construction of the quantum effective action. I conclude in Sect. 6 with a short summary of the discussion.

2 Inflation

In this section, I briefly summarize the essential assumptions and observable consequences of inflation. I discuss the dynamics of the homogeneous and isotropic background and the observational implications following from the linear perturbations propagating on this background.

2.1 Background Dynamics

Observations support the underlying theoretical assumption that on large scales ($\gtrsim$ 200 Mpc), the universe is spatially flat and statistically homogeneous and isotropic, leading to the Friedman-Lemaître-Robertson-Walker (FLRW) line-element

$$\mathrm{d}s^2 = -\mathrm{d}t^2 + a^2(t)\delta_{ij}\mathrm{d}x^i\mathrm{d}x^j. \tag{1}$$

Here, t and x^i, $i = 1, 2, 3$ are comoving coordinates and $a(t)$ is the scale factor. A consistent explanation of the cosmic history requires a *quasi de Sitter stage*, in which the early universe underwent a phase of accelerated expansion—inflation. In most theoretical models, the inflationary stage is realized dynamically by one or more scalar fields. In contrast to a de Sitter stage, corresponding to a cosmological constant Λ, the dynamics of the scalar inflaton field allows inflation to last sufficiently long to explain the high degree of isotropy of the Cosmic Microwave Background (CMB) radiation, and, at the same time, to end the stage of accelerated expansion, necessary to explain the formation of structure we observe today, see [3, 8, 69, 70, 73, 96, 106, 107, 128–130]. In its simplest incarnation, inflation is described by a single minimally coupled scalar field, the *inflaton* φ with the action

$$S[g,\varphi] = \int \mathrm{d}^4x\sqrt{-g}\left[\frac{M_{\mathrm{P}}^2}{2}R - \frac{1}{2}\partial_\mu\varphi\partial^\mu\varphi - V(\varphi)\right]. \tag{2}$$

Here, $M_{\mathrm{P}} = 1/\sqrt{8\pi G_{\mathrm{N}}} \approx 2.4 \times 10^{18}$ GeV (in units $c = \hbar = 1$) is the reduced Planck mass, $g = \det(g_{\mu\nu})$ the determinant of the metric field $g_{\mu\nu}$, R the Ricci scalar and $V(\varphi)$ the scalar field potential. For the symmetric FLRW background (1) and a homogeneous scalar field $\varphi(t)$, the field equations for $g_{\mu\nu}$ and φ reduce to

$$H^2 = \frac{1}{3M_{\mathrm{P}}^2}\left(\frac{1}{2}\dot{\varphi}^2 + V\right), \qquad \ddot{\varphi} + 3H\dot{\varphi} + V_{,\varphi} = 0. \tag{3}$$

Here $H(t) := \dot{a}(t)/a(t)$ is the Hubble parameter and the dot denotes a derivative with respect to Friedman time t. A sufficiently long phase of inflation is realized for a slowly rolling inflaton $\ddot{\varphi}/H^2 \ll \dot{\varphi}^2/H^2 \ll 1$, for which the equations (3) reduce to

$$H^2 \approx \frac{1}{3}\frac{V}{M_{\mathrm{P}}^2}, \qquad 3H\dot{\varphi} \approx -V_{,\varphi}. \tag{4}$$

In the slow-roll approximation, the inflationary dynamics is largely dictated by the properties of the inflaton potential $V(\varphi)$. This motivates the introduction of the dimensionless slow-roll parameters which measure the slope and the curvature of V,

$$\epsilon_{\mathrm{v}} := \frac{M_{\mathrm{P}}^2}{2}\left(\frac{V_{,\varphi}}{V}\right)^2, \qquad \eta_{\mathrm{v}} := M_{\mathrm{P}}^2\left(\frac{V_{,\varphi\varphi}}{V}\right). \tag{5}$$

During slow-roll inflation, $\epsilon_{\mathrm{v}} \ll 1$ and $|\eta_{\mathrm{v}}| \ll 1$ quantify the deviation from de Sitter space ($\epsilon_{\mathrm{v}} = \eta_{\mathrm{v}} = 0$). More precisely, $\varepsilon_{\mathrm{v}} < 1$ is required for inflation, i.e. accelerated expansion $\ddot{a} > 0$, to be realized in the first place and $|\eta_{\mathrm{v}}| \ll 1$ ensures that the inflationary phase lasts sufficiently long to solve the horizon problem.

2.2 Perturbations and Inflationary Observables

Inflation amplifies the quantized inhomogeneous perturbations around the FLRW background, which provide the seeds for the density perturbations that clump under the influence of gravity and ultimately give rise to the CMB and the structure we observe today. The main inflationary observables are the power spectra of the scalar and tensor perturbations, which, due to their weak logarithmic k-dependence, are parametrized by the power law ansatz, see e. g. [1],

$$\mathcal{P}_{\rm t} := A_{\rm t}\left(\frac{k}{k_*}\right)^{n_{\rm t}+\cdots}, \qquad \mathcal{P}_{\rm s} := A_{\rm s}\left(\frac{k}{k_*}\right)^{n_{\rm s}-1+\cdots}. \tag{6}$$

The reference scale k_* in the observable window $10^{-4}\ \text{Mpc}^{-1} \leq k_* \leq 10^{-1}\ \text{Mpc}^{-1}$, first crosses the horizon at the moment $k_* = a_*\, H_*$ chosen to correspond to $N = 60$, where $N := \ln a$ is the number of e-folds. The tensor and scalar amplitudes $A_{\rm t}$ and $A_{\rm s}$ characterize the strengths of the power spectra, while the tensor and scalar spectral indices $n_{\rm t}$ and $n_{\rm s}$ characterize their tilts, i.e. their weak scale dependence. The ellipsis indicate higher order terms in the expansion, which I neglect. To first order in the slow-roll approximation, these quantities are assumed to be constant and can be expressed in terms of V, $\epsilon_{\rm v}$ and $\eta_{\rm v}$,

$$A_{\rm t} = \frac{2\,V}{3\,\pi^2\,M_{\rm P}^4}, \qquad A_{\rm s} = \frac{V}{24\,\pi^2\,M_{\rm P}^4\,\epsilon_{\rm v}}, \tag{7}$$

$$n_t = -\,2\,\epsilon_{\rm v}, \qquad n_{\rm s} = 1 + 2\,\eta_{\rm v} - 6\,\epsilon_{\rm v}\,. \tag{8}$$

All observables (7) and (8) are to be evaluated at φ_*, which can be expressed in terms of the number of e-folds N via the integral relation

$$N_* = \int_{t_*}^{t_{\rm end}} {\rm d}t\, H \simeq \int_{\varphi_{\rm end}}^{\varphi_*} \frac{{\rm d}\varphi}{M_P^2}\frac{V}{V_{,\varphi}}\,. \tag{9}$$

The value $\varphi_{\rm end}$ is defined by the breakdown of the slow-roll approximation

$$\epsilon_{\rm v}(\varphi_{\rm end}) := 1\,. \tag{10}$$

To first order in the slow-roll approximation, the scalar-to-tensor ratio is given by

$$r := \frac{A_{\rm t}}{A_{\rm s}} = 16\,\epsilon_{\rm v} = -8\,n_{\rm t}\,. \tag{11}$$

The last equality is a consistency equation valid in single field models. There only exists an observational upper bound on r, since the tensor power spectrum has not been measured. Recent observational constraints from the CMB at $k_* = 0.05\ \text{Mpc}^{-1}$ are provided in [2],

$$A_s^* = (2.099 \pm 0.014) \times 10^{-9} \qquad 68\%\ \text{CL}, \tag{12}$$

$$n_{s,*} = 0.9649 \pm 0.0042 \qquad 68\%\ \text{CL}, \tag{13}$$

$$r_* < 0.11 \qquad 95\%\ \text{CL}. \tag{14}$$

For any consistent inflationary model, these observational constraints have to be in agreement with the predictions for (7) and (8).

3 Standard Model of Particle Physics

The Higgs boson h is an integral part of the SM of particle physics and provides a mechanism by which the SM particles acquire their mass. The Higgs field is a complex $SU(2)$ doublet Φ, which in the real parametrization consists of the radial massive Higgs component h and the three angular Goldstone bosons θ_i, $i = 1, 2, 3$. So far, the Higgs boson is the only fundamental scalar particle which has been detected. It is an interesting question whether there are more (maybe many more) scalar fields in nature. The Higgs sector of the SM is described by the Lagrangian density

$$\mathcal{L}_{\text{Higgs}}^{\text{SM}} = -\frac{1}{2}|\partial\Phi|^2 - \frac{\lambda}{4}\left(|\Phi|^2 - v^2\right)^2, \qquad |\Phi|^2 = \Phi^\dagger\Phi. \tag{15}$$

Here λ is the quartic Higgs self-coupling and $v \approx 246$ GeV the EW scale. In unitary gauge, with $|\Phi|^2 = h^2$, the Lagrangian describing the non-derivative interaction of the Higgs boson h with the other SM particles schematically reads

$$\mathcal{L}_{\text{int}}^{\text{SM}} = -\sum_\chi \frac{1}{2}\lambda_\chi \chi^2 h^2 - \sum_A \frac{1}{2} g_A^2 A_\mu^2 h^2 - \sum_\psi y_\psi \bar{\psi}\psi h. \tag{16}$$

The sum extends over scalar fields χ, vector gauge fields A_μ and Dirac spinors ψ with the corresponding scalar, gauge and Yukawa couplings λ_χ, g_A and y_ψ. The interaction sector is dominated by the heaviest particles in the SM: the Yukawa top-quark, the Higgs boson, the Z boson and the $W^\pm$ bosons, with masses

$$M_t^2 = \frac{1}{2} y_t^2 h^2, \qquad M_h^2 = 2\lambda h^2, \qquad M_Z^2 = \frac{\left(g + g'\right)^2}{4} h^2, \qquad M_W^2 = \frac{1}{4} g^2 h^2. \tag{17}$$

Data from collider experiments constrain these masses, see [138],

$$M_t = 173.0 \pm 0.4\ \text{GeV}, \qquad M_h = 125.18 \pm 0.16\ \text{GeV}, \tag{18}$$

$$M_Z = 91.1876 \pm 0.0021\ \text{GeV}, \qquad M_W = 80.379 \pm 0.012\ \text{GeV}. \tag{19}$$

The values for (17) and (18) constrain the quartic Higgs self-coupling $\lambda \approx 0.1$ at the EW scale $h \simeq v$.

4 Higgs Inflation

The basic idea of Higgs inflation is to identify the SM Higgs boson h with the cosmic inflaton φ, thereby establishing a direct connection between elementary particle physics and inflationary cosmology,

$$h \equiv \varphi. \tag{20}$$

Such a unified scenario is not only very appealing from a theoretical point of view, but also very predictive, as it requires to simultaneously match observational constraints from particle physics and cosmology. I first discuss the difference between a minimally and non-minimally coupled SM Higgs to gravity at tree-level and then extend the discussion to the inclusion of important and unavoidable quantum corrections.

4.1 Minimal Higgs Inflation

The most direct approach to construct the unified scenario of Higgs inflation is to embed the SM in curved spacetime and to analyse the inflationary consequences of a SM Higgs boson minimally coupled to gravity. In this case, the graviton-Higgs sector is described by the action

$$S[g, \varphi] = \int \mathrm{d}^4x\sqrt{-g}\left[\frac{M_{\mathrm{P}}^2}{2}R - \frac{1}{2}\partial_\mu\varphi\partial^\mu\varphi - \frac{\lambda}{4}\left(\varphi^2 - v^2\right)^2\right]. \tag{21}$$

The action (21) was already investigated in one of the earliest models of inflation, formulated in [97]. From the point of view of the inflationary slow-roll analysis, (21) leads to the chaotic inflation scenario with the monomial potential (for $\varphi/v \gg 1$),

$$V(\varphi) = \frac{\lambda}{4}\varphi^4. \tag{22}$$

To leading order in $N_* \gg 1$, the model (21) with the potential (22) predicts for the inflationary observables (7), (8), and (11),

$$A_{\mathrm{s}} = \frac{2\lambda N_*^3}{3\pi^2}, \qquad n_{\mathrm{s}} = 1 - \frac{3}{N_*} = 0.95, \qquad r = \frac{16}{N_*} \approx 0.2667. \tag{23}$$

The numerical values are obtained for $N_* = 60$. Combining the observational constraint on A_s given in (12) with the predicted value in (23), directly translates into a constraint for the quartic Higgs self-coupling $\lambda \approx 10^{-13}$. Such a tiny value for λ is clearly incompatible with the value $\lambda \approx 0.1$, required by the observational SM constraint on the Higgs mass (17) and therefore spoils this first approach to identify the inflaton with the SM Higgs boson.

Note, however, that the constraint $\lambda \approx 0.1$ only has to be satisfied at the EW scale $E_{\mathrm{EW}} \approx 10^2$ GeV, while at the inflationary energy scale $E_{\mathrm{inf}} \approx 10^{15}$ GeV, λ might attain different values. As I discuss in Sect. 4.3, the RG flow of the SM drives the running $\lambda(t)$ to very small values at high energy scales. But even if values as small as $\lambda(t_{\mathrm{inf}}) \approx 10^{-13}$ could be attained dynamically by the RG flow, such that the CMB normalization condition (12) would be satisfied at the energy scale of inflation, at the same time, the RG corrections would also have to improve the situation with the spectral observables, as the tree-level chaotic inflationary model with potential (22) predicts a scalar spectral index and a tensor-to-scalar ratio (23) incompatible with the observational constraints (13) and (14).

4.2 *Non-minimal Higgs Inflation*

The central assumption for a successful identification of the inflaton with the SM Higgs boson is to include a non-minimal coupling of φ to gravity.

Early ideas to incorporate a non-minimal coupling of an abstract inflaton field to gravity were formulated in [58, 117] in order to improve the situation with the observational constraints on λ and on the spectral observables (23). The identification of a non-minimally coupled inflaton with the SM Higgs boson was proposed in [21]. Independent of its phenomenological impact on the observational constraints, a non-minimal coupling might be motivated by several theoretical reasons: First, a non-minimal coupling might, to some extent, be viewed as incorporating the Machian idea of a variational gravitational constant. Second, its presence is required for technical reasons. Even in the absence of a non-minimal coupling, already the first quantum corrections for a self-interacting scalar field induce a non-minimal coupling term and the consistency of the renormalization procedure requires that this term must be included in the action. Third, from an effective field theory point of view, the non-minimal coupling term in the action corresponds to a marginal operator, which is on equal footing with the Einstein-Hilbert term in a derivative expansion and should therefore be included in the defining low energy limit of the theory. Fourth, the inclusion of such a term leads to an asymptotic scale invariance for large values of the scalar field, which realizes inflation in a natural way and ends it when the scale invariance is explicitly broken by the Einstein-Hilbert operator. Fifth, in the context of effective string theory inspired models, a non-minimal coupling unavoidably arises in the form of a dilaton or moduli field. Irrespectively of these theoretical motivations, in the following sections, I discuss the phenomenological consequences of a non-minimal coupling in the context of Higgs inflation, for which the action of the graviton-Higgs sector acquires an additional non-minimal coupling term,

$$S[g,\varphi] = \int \mathrm{d}^4x\sqrt{-g}\left[\frac{1}{2}\left(M_{\mathrm{P}}^2 + \xi\varphi^2\right) R - \frac{1}{2}\partial_\mu\varphi\partial^\mu\varphi - \frac{\lambda}{4}\left(\varphi^2 - v^2\right)^2\right]. \quad (24)$$

The formalism of Sect. 2 is directly applicable by performing the field redefinitions

$$\hat{g}_{\mu\nu} = \left(1 + \xi\frac{\varphi^2}{M_{\mathrm{P}}^2}\right) g_{\mu\nu}, \qquad \left(\frac{\partial\hat{\varphi}}{\partial\varphi}\right) = \frac{1 + \xi\left(1 + 6\xi\right)\frac{\varphi^2}{M_{\mathrm{P}}^2}}{\left(1 + \xi\frac{\varphi^2}{M_{\mathrm{P}}^2}\right)^2}. \quad (25)$$

In this way, the action (24), originally formulated in the Jordan frame (JF) variables $(g_{\mu\nu}, \varphi)$, is mapped to the action in the Einstein frame (EF) variables $(\hat{g}_{\mu\nu}, \hat{\varphi})$,[1]

$$\hat{S}[\hat{g},\hat{\varphi}] = \int \mathrm{d}^4x\sqrt{-\hat{g}}\left[\frac{1}{2}M_{\mathrm{P}}^2\hat{R} - \frac{1}{2}\hat{g}^{\mu\nu}\partial_\mu\hat{\varphi}\partial_\nu\hat{\varphi} - \hat{V}(\hat{\varphi})\right], \quad (26)$$

with the EF potential defined as

$$\hat{V}(\hat{\varphi}) := \frac{M_{\mathrm{P}}^4\lambda}{4\xi^2}\left(1 - e^{-\sqrt{\frac{2}{3}}\frac{\hat{\varphi}}{M_{\mathrm{P}}}}\right)^2. \quad (27)$$

The main purpose of this transformation is to remove the non-minimal coupling by a Weyl transformation of the metric field $g_{\mu\nu} \rightarrow \hat{g}_{\mu\nu} = \Omega^2(\varphi)g_{\mu\nu}$, with a field-dependent conformal factor $\Omega^2(\varphi)$, explicitly given in the first equation of (25). Since the derivatives of $\Omega(\varphi)$, which arise in this transformation, also induce a contribution to the kinetic term of the scalar field, a reparametrization $\varphi \rightarrow \hat{\varphi}$ is required to obtain a canonically normalized kinetic term. Ultimately, in this way the complexity associated with the non-minimal coupling is shifted to the scalar potential. While the EF action formally resembles that of the minimally coupled scalar field, the potentials (27) and (22) differ and matter fields also feel the coupling to the scalar field-dependent EF metric.[2]

Applying the inflationary formalism of Sect. 2 to the EF action (26), to leading order in $\hat{N}_* \gg 1$, the inflationary observables for the potential (27) are given by

$$\hat{A}_{\mathrm{s}}^* = \frac{\hat{N}_*^2}{72\pi^2}\frac{\lambda}{\xi^2}, \qquad \hat{n}_{\mathrm{s}}^* = 1 - \frac{2}{\hat{N}_*} \approx 0.9667, \qquad \hat{r}^* = \frac{12}{\hat{N}_*^2} \approx 0.0033. \quad (28)$$

[1] Since the form of the EF action (26) formally resembles that of the Einstein-Hilbert action (21), the formulation in terms of the variables $(\hat{g}_{\mu\nu}, \hat{\varphi})$ is called "Einstein frame". The formulation in terms of the "Jordan frame" variables $(g_{\mu\nu}, \varphi)$, for which the scalar-tensor character of the action is manifest, derives from the early work of Pascual Jordan on such models.

[2] Note however, that matter fields which do not directly couple to φ and whose action is invariant under Weyl transformations are insensitive to the transformation (25)—at least at the classical level.

The numerical values are again presented for $N_* = 60$. Comparing the potential (22) of the minimally coupled action (21) with the EF potential (27), there are two main effects which improve the situation with the observational constraints: First, for the non-minimally coupled Higgs boson, the normalization of the EF potential (27) depends on the ratio λ/ξ^2 in contrast to the pure λ dependence for the minimally coupled Higgs boson. Therefore, by making ξ sufficiently large ($\xi \approx 10^4$), the quartic coupling can be tuned to $\lambda \approx 0.1$, such that the CMB constraint (12) as well as the Higgs mass constraint (18) are satisfied simultaneously. The second effect is that, in contrast to the quartic chaotic inflation potential (22), the EF potential (27) becomes field independent and almost flat for large field values of $\hat{\varphi}$, thereby improving the situation with the constraints on the spectral observables in (28)—both the spectral index as well as the tensor-to-scalar ratio are in perfect agreement with the observed value (13) and the observational bound (14).

4.3 Quantum Corrections and the Renormalization Group

When identifying the non-minimally coupled SM Higgs boson with the inflaton at tree-level, the main compatibility requirement is that the observational restrictions on the CMB normalization (12) and the Higgs mass (18) are satisfied simultaneously, requiring $\xi \approx 10^4$ and $\lambda \approx 10^{-1}$. As can be seen from (28), the scalar-tensor sector of the non-minimally coupled Higgs field belongs to a more general class of inflationary models, for which the predictions of the scalar spectral index and the tensor-to-scalar ratio are independent of the model parameters. Therefore, provided the constraints (12) and (18) are satisfied, the inflationary predictions of the tree-level non-minimal Higgs inflation model are insensitive to the detailed properties of the SM particles.

It is clear that such a tree-level consideration is incomplete and cannot be correct. The quantum loop corrections of the SM particles have to be taken into account. In particular, the quantum corrections to the EF effective potential, which are dominated by the heaviest particles of the SM, ultimately induce a dependence of the spectral observables n_s and r on the particle content of the SM. In [14], the one-loop effective Coleman-Weinberg potential in the EF (expressed in terms of the JF field φ) was obtained as

$$\hat{V} = \frac{\lambda M_P^4}{4\xi^2}\left(1 - \frac{2M_P^2}{\xi\varphi^2} + \frac{\mathbf{A}_I}{16\pi^2}\ln\frac{\varphi}{\mu_0}\right), \tag{29}$$

with arbitrary renormalization point μ_0 and the inflationary anomalous scaling

$$\mathbf{A}_I = \mathbf{A} - 12\lambda = \frac{3}{8\lambda}\left[2g^4 + \left(g^2 + g'^2\right)^2 - y_t^4\right] - 6\lambda. \tag{30}$$

Here g and g' are the EW gauge couplings and y_t the Yukawa top-quark coupling, cf. (17). As shown in [14], the impact of these quantum corrections leads to essential modifications of the shape of the inflationary potential, as during inflation $\varphi \gg M_{\mathrm{P}}/\sqrt{\xi}$, the second term in (29) is negligible and the logarithmic quantum corrections dominate over the flat tree-level part. The quantum contribution can be parametrized by the dimensionless quantity

$$x := \frac{\hat{N}\mathbf{A}_I}{48\pi^2}, \tag{31}$$

which enters the inflationary observables (28) in the form of correction factors

$$\hat{A}_{\mathrm{s}}^* = \frac{\hat{N}_*^2}{72\pi^2}\frac{\lambda}{\xi^2}\left(\frac{e^x-1}{xe^x}\right)^2, \qquad \hat{n}_{\mathrm{s}}^* = 1 - \frac{2}{\hat{N}_*}\frac{x}{e^x-1}, \qquad \hat{r}^* = \frac{12}{\hat{N}_*^2}\left(\frac{xe^x}{e^x-1}\right). \tag{32}$$

As demonstrated in [14], the impact of these quantum corrections could render the identification of the SM Higgs with the inflaton invalid.

However, taking into account the quantum corrections at the EW scale is not sufficient. The coupling constants depend on the energy scale of the underlying physical process and their change is determined by the system of RG equations

$$\frac{\mathrm{d}g_i}{\mathrm{d}t} = \beta_i, \qquad \frac{\mathrm{d}Z}{\mathrm{d}t} = \gamma Z. \tag{33}$$

The beta functions β_i of the couplings $g_i = \{\lambda,\ y_{\mathrm{t}},\ \xi,\ g,\ g',\ g_{\mathrm{s}}\}$ include the quartic Higgs self-coupling λ, the Yukawa top-quark coupling y_t, the non-minimal coupling ξ, the EW gauge couplings g, g' and the strong gauge coupling g_{s}, which all depend on the logarithmic RG scale $t := \ln(\varphi/M_{\mathrm{t}})$. The arbitrary renormalization point μ_0 has been fixed to match the highest mass scale M_{t} in the SM. In addition, the wave function renormalization Z of the Higgs boson determined by its anomalous dimension γ has been taken into account. However, since it was found in [11] that the running of Z is very slow, I neglect the running of Z and assume $Z = 1$ in the following discussion. Although the dimensionless couplings g_i only change logarithmically with the energy scale, the effects of the RG improvement become sizeable as the EW scale and the inflationary scale are separated by many orders of magnitude, as illustrated in Fig. 1.

The dominant one-loop RG improvement of the model was investigated in [11, 20, 42] and shown to be the essential mechanism by which the Higgs inflation

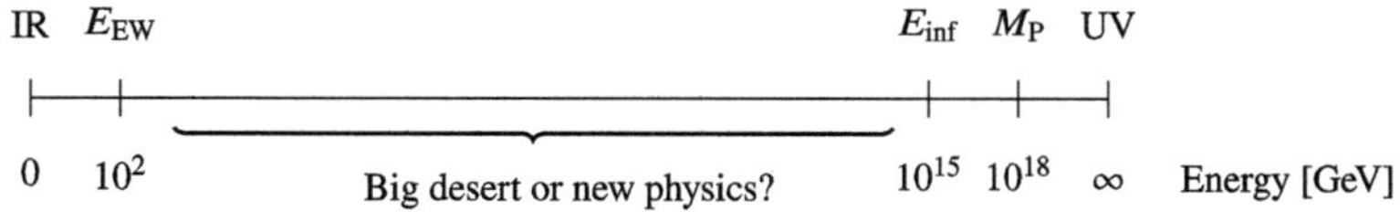

Fig. 1 Different energy scales and their connection to the model of non-minimal Higgs inflation

compatibility constraints can be satisfied. The subleading two-loop contributions to the running, first considered in [23], are significant and, compared to the one-loop running, reduce the bound of the cosmologically compatible Higgs mass about 10 GeV down to the observed value (18).[3]

In general, the functional shape of the tree-level EF potential (27) is changed by the RG improvement, which in turn leads to modified predictions for the inflationary observables (28). Moreover, since the RG flow of the SM is very sensitive to the initial conditions of $\lambda(t)$ and $y_t(t)$ at the EW scale $t_{\rm EW} \approx 0$, and since these initial conditions are related to the observed masses $M_{\rm h}$ and $M_{\rm t}$, the RG improved spectral observables can induce a strong dependence on the precise values of these masses and therefore on the details of the SM.

The non-minimal coupling in the JF directly couples the Higgs boson h to derivatives of the metric field and thereby mixes gravitational and Higgs degrees of freedom. In contrast, the field content of the graviton-scalar sector in the EF is diagonal and leads to a Higgs propagator

$$\langle h(x), h(0)\rangle \propto \frac{s(\varphi)}{\hat{g}^{1/2}\left(\hat{\Box} - M_{\rm h}^2\right)}. \tag{34}$$

In [11, 42], it was demonstrated that the impact of the non-minimal coupling on the SM beta functions can effectively be incorporated by weighting each internal Higgs propagator in the corresponding Feynman diagrams by a power of the suppression function

$$s(\varphi) = \frac{U}{GU + 3(U_1)^2} \simeq \begin{cases} 1 & \text{for } \varphi \ll \frac{M_{\rm P}}{\sqrt{\xi}}, \\ \frac{1}{6\xi} & \text{for } \varphi \gg \frac{M_{\rm P}}{\sqrt{\xi}}. \end{cases} \tag{35}$$

Here, the functions $U(\varphi)$ and $G(\varphi)$ are defined in (65) below, and, upon comparison with the model of non-minimal Higgs inflation (24), are given by $U = (M_{\rm P}^2 + \xi\varphi^2)/2$ and $G(\varphi) = 1$. Since for large field values $\varphi \gg M_{\rm P}/\sqrt{\xi}$ the suppression function (35) behaves like $s \approx 1/6\xi$, a large non-minimal coupling $\xi \approx 10^4$ leads to a strong suppression of the Higgs contributions in the beta functions. Prior to the Higgs discovery, the Higgs mass was expected to lie in the interval 118 GeV $\lesssim M_{\rm h} \lesssim$ 180 GeV. The suppression phenomenon was termed "asymptotic freedom" in [11], as the suppression of Higgs contribution in the beta function β_λ essentially prevents $\lambda(t)$ to run into a Landau pole before the energy scale of inflation and allows a perturbative treatment up to the inflationary energy scale—even for large Higgs mass values.

From Fig. 2, it is clear that the suppression is strongest for large Higgs masses. Since the discovery of a light Higgs with $M_{\rm h} \approx 125$ GeV, the suppression mechanism

[3] A non-perturbative treatment in the context of the asymptotic safety paradigm, which includes the running of Newton's constant, has been investigated in [125]. Imposing that λ and β_λ vanish at the Planck scale and evolving the flow towards the IR leads to a Higgs mass prediction of $M_{\rm h} \approx 126$ GeV.

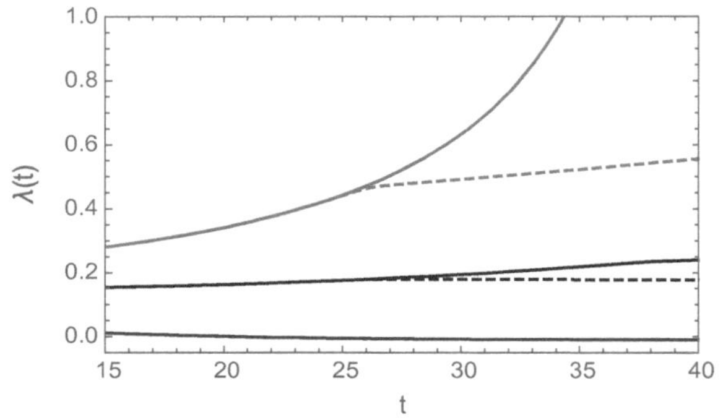

Fig. 2 The RG running of $\lambda(t)$ for fixed top-quark mass $M_t = 173$ GeV and different Higgs masses $M_h = 180$ GeV (gray), $M_h = 160$ GeV (purple) and $M_h = 125$ GeV (blue) including the effect of the non-minimal coupling due to the Higgs-propagator weighting with the suppression function $s(\varphi)$ (dashed) and without the Higgs-propagator weighting (solid). For large Higgs masses, the unsuppressed beta functions would drive $\lambda(t)$ into a Landau pole for scales below the energy scale of inflation $t < t_{\text{inf}}$, c.f. the gray solid line. The numerically integrated two-loop beta functions with the weighting were taken from [4] and those without weighting from the two-loop truncation of the beta functions in the appendix of [34]

is no longer that relevant.[4] Instead, the RG flow of the SM drives the running $\lambda(t)$ to very small values at high energy scales and λ-dependent contributions in the beta functions are anyway small. Therefore, the influence of the non-minimal coupling ξ on the SM beta functions is very weak. Moreover, given that the running of the non-minimal coupling ξ itself is rather slow, I neglect its running $\beta_\xi \approx 0$ and consider only the system of unmodified pure SM beta functions, shown in Fig. 3, in the following discussion.

Even if the running of the non-minimal coupling and its impact on the running of the SM beta functions is rather mild for a light SM Higgs boson, the presence of the non-minimal coupling is nevertheless crucial, as it ensures that the EF potential (27) is almost field independent (i.e. flat) with the overall normalization factor $\lambda(t)/\xi^2$. Before I address the (in)stability of the RG improved EF potential, I briefly discuss three qualitatively different scenarios under the assumption that $\lambda(t) > 0$ for all t. For a light Higgs boson, the RG flow of the SM drives $\lambda(t)$ to very small values at high energies. At the same time its flow is also very slow at high energies, $\beta_\lambda \ll 1$. Moreover, $\lambda(t)$ develops a minimum $\lambda_0 := \lambda(t_0)$ at t_0, defined by $\partial_t \lambda(t)|_{t=t_0} = \beta_\lambda(t)|_{t=t_0} = 0$. Hence, the running in the vicinity of λ_0 might be described by the Taylor expansion

[4]It might however become relevant for strong non-minimal couplings $\xi \gg 10^4$ arising e.g. in induced inflation—not because of the stronger suppression with $s = 1/6\xi$, but because the scale of the regime $M_P/\xi \lesssim \varphi \lesssim M_P/\sqrt{\xi}$, during which the suppression mechanism becomes effective, is essentially lowered for large ξ and might therefore lead to a strong suppression of λ-dependent terms already at energy scales at which the RG flow has not yet driven λ to very small values $\lambda \ll 0.1$.

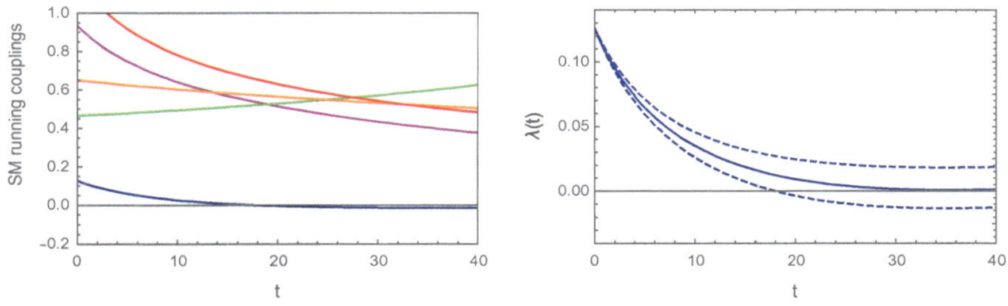

Fig. 3 Left: The pure SM running of the quartic Higgs self-coupling λ (blue), the Yukawa top-quark coupling y_t (pink) and the EW and strong gauge couplings g (green line), g' (orange line) and g_s (red line). Right: Zoomed in plot of the pure SM running of the quartic Higgs self-coupling λ for fixed Higgs mass $M_{\rm h} = 125$ GeV and three different values of the Yukawa top quark mass $M_{\rm t} = 168$ GeV (upper blue dashed line), $M_{\rm t} = 170.8$ GeV (middle blue solid line), and $M_{\rm t} = 173$ GeV (lower blue dashed line), illustrating that the Higgs coupling might be driven to negative values, depending on the precise value of $M_{\rm t}$. In both plots the two-loop approximation of the beta functions presented in [34] were numerically integrated with `Mathematica`

$$\lambda(t) = \lambda_0 + \frac{\lambda_2}{(16\pi^2)^2}\, t^2 + \mathcal{O}(t^3). \tag{36}$$

In general, the values of λ_0 and λ_2 are functions of the SM input and predominately depend on the values of $M_{\rm h}$ and $M_{\rm t}$. However, numerical integration of the RG flow reveals that λ_2 is rather insensitive to changes in $M_{\rm h}$ and $M_{\rm t}$ and is well approximated by a constant $\lambda_2/(16\pi^2)^2 \approx 4 \times 10^{-5}$. In contrast, the value of λ_0 varies between 10^{-2} and 10^{-6} and its dependence on $M_{\rm h}$ and $M_{\rm t}$ can be parametrized by a fitting formula as e.g. discussed in [24, 71]; see also the discussion in the recent review article of [113]. The value t_0 is related to a value $t_{\rm crit}$ at which the RG improved effective EF potential has an inflection point $\partial_t^2 \hat{V}(t)|_{t=t_{\rm crit}} := 0$. There are three qualitatively different scenarios, depending on the sign of the slope $\partial_t \hat{V}(t)|_{t=t_{\rm crit}}$ at that point :

I. **Universal**: $\partial_t \hat{V}(t)|_{t=t_{\rm crit}} \gg 0$

The positive slope of the RG improved EF potential cannot be too large (such as exaggeratedly shown in Fig. 4 for illustrative purposes) and must be cut off by a sufficiently strong non-minimal coupling ξ in order not to spoil the flatness of the potential required for slow-roll inflation. Since the shape of the RG improved potential is almost unchanged compared to the shape of the tree-level EF potential, the RG improved spectral observables (28), which depend on derivatives of the potential, are almost identical to the tree-level predictions (27). Therefore, the cosmological predictions are largely insensitive to the details of the SM in this scenario.

II. **Critical**: $\partial_t \hat{V}(t)|_{t=t_{\rm crit}} \geq 0$

The exact relation $\partial_t \hat{V}(t)|_{t=t_{\rm crit}} = 0$ would lead to a strictly constant plateau of the RG improved EF potential, as shown in Fig. 5, preventing any inflation-

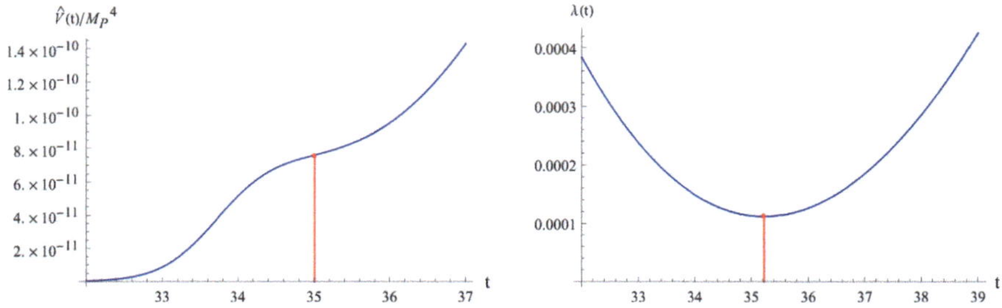

Fig. 4 Left: The RG improved EF potential is a monotonically increasing function. The red line connects t_{crit} with $\partial_t^2 \hat{V}(t)|_{t=t_{\text{crit}}}$. Right: The running of λ, where the red line connects t_0 with λ_0

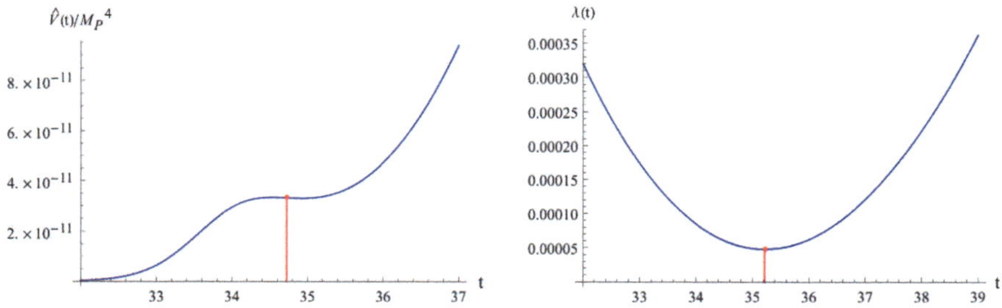

Fig. 5 Left: The inflection point of the RG improved EF potential coincides with its extremum. The red line connects t_{crit} with $\partial_t^2 \hat{V}(t)|_{t=t_{\text{crit}}}$. Right: The RG running of $\lambda(t)$. The red line connects t_0 with the minimum λ_0

ary dynamics. In contrast a slight violation of this strict condition by allowing $\partial_t \hat{V}(t)|_{t=t_{\text{crit}}} \geq 0$ with $\left|\partial_t \hat{V}(t)|_{t=t_{\text{crit}}}\right| \ll 1$ leads to an extremely flat plateau on which (ultra) slow-roll inflation can take place. Since this configuration can only be obtained by a highly fine-tuned combination of parameters M_{h}, M_{t} and ξ, the RG improved cosmological predictions of non-minimal Higgs inflation in the critical regime strongly depend on the details of the SM at the EW scale, in particular on the values of M_{t}. Compared to its value at the EW scale $\lambda_{\text{EW}} \approx 0.1$, the running λ can be as small as $\lambda_{\text{inf}} \approx 10^{-6}$ during inflation, such that the CMB normalization condition (28) for A_{s} allows for a significant smaller $\xi = \mathcal{O}(10)$ as found in [4, 24, 71]. During the (ultra) slow-roll dynamics on the (ultra) flat plateau, the background dynamics of the inflaton field might no longer be dominated by the overall classical slow-roll drift but by quantum fluctuations, a scenario which can be consistently described within the stochastic approach, see [131]. In contrast to the universal regime, the slow-roll parameter $\varepsilon_{\text{V}}(\varphi)$ defined in (5) is no longer a monotonic function of φ, but changes the sign of its slope in accordance with the change of slope of the RG improved EF potential at the inflection point. In particular, the tree-level consistency condition (11) implies that, in contrast to the small universal tree-level prediction $r \approx 3 \times 10^{-3}$, in the critical scenario the tensor-to-scalar ratio r can attain larger values up to

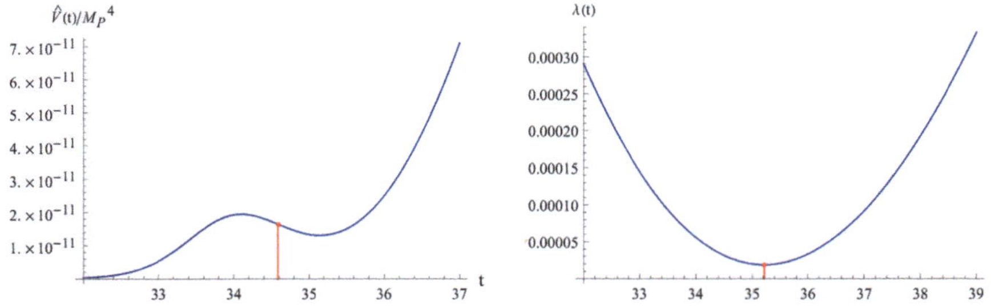

Fig. 6 Left: The RG improved EF potential forms a second minimum at high energies. The red lie connects t_{crit} with the inflection point $\partial_t^2 \hat{V}(t)|_{t=t_{\text{crit}}}$. Right: The RG running of $\lambda(t)$. The red line connects t_0 with the minimum λ_0

$r = \mathcal{O}(10^{-1})$, c.f. [4, 24, 71]. In general, the non-monotonic behaviour of the slow-roll parameter ε_{v} also leads to a rather strong change (k-dependence) of n_{s} where the simple power-law parametrization of the primordial power spectra (6) is no longer appropriate.

III. **Hilltop**: $\partial_t \hat{V}(t)|_{t=t_{\text{crit}}} < 0$

The RG improved EF potential develops a second local minimum at high energy scales, shown in Fig. 6. By lowering the values of λ_0 (increasing M_{t} for fixed M_{h}), the second local minimum can be continuously lowered, up to the point where $\lambda_0 = 0$ and its height degenerates with the EW vacuum. The two minima are separated by a local maximum at which hilltop inflation can take place. Such a behaviour of the RG improved EF potential has been found in [11], see also [53]. In order to realize a successful phase of inflation, it must be ensured that the inflaton field can roll down all the way to the EW vacuum and does not get trapped in the second (false) vacuum when rolling down the hilltop in the opposite direction. In general, this scenario would require a rather strong fine tuning to arrange for the correct initial conditions of inflation. However, as I discuss in Sect. 4.6, the formation of the initial conditions for non-minimal Higgs inflation might be consistently derived from more fundamental quantum cosmological considerations.

4.4 Instability of the Electroweak Vacuum

The stability of the EW vacuum together with the associated restrictions on the SM masses has already been investigated in [5, 6, 57, 63, 81, 127]. The RG flow of the SM is known to high precision and, for the central values of the Higgs mass and the top-quark mass at the EW scale (18), the RG flow of the SM drives λ to negative values $\lambda(t) < 0$ at high energies $t_{\text{inst}} < t < t_{\text{inf}}$, see [16, 18, 34, 43]. As illustrated in

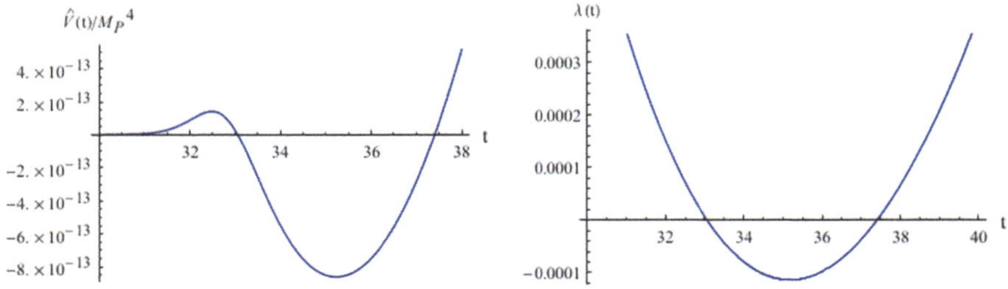

Fig. 7 Left: The RG improved EF potential develops a negative vacuum at high energy scales but below the energy scale of inflation. Right: The running Higgs self-coupling λ turns negative for $t_{\text{inst}} < t < t_{\text{inf}}$

Fig. 7, a negative λ leads to the formation of a negative global minimum at an energy scale below the energy scale of inflation and therefore to a non-stable RG improved EF potential. A scenario which could have disastrous consequences for the universe. Various ways to stabilize the EW vacuum have been proposed, among which are thermal effects discussed in [17], additional heavy scalar fields suggested in [49], the inclusion of higher dimensional operators considered in [27, 47, 66], the extended scalaron-Higgs model analysed in [35, 50–52, 54, 65, 67, 68, 75, 76, 89, 119, 141], or the coupling of a quintessence field to the SM Higgs sector investigated in [72].

Aside from the division of the parameter space into a "stable" regime ($\lambda(t) \geq 0$) and a "non-stable" regime ($\lambda(t_{\text{inst}}) < 0$) with t_{inst} corresponding to an energy scale $E_{\text{inst}} \approx 10^{11}$ GeV, the "non-stable" parameter region can be further subdivided into a "metastable" region and an "unstable" region by calculating the probability Γ_{EW} for the quantum tunnelling from the EW vacuum into the negative false vacuum. Comparing the lifetime of the EW vacuum

$$\tau_{\text{EW}} \sim \Gamma_{\text{EW}}^{-1}, \tag{37}$$

with the lifetime τ_{U} of the universe, metastability (instability) implies $\tau_{\text{EW}} > \tau_{\text{U}}$ ($\tau_{\text{EW}} < \tau_{\text{U}}$). This scenario and its cosmological implications has been studied in many works, see e.g. [28–30, 48, 55, 56, 78, 80, 84, 93–95, 101, 118, 120].

In contrast to the firm and clear distinction between a "stable" region and a "non-stable" region in parameter space, which only depends on the values for the SM couplings at the EW scale and the precision of the perturbatively calculated beta functions, the further subdivision of the non-stable region into a "meta-stable" and an "unstable" region strongly depends on the details of the non-perturbative tunnelling scenario such as e.g. the decay via bubble formation discussed in [40] or via the Hawking-Moss instanton proposed in [74]. Summarizing, various tunnelling probabilities have been calculated which, depending on the concrete realization, lead to very different results. Therefore, the conclusion about the ultimate fate of our universe seems to remain obscure.

4.5 Validity of the Effective Field Theory

When extrapolating the SM as perturbative quantum field theory up to the inflationary energy scale, it is important to ensure that the effective field theory expansion is well under control and does not break down below or at the energy scale of inflation. The validity of the effective field theory depends on whether irrelevant higher-dimensional operators are sufficiently suppressed by the associated cutoff Λ. In the absence of any new physics in between the EW scale and the Planck scale, the natural cutoff in the context of quantum gravity is $\Lambda = \mathrm{M_P}$. However, from a tree-level unitary analysis, it was found in [7, 31, 32] that the cutoff in non-minimal Higgs inflation is essentially lowered to

$$\Lambda = \frac{4\pi M_{\mathrm{P}}}{\xi}. \tag{38}$$

In view of the strong non-minimal coupling $\xi \approx 10^4$, the cutoff (38) corresponds to a significantly lower scale than the typical field values during inflation

$$\varphi_{\mathrm{inf}} \approx \frac{M_{\mathrm{P}}}{\sqrt{\xi}}. \tag{39}$$

If true, this would suggest that the predictions based on the low energy approximation (24) are not valid during inflation, unless an unnatural suppression of higher-dimensional operators is assumed. However, in the context of small gradient terms and large background fields during inflation, the gravitational interaction strength in the JF parametrization (24) is not given by M_{P}, but by the effective Planck mass

$$M_{\mathrm{P}}^{\mathrm{eff}}(\varphi) := \sqrt{M_{\mathrm{P}}^2 + \xi\varphi^2} \geq \sqrt{\xi}\varphi. \tag{40}$$

In [12, 22] it was shown that the cutoff itself is running $\Lambda \to \Lambda(\varphi)$. The power counting method of [31, 32], which leads to (38), remains valid if the Planck mass M_{P} is replaced by $M_{\mathrm{P}}^{\mathrm{eff}}$ and leads to a *running* cutoff, which, in view of (40), is bounded from below by

$$\Lambda(\varphi) = \frac{4\pi\varphi}{\sqrt{\xi}}. \tag{41}$$

This cutoff, which controls the gradient and curvature expansion, can also be derived from the leading contribution to the quadratic one-loop operator R^2 in the large ξ expansion $\sim \xi^2 R^2/(4\pi)^2$ calculated in [11, 133]. Comparing this to the tree-level interaction term $\sim (M_{\mathrm{P}}^2 + \xi\varphi^2)R$, the one-loop contribution is suppressed by a cutoff $\Lambda^2(\varphi)$, in agreement with the running cutoff (41).

As demonstrated in [12], the gradient and curvature expansion with respect to the running cutoff (41) is efficient. Indeed, making use of the JF on-shell relation $R \sim V/U \sim \lambda\varphi^2/\xi$, the curvature expansion runs in powers of

$$\frac{R}{\Lambda^2(\varphi)} \sim \frac{\lambda}{16\pi^2} \ll 1. \tag{42}$$

The smallness in (42) holds in the range in which the SM remains perturbative ($\lambda \ll 1$). Likewise, as shown in [12], the gradient terms is suppressed even stronger

$$\frac{\partial}{\Lambda} \sim \frac{\lambda}{48\pi}\sqrt{\hat{\varepsilon}_{\rm V}}. \tag{43}$$

According to (9), the additional factor $\sqrt{2\hat{\varepsilon}_v}$ scales between $1/\hat{N}$ at the beginning of inflation and $\mathcal{O}(1)$ at the end of it. Having established the efficiency of the curvature and gradient expansion (42) and (43), the only higher dimensional operators which could potentially spoil the truncation (24) of the effective field theory expansion are multi-loop corrections in the form of monomial operators

$$\frac{\varphi^n}{\Lambda^{n-4}(\varphi)}. \tag{44}$$

While these contributions might be large, they do not spoil the flatness of the inflationary potential as, in view of the running cutoff (41), the ratio $\varphi/\Lambda(\varphi)$ is field independent. Moreover, as noted in [12, 22, 64], the asymptotic tree-level scale invariance of the model (24) in the JF is only weakly broken by the quantum logs and leads to an asymptotic shift invariance in the EF, protecting the EF inflaton potential from large quantum corrections. See also [102, 103, 122–124, 126, 144] for a recent discussion on classical and quantum scale (conformal, respectively) invariance.

4.6 *Quantum Cosmology and Initial Conditions for Higgs Inflation*

While the model of non-minimal Higgs inflation provides an answer to the question about the fundamental nature of the inflaton field, which largely remains an open question in many other models of inflation, it does not answer the question about the initial conditions for inflation or the origin of the universe itself.

An attempt to derive the initial conditions for the inflationary background evolution from quantum cosmology was undertaken in [13] by reviving the old idea that the universe tunnelled from nothing to existence, see [98, 112, 139, 146]. The nucleation process of the universe in the tunnelling scenario is described by a gravitational instanton—the solution to the Euclidean version of Einstein's equation. Starting from the Euclidean path integral

$$\exp(-W) := \int \mathcal{D}g \exp(-S_{\text{eff}}[g]), \tag{45}$$

the effective gravitational action $S_{\text{eff}}[g]$ is obtained by integrating out all matter fields, collectively denoted by ψ,

$$\exp(-S_{\text{eff}}[g]) := \int \mathcal{D}\psi \exp(-S[g;\psi]). \tag{46}$$

For the large and slowly varying background fields during inflation, the effective action $S_{\text{eff}}[g]$ admits a local expansion in gradients and curvatures

$$S_{\text{eff}}[g] = \frac{M_{\text{P}}^2}{2}\int \mathrm{d}^4x\sqrt{g}\,[2\Lambda_{\text{eff}} - R(g) + \cdots]. \tag{47}$$

The ellipsis indicate, that I have only kept the leading orders and neglected gradient terms and higher curvature invariants. In the inflationary slow-roll approximation, the quantum effective scalar field potential might be identified with an effective cosmological constant $M_{\text{P}}^2\Lambda_{\text{eff}} := V_{\text{eff}}(\varphi)$. The line element reduces to the Euclidean version of the homogeneous and isotropic FLRW universe

$$\mathrm{d}s^2 = N(\tau)\mathrm{d}\tau^2 + a(\tau)^2\mathrm{d}\Omega_{(3)}^2, \tag{48}$$

with the Euclidean time τ, the volume element of the unit three-sphere $\mathrm{d}\Omega_{(3)}^2$, the Euclidean lapse function $N(\tau)$ and the Euclidean scale factor $a(\tau)$. The "matter" fields $\psi(\tau, \mathbf{x})$ are associated with all inhomogeneous degrees of freedom, including metric perturbations. On the background (48), the Euclidean action (47) reduces to

$$S_{\text{eff}}[a, N] = 12\pi^2 M_{\text{P}}^2 \int \mathrm{d}\tau N\left[-a + (a')^2 a + H^2 a^3\right]. \tag{49}$$

In (49), I have defined $a' := N^{-1}\mathrm{d}/\mathrm{d}\tau$. The tunnelling instanton is obtained for the gauge choice $N = -1$ as stationary configuration with respect to variations of the lapse function in the saddle point approximation

$$\left(a'\right)^2 = 1 - H^2 a^2. \tag{50}$$

The Euclidean Friedman equation (50) has a turning point at a_+ corresponding to the equator of the Euclidean S^4 half-sphere. The positive solution of (50) is

$$a(\tau) = H^{-1}\sin(H\tau), \tag{51}$$

where I have fixed the constant of integration by the condition $da/d\tau|_{a=0} = 1$, which follows from (50) at $a_- = 0$. The tunnelling probability distribution function (PDF) for $H^2 = \Lambda_{\rm eff}/3$ and $\Lambda_{\rm eff} > 0$ is obtained from (49) by integrating the configurations (51) from the pole of the Euclidean half-sphere at $a_- = a(\tau_-) = 0$ to the nucleation point $a_+ = a(\tau_+) = 1/H$ at the equator [5]

$$P(\varphi) = \exp[-W(\varphi)] = \exp\left[-\frac{24\pi^2 M_{\rm P}^4}{V_{\rm eff}(\varphi)}\right]. \tag{52}$$

At the moment of nucleation $\tau_+ = \pi/2H$, the solution (51) can be analytically continued $\tau \to it$ to the Lorentzian regime

$$a_{\rm L}(t) = \frac{1}{H}\cosh(Ht). \tag{53}$$

The tunnelling instanton (52) can be interpreted as representing the PDF of scale factors (53) in the quantum ensemble of de Sitter models after nucleation, i.e. the realizations of scale factors a_L (with different $H(\varphi) = \sqrt{\Lambda_{\rm eff}(\varphi)/3}$) are distributed according to (52). The maximum $\varphi_{\rm max}$ of the potential $V_{\rm eff}(\varphi)$ corresponds to a peak in the distribution (52) and can be interpreted as the most probable value for the inflationary trajectory to start, i.e. provides the initial condition for inflation.

We note that the quantum cosmological scenario could in principle be falsified by observations, as it must satisfy a consistency condition—the value $\varphi_{\rm max}$ determined by the tunnelling scenario has to be compatible with the value $\varphi_{\rm inf}$ derived from the energy scale of inflation, i.e. $\varphi_{\rm max} \geq \varphi_{\rm inf}$. The energy scale of inflation could be inferred from the detection of primordial gravitational waves,

$$E_{\rm inf}^{\rm obs} = M_{\rm P}\left(\frac{3}{2}\pi^2 A_{\rm t,*}\right)^{1/4}. \tag{54}$$

The observational data (7), (12) and (14) imply the upper bound $E_{\rm obs}^{\rm inf} \lesssim 10^{16}$ GeV.

When this general formalism is applied to the model of non-minimal Higgs inflation, the RG improved EF potential (27) enters the tunnelling PDF (52),

$$P(\varphi) \simeq \exp\left[-96\pi^2\frac{\xi^2}{\lambda}\left(1+\frac{2M_{\rm P}^2}{\xi\varphi^2}\right)\right]. \tag{55}$$

Here, the couplings $\lambda(t)$ and $\xi(t)$ are function of the logarithmic RG scale $t = \log(\varphi/M_{\rm t})$ and I have neglected the wave function renormalization of the Higgs boson, i.e. set $Z = 1$. The distribution (55) has a sharply peaked maximum at

---

[5] Note that (52) coincides with the PDF obtained from the semiclassical expansion of the Wheeler-DeWitt equation $e^{-W} = |\Psi|^2$, with the tunnelling wave function of the universe Ψ. Corrections from canonical quantum gravity to the inflationary power spectra have first been derived for the minimally coupled theory [92] and later for a general scalar-tensor theory in [135, 136].

$$\varphi_{\max}^2 = -\left.\frac{64\pi^2 M_P^2}{\xi \mathbf{A}_I}\right|_{t_{\max}}. \tag{56}$$

Here, $\mathbf{A}_I$ is the inflationary anomalous scaling defined in (30). The value (56) for $\varphi_{\max}$ satisfies the consistency condition $\varphi_{\max} \geq \varphi_{\inf}$ as in [13], the value $\varphi_{\inf}$ at horizon crossing was found to be related to $\varphi_{\max}$ by

$$\frac{\varphi_{\inf}^2}{\varphi_{\max}^2} = 1 - \exp\left(-\hat{N}\frac{\mathbf{A}_I(t_{\text{end}})}{48\pi^2}\right). \tag{57}$$

Thus, for wavelengths longer than the pivotal one, the moment of horizon crossing $\varphi_{\inf}$ comes closer to the moment of nucleation $\varphi_{\max}$, but always chronologically stays behind it $\varphi_{\max} > \varphi_{\inf}$ and approaches it in the limit $N \to \infty$.

The quantum cosmological analysis provides a complete picture, as it suggests that non-minimal Higgs inflation might even predict its own initial conditions in a self-consistent way, followed by a successful inflationary phase and a subsequent transition to the SM at low energies.

5 Quantum Field Parametrization Dependence in Cosmology

In this section, I discuss another aspect of Higgs inflation, which is connected to a more general field theoretical problem. The predictions (28) were derived by transforming between the JF parametrization (24) and the EF parametrization (26). In fact, the class of models leading to inflationary predictions equivalent to those of non-minimal Higgs inflation (28) is much larger and also includes geometrical modifications of general relativity.

5.1 *Starobinsky Inflation*

The first model of inflation, proposed in [129], is favoured by current data, see [2] and the analysis in [100]. In addition to the Einstein-Hilbert action, which is linear in the scalar curvature, the action of Starobinsky's model includes the square of the Ricci scalar

$$S[g] = \frac{M_P^2}{2}\int d^4x\sqrt{-g}\left[R + \frac{R^2}{6M^2}\right], \tag{58}$$

Aside from the Planck mass M_P, the mass scale M is the only new scale in the model. Due to the fourth-order derivatives implicit in the R^2 term, the theory effectively propagates a massive scalar particle—the *scalaron*. In contrast to other higher-derivative modifications of Einstein's theory, which involve quadratic invariants built from the Ricci tensor and the Riemann tensor, $f(R)$ gravity does not suffer from the Ostrogradski instability and the associated problem of the higher derivative massive spin-two ghost, which spoils the unitarity of the corresponding quantum theory, see [137] and [145]. The additional scalar degree of freedom can be made explicit by transforming the action (58) into its scalar-tensor representation. Performing the field redefinitions

$$\hat{g}_{\mu\nu} = \exp\left(\sqrt{\frac{2}{3}}\frac{\hat{\chi}}{M_{\mathrm{P}}}\right) g_{\mu\nu}, \qquad \hat{\chi}(R) = \sqrt{\frac{2}{3}} M_{\mathrm{P}} \ln\left(1 + \frac{R}{3M}\right), \tag{59}$$

the scalaron $\hat{\chi}$ becomes manifest in the EF formulation

$$S[\hat{g}, \hat{\chi}] = \int \mathrm{d}^4x \sqrt{-\hat{g}} \left[\frac{M_{\mathrm{P}}^2}{2}\hat{R} - \frac{1}{2}\hat{g}^{\mu\nu}\partial_\mu\hat{\chi}\partial_\nu\hat{\chi} - \hat{V}(\hat{\chi})\right]. \tag{60}$$

The Starobinsky EF potential has the same structure as the potential (27),

$$\hat{V}(\hat{\chi}) = \frac{3}{4} M_{\mathrm{P}}^2 M^2 \left(1 - e^{-\sqrt{\frac{2}{3}}\frac{\hat{\chi}}{M_{\mathrm{P}}}}\right)^2. \tag{61}$$

Evaluating (7)-(11) for (61) to leading order in $N_* \gg 1$ results in

$$A_{\mathrm{s}}^* = \frac{N_*^2}{24\pi^2}\frac{M^2}{M_{\mathrm{P}}^2}, \qquad n_{\mathrm{s}}^* = 1 - \frac{2}{N_*} \approx 0.9667, \qquad r^* = \frac{12}{N_*^2} \approx 0.0033. \tag{62}$$

The numerical values are again obtained for $N_* = 60$. The predictions for $n_{\mathrm{s}*}$ and r^* are not only in excellent agreement with the current observational bounds (13) and (14), they are identical to the tree-level predictions of non-minimal Higgs inflation (28). The normalization condition (12) fixes the only free parameter $M \approx 10^{-5} M_{\mathrm{P}}$. Identifying $M = \sqrt{\lambda/3\xi^2} M_{\mathrm{P}}$ shows that non-minimal Higgs inflation and Starobinsky inflation are indistinguishable regarding their inflationary observables. However, the couplings to other fields, the predictions for reheating and the inclusion of quantum corrections makes them at least in principle observationally distinguishable, see [19]. The degenerate inflationary predictions of these two models can be explained naturally, as both are part of a common universality class of inflationary models, see [105].

Before discussing in more detail the general question of equivalence between different field parametrizations, another interesting point regarding the field transformations (25) and (59) is the following: In the geometric formulation of Starobinsky's model, the quadratic R^2 term dominates over the linear R term for high energies $R/6M^2 \gg 1$ and leads to an asymptotic scale invariance of the action (58). Likewise, for high energies $\varphi \gg M_{\mathrm{P}}/\sqrt{\xi}$, the non-minimally coupled term in the action (24) dominates and leads to the aforementioned asymptotic scale invariance. In both cases, inflation is realized for the approximately scale-invariant quasi de Sitter phase and ends when the scale invariance is explicitly broken by the Einstein-Hilbert term. Moreover, the transformations (25) and (59) show that the asymptotic invariance under the scale transformations with constant scaling parameter α,

$$g_{\mu\nu} \to g'_{\mu\nu}\alpha^{-2}g_{\mu\nu}, \qquad \varphi \to \varphi' = \alpha\varphi, \tag{63}$$

translates into an asymptotic shift symmetry of the scalar field in the EF formulation

$$\hat{\varphi} \to \hat{\varphi}' = \hat{\varphi} + \ln\alpha. \tag{64}$$

An approximate shift symmetry, in turn, naturally explains why the inflationary quasi de Sitter phase is realized by an almost flat, quasi-constant potential (27) and (61) in the EF formulation.

5.2 *Classical and Quantum Equivalence in Cosmology*

The equivalence between the JF and EF formulations in non-minimal Higgs inflation as well as the equivalence between non-minimal Higgs inflation and Starobinsky inflation are just particular realizations of a more general equivalence. In fact, the gravity-Higgs sector (24) is a particular case of the more general scalar-tensor theory

$$S[g,\varphi] = \int \mathrm{d}^4x\sqrt{-g}\left[U(\varphi)R - \frac{1}{2}G(\varphi)\partial_\mu\varphi\partial^\mu\varphi - V(\varphi)\right], \tag{65}$$

with general functions $U(\varphi)$, $G(\varphi)$ and $V(\varphi)$, which parametrize the non-minimal coupling to gravity, the non-standard kinetic term and the arbitrary potential. The action (65) covers almost all single-field models of inflation. Likewise, Starobinsky's model (58) is just a specific case of the more general class of $f(R)$ models

$$S[g] = \int \mathrm{d}^4x\sqrt{-g}f(R). \tag{66}$$

Note that since $f(R)$ is only a function of the undifferentiated Ricci scalar R, apart from the scalaron, there are no additional propagating degrees of freedom despite the arbitrariness of f.

Provided $U \neq 0$, the transition of the general scalar-tensor theory (65) to the EF action (26) is achieved by performing the field redefinitions with the EF potential

$$\hat{g}_{\mu\nu} = \frac{2U}{M_\mathrm{P}^2}g_{\mu\nu}, \qquad \left(\frac{\partial\hat{\varphi}}{\partial\varphi}\right)^2 = \frac{M_\mathrm{P}^2}{2}\frac{GU + 3U_{,\varphi}^2}{U^2}, \qquad \hat{V}(\hat{\varphi}) = \left(\frac{M_\mathrm{P}^2}{2}\right)^2\frac{V(\hat{\varphi})}{U^2(\hat{\varphi})}. \tag{67}$$

Similarly, provided $f_{,R} \neq 0$ and $f_{,RR} \neq 0$, the $f(R)$ theory (66) is mapped to the EF action (26) by performing the field redefinitions with the EF potential

$$\hat{g}_{\mu\nu} = \frac{2f_{,R}}{M_{\rm P}^2} g_{\mu\nu}, \qquad \hat{\varphi}(R) = M_{\rm P}\sqrt{\frac{3}{2}} \ln f_{,R}, \qquad \hat{V}(R) = \left(\frac{M_{\rm P}^2}{2}\right)^2 \frac{Rf_{,R} - f}{f_{,R}^2}. \tag{68}$$

Comparing the transformations (67) and (68) implies the identifications

$$U \leftrightarrow f_{,R}, \qquad V \leftrightarrow Rf_R - f, \qquad GU + 3U_{,\varphi}^2 \leftrightarrow 3f_{,RR}. \tag{69}$$

At the level of the classical action, the classical equations of motion and the linear perturbations propagating on the background (which is a solution of the classical equations of motion), the formulations in different parametrizations related by non-singular but arbitrary non-linear transformations are mathematically equivalent. All parametrizations correspond to one and the same theory, just expressed in terms of different field variables. In contrast to the non-linear but ultra-local transformation (67) relating the JF and EF parametrizations of the scalar-tensor theory, the transformations (68) relating the $f(R)$ theory with the EF scalar-tensor parametrization involves in addition derivatives—reflecting the presence of the additional propagating scalar degree of freedom.

5.3 Quantum Equivalence and Renormalization

In Sect. 5.2, the identical inflationary predictions of the model of non-minimal Higgs inflation (24) and Starobinsky's model of inflation (58) has been traced back to the more general classical equivalence between scalar-tensor theories (65) and $f(R)$ gravity (66). In this section, I discuss the equivalence between different field parametrizations at the quantum level.

As emphasized in Sect. 4.3, quantum corrections and the RG improvement are crucial. In the RG improved treatment, the inflationary observables are expressed in terms of the running couplings evaluated at the energy scale of inflation. The running couplings are solutions to the RG equations. The RG system is determined by the beta functions. The beta functions are derived order-by-order in perturbation theory from the ultraviolet divergences of the theory, or, in non-perturbative approaches, such as the asymptotic safety program proposed in [142], from solving the Wetterich equation for the functional RG flow of the averaged effective action within a given truncation, cf. [143].

In the context of the field transformation, relating different representations of the same classical theory, an important question arises: does the RG improvement derived in different field parametrizations lead to the same results for the inflationary observables, or, in other words, does the classical equivalence between the differ-

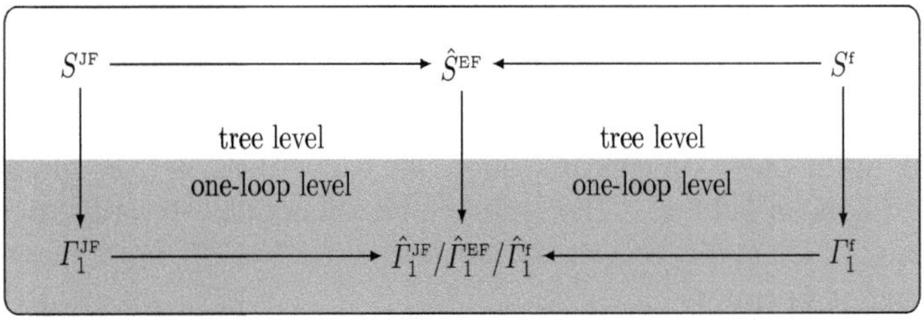

Fig. 8 Transition between formulations in terms of different field parametrizations at the classical and quantum level. If the classical equivalence extends to the one-loop quantum level, the diagram must commute

ent formulations extend to the quantum level? In order to answer this question to leading order in the perturbative loop expansion, the one-loop divergences are explicitly calculated in different parametrizations and the results are compared. Phrased diagrammatically, the question of quantum equivalence reduces to the question of whether the diagram shown in Fig. 8 commutes or not.

The combination of the background field method with heat-kernel techniques provides an efficient and manifest covariant tool to calculate the ultraviolet divergences in curved spacetime. The Euclidean quantum effective action $\Gamma[\phi_0]$ for a theory with bare action functional $S[\phi]$ and generalized field ϕ^i can be defined by the functional integro-differential equation[6]

$$\exp(-\Gamma) = \int \mathcal{D}\phi\mathcal{M}(\phi)\exp\left\{-S[\phi] - (\phi_0^i - \phi^i)\Gamma_{,i}[\phi_0]\right\}, \tag{70}$$

Here, ϕ_0^i is the mean field (one-point correlation function) and $\mathcal{M}(\phi)$ is the functional measure. Expanding the action $S[\phi]$ around the mean field ϕ_0^i, this equation serves as starting point for a perturbative expansion of Γ in powers of $\hbar$, corresponding to the number of closed loops in terms of a Feynman diagrammatic representation

$$\Gamma[\phi_0] = S[\phi_0] - \hbar\ln\mathcal{M}(\phi_0) + \hbar\Gamma_1[\phi_0] + \mathcal{O}(\hbar^2). \tag{71}$$

Neglecting the contribution from the (ultra) local measure $\mathcal{M}(\phi_0) \propto 1 + \delta(0)(\ldots)$, the one-loop effective action Γ_1 is given in terms of the functional trace

$$\Gamma_1 = \frac{1}{2}\mathrm{Tr}\ln F_{ij}, \tag{72}$$

with the fluctuation operator defined by the Hessian of the classical action

[6]The compact DeWitt notation combines the collection of discrete bundle indices $A, B, \ldots$ with the continuous spacetime arguments $x, y, \ldots$ into the generalized DeWitt index $i = (A, x)$, where it is understood that the Einstein summation convention for the DeWitt indices $i, j \ldots$ includes integration over the continuous spacetime arguments, i.e. $\phi^i\phi_i = \int \mathrm{d}^4x\phi^A(x)\phi_A(x)$.

$$F_{ij}(\nabla) := \partial_i \partial_j S = \left.\frac{\delta^2 S[\phi]}{\delta\phi^i \delta\phi^j}\right|_{\phi=\bar{\phi}} . \tag{73}$$

At the level of the one-loop approximation, the background field $\bar{\phi}^i$ might be identified with the mean field ϕ_0^i.[7] The one-loop beta functions are proportional to the counterterms, which in the $\overline{\text{MS}}$ scheme are determined by Γ_1^{div}. For minimal second-order fluctuation operators

$$\mathbf{F}(\nabla) = \mathbf{1}\Delta + \Pi, \tag{74}$$

the Schwinger-DeWitt technique, originally developed in [45], allows to calculate the one-loop divergences in a closed form. Here, bundle indices are suppressed and operator valued quantities such as e.g. $\mathbf{F} = F^A_{\ B} = \mathcal{G}^{AC} F_{CB}$ are written in bold face, $\Delta = -g^{\mu\nu}\nabla_\nu\nabla_\nu$ is the positive definite Laplacian and Π is the potential part acting multiplicatively. The one-loop UV divergences in dimensional regularization $d = 4 - 2\varepsilon$ are isolated as poles $1/\varepsilon$ in the limit $\varepsilon \to 0$. For the operator (74), the divergent part of the one-loop effective action in curved spacetime reads

$$\Gamma_1^{\text{div}} = -\frac{1}{32\pi^2\varepsilon} \int \mathrm{d}^4x \sqrt{g}\, \mathrm{tr}\, \mathbf{a}_2(x, x). \tag{75}$$

It involves the bundle trace tr of the coincidence limit of the second Schwinger-DeWitt coefficient, which, up to total derivative terms, is given by

$$\mathbf{a}_2(x, x) = \frac{1}{180}\left(R_{\mu\nu\rho\sigma}R^{\mu\nu\rho\sigma} - R_{\mu\nu}R^{\mu\nu}\right)\mathbf{1} + \frac{1}{2}\mathcal{R}_{\mu\nu}\mathcal{R}^{\mu\nu} + \frac{1}{2}\left(\Pi - \frac{R}{6}\mathbf{1}\right)^2 . \tag{76}$$

The bundle curvature (suppressing bundle indices) is defined as $\mathcal{R}_{\mu\nu} := [\nabla_\mu, \nabla_\nu]\phi$. For higher-order and non-minimal operators $\mathbf{F}$, the Schwinger-DeWitt algorithm is not directly applicable and more general techniques developed in [15] are required. For operators with a degenerate principal part or with an effective Laplacian, even these generalized techniques cannot be applied directly and other methods discussed in [77, 114–116] must be employed.

In order to perform the explicit comparison illustrated in Fig. 8, first the individual calculations in the different field parametrizations have to be performed. The derivation of the one-loop divergences for the EF action (26) with arbitrary EF potential $\hat{V}$ has been performed in the EF field parametrization in [10, 133]. The corresponding calculation for the general scalar-tensor action (65) in the JF field parametrization has been carried out in [121] and has been generalized in [133] for a $O(N)$ symmetric multiplet of scalar fields. Finally, the one-loop divergences for the $f(R)$ action were

[7] A systematic order-by-order renormalization procedure for arbitrary loop order, which ensures the gauge invariant structure of the counterterms and keeps track of the background field and the mean field separately, has been proposed in [9].

obtained in [114]. Note that in all cases the off-shell divergences were calculated on a general background, which is crucial in order to uniquely ascribe the individual coefficients to different operator structures.[8]

Using the transition formulas (67) and (68), the one-loop results for the JF scalar-tensor theory and for $f(R)$ gravity were transformed back to the EF and compared with the one-loop divergences directly obtained in the EF to explicitly check the commutativity of the diagram Fig. 8. The comparison between the EF and JF parametrizations has been performed in [88], while the comparison between the $f(R)$ and scalar-tensor formulations has been carried out in [116]. In both cases, the one-loop comparison showed that the divergent part of the off-shell one-loop effective action, calculated in different parametrizations, does not coincide, implying that the classical equivalence is lost at the quantum level. However, in both cases it was also found that the quantum equivalence is restored for the on-shell reduction. The on-shell equivalence is in agreement with formal S-matrix equivalence theorems formulated in [39, 41, 85, 86].[9]

A naive application of the of the RG improvement by replacing the coupling constants in cosmological observables by the running constants leads to ambiguous results. The couplings are solutions of the system of RG equations which are in turn controlled by the off-shell beta function. Since the off-shell beta functions are derived from the off-shell divergences, they inherit the off-shell parametrization ambiguity.

5.4 Geometry of Field Space and Field Covariant Formalism

From a quantum field theoretical point of view, the configuration space of fields (including the spacetime metric) might formally be viewed as a differentiable manifold. In this geometric setup, different field parametrizations simply correspond to different local coordinates and the quantum off-shell parametrization dependence is revealed as a failure of the non-field covariant mathematical formalism underlying the ordinary definition of the quantum effective action. In more detail, the analysis shows that the ordinary quantum effective action is not a true scalar with respect to configuration space diffeomorphisms (although it is of course a scalar with respect to spacetime diffeomorphisms). Within a field covariant formalism, it becomes meaningless to talk about a preferred physical field parametrization—any frame is as good as any other, see e.g. [88, 132, 134]. In this sense, the off-shell quantum frame ambiguity finds a natural resolution in terms of the geometrically defined effective action, introduced in [140]. Instead of trying to select a preferred parametrization on physical grounds, the quantum effective action is defined in a manifestly field parametrization invariant way. The fact that the ordinary effective action is not invariant under

[8] On particular symmetric backgrounds such as e.g. spaces of constant curvature, the different curvature invariants degenerate and their individual contributions are no longer resolvable.

[9] Note however, that some of the propositions in the formal theorems, such as e.g. the assumption of asymptotically free states, are in general not satisfied in the context of gravity and cosmology.

Fig. 9 Synge's world function is a measure of the (squared) geodesic distance between the two points ϕ_0 and ϕ in configuration space

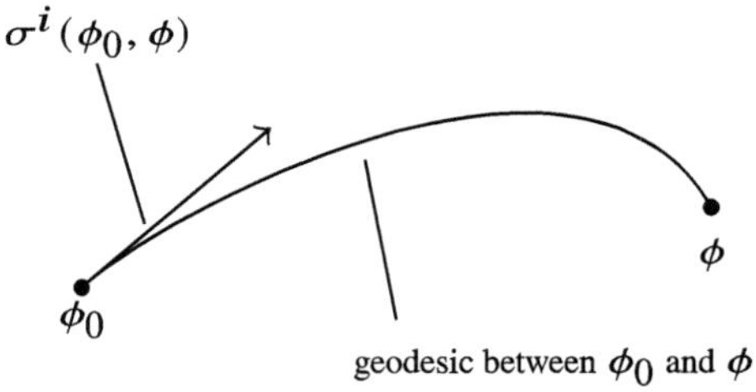

field reparametrizations can be traced back to the geometrically meaningless coordinate difference $(\phi_0^i - \phi^i)$, which enters in the exponent of the defining equation (70). According to Vilkovisky's proposal in [140], it should be replaced by the geometrical meaningful two-point quantity $\sigma^i(\phi_0, \phi)$,

$$(\phi_0^i - \phi^i) \to \sigma^i(\phi_0, \phi) := \mathcal{G}^{ij}\nabla_j\sigma(\phi_0, \phi). \tag{77}$$

The bi-scalar $\sigma(\phi_0, \phi)$ is Synge's world function on the configuration space manifold.

Here, $\mathcal{G}^{ij}(\phi)$ is the inverse of the configuration space metric $\mathcal{G}_{ij}(\phi)$ and ∇_i is the field covariant derivative, which defines the configuration connection $\Gamma^k_{ij}(\phi)$. The derivative (77) is a vector at ϕ_0 and a scalar at ϕ as depicted in Fig. 9. In the one-loop approximation, the identification of the functional measure with $\mathcal{M} = \sqrt{\mathrm{Det}(\mathcal{G}_{ij})}$ and the replacement of the coordinate difference $(\phi_0^i - \phi^i) \to \sigma^i(\phi_0, \phi)$ in (70), leads to a replacement of partial derivatives by covariant ones in the definition of the fluctuation operator (73),

$$\partial_i\partial_j S \to \nabla_i\nabla_j S = \partial_i\partial_j S - \Gamma^k_{ij}\partial_k S. \tag{78}$$

Independently of the concrete prescription for the explicit construction of the configuration space metric $\mathcal{G}_{ij}$ and the Vilkovisky connection Γ'^i_{jk}, it is clear that the additional term in the covariant construction (78) is proportional to $\partial_i S$ and vanishes on-shell $\partial_i S = 0$. Hence, on-shell the one-loop approximation of the covariant Vilkovisky-DeWitt effective action reduces to the one-loop approximation of the ordinary effective action.

As first discussed in [88, 132, 134], the problem of quantum frame dependence in cosmology is just a particular manifestation of this more general field theoretic problem of field parametrization dependence and the construction of a unique off-shell extension of the quantum effective action. For related work on classical and quantum frame dependence in cosmology, see also [25, 26, 33, 36–38, 44, 46, 59–62, 79, 82, 83, 87, 90, 91, 99, 104, 108–111]. However, besides the technical difficulties in the explicit construction of the Vilkovisky-DeWitt effective action, it is questionable whether such a construction is actually required in the context of cosmological observables. A direct construction of manifestly reparametrization and gauge invariant cosmological observables (n-point correlation functions) would most likely be more efficient and would also lead to unique results.

6 Conclusion

In this contribution, I reviewed and discussed various classical and quantum aspects of the model of non-minimal Higgs inflation. Summarizing, this model offers a theoretically well motivated and phenomenologically successful unified description of particle physics with inflationary cosmology. The identification of the inflaton field with the SM Higgs boson offers an explanation for the origin and fundamental nature of the scalar field which drives the accelerated expansion of the early universe. The RG flow connects the EW scale with the energy scale of inflation and supports the scenario that the SM could be a perturbative quantum field theory all the way up to the Planck scale. The effective field theory expansion of the model can be arranged in a controlled way upon the introduction of a field dependent cutoff. Moreover, within the quantum cosmological tunnelling scenario, the model predicts its own initial conditions for the onset of inflation in a self-consistent way.

In spite of all these appealing properties, I also highlighted several questions and open problems. For the central values of the SM Higgs boson mass and the Yukawa top quark mass as measured by collider experiments at the EW scale, the RG flow of the SM drives the quartic Higgs self-coupling to negative values, resulting in the formation of a negative vacuum at a high energy scale, which in turn triggers an instability of the EW vacuum. This instability has to be cured, either by finding ways to directly stabilize the RG improved Higgs potential or by invoking a mechanism which prevents the EW vacuum to decay within the lifetime of our universe.

Besides, I discussed the ambiguity in the definition of the perturbative off-shell beta function, which results from a quantization in different field parametrizations related by a Weyl transformation of the metric field and a highly non-linear transformation of the scalar field. I also illustrated the classical equivalence between scalar-tensor sector of non-minimal Higgs inflation and Starobinsky's model of inflation, which corresponds to a subclass of geometric $f(R)$ modifications of General Relativity. I discussed how, within the ordinary definition of the quantum effective action, off-shell quantum corrections break this equivalence and how this equivalence is restored on-shell. I gave a natural explanation of these results in terms of Vilkovisky's geometrically defined parametrization invariant off-shell extension of the quantum effective action and emphasized the importance of and the need for gauge and parametrization invariant cosmological quantum observables. Finally, coming back to Hermann Weyl, I hope that already from the very limited discussion in this contribution, it became evident how his original ideas introduced 100 years ago, still strongly influence active research in theoretical physics today.

Acknowledgements It is a pleasure to thank Silvia De Bianchi and Claus Kiefer for inviting me to this very stimulating interdisciplinary conference and the physics centre in Bad Honnef for the warm hospitality. I also thank the other participants for many interesting after-dinner discussions.

References

1. P.A.R. Ade et al. Planck 2015 results. XX. Constraints on inflation, Astron. Astrophys. **594**, A20 (2016)
2. Y. Akrami et al. Planck 2018 results. X. Constraints on inflation (2018). arXiv:1807.06211v2 [astro-ph.CO]
3. A. Albrecht, P.J. Steinhardt, Cosmology for Grand Unified Theories with radiatively induced symmetry breaking. Phys. Rev. Lett. **48**, 1220–1223 (1982)
4. K. Allison, Higgs xi-inflation for the 125–126 GeV Higgs: a two-loop analysis. J. High Energy Phys. **02**, 040 (2014)
5. G.W. Anderson, New cosmological constraints on the higgs boson and top quark masses. Phys. Lett. B **243**, 265–270 (1990)
6. P.B. Arnold, Can the electroweak vacuum be unstable? Phys. Rev. D **40**, 613 (1989)
7. J.L.F. Barbon, J.R. Espinosa, On the naturalness of higgs inflation. Phys. Rev. D **79**, 081302 (2009)
8. J.M. Bardeen, P.J. Steinhardt, M.S. Turner, Spontaneous creation of almost scale-free density perturbations in an inflationary universe. Phys. Rev. D **28**, 679 (1983)
9. A.O. Barvinsky, D. Blas, M. Herrero-Valea, S.M. Sibiryakov, C.F. Steinwachs, Renormalization of gauge theories in the background-field approach. J. High Energy Phys. **07**, 035 (2018)
10. A.O. Barvinsky, A.Y. Kamenshchik, I.P. Karmazin, The Renormalization group for nonrenormalizable theories: Einstein gravity with a scalar field. Phys. Rev. D **48**, 3677–3694 (1993)
11. A.O. Barvinsky, A.Y. Kamenshchik, C. Kiefer, A.A. Starobinsky, C.F. Steinwachs, Asymptotic freedom in inflationary cosmology with a non-minimally coupled Higgs field. J. Cosmol. Astropart. Phys. **09**(12), 003 (2009)
12. A.O. Barvinsky, A.Y. Kamenshchik, C. Kiefer, A.A. Starobinsky, C.F. Steinwachs, Higgs boson, renormalization group, and naturalness in cosmology. Europ. Phys. J. C **72**, 2219 (2012)
13. A.O. Barvinsky, A.Y. Kamenshchik, C. Kiefer, C.F. Steinwachs, Tunneling cosmological state revisited: Origin of inflation with a non-minimally coupled Standard Model Higgs inflaton. Phys. Rev. D **81**, 043530 (2010)
14. A.O. Barvinsky, A.Y. Kamenshchik, A.A. Starobinsky, Inflation scenario via the Standard Model Higgs boson and LHC. J. Cosmol. Astropart. Phys. **08**(11), 021 (2008)
15. A.O. Barvinsky, G.A. Vilkovisky, The generalized Schwinger-Dewitt technique in gauge theories and quantum gravity. Phys. Reports **119**, 1–74 (1985)
16. A.V. Bednyakov, B.A. Kniehl, A.F. Pikelner, O.L. Veretin, Stability of the electroweak vacuum: gauge independence and advanced precision. Phys. Rev. Lett. **115**, 201802 (2015)
17. F. Bezrukov, D. Gorbunov, M. Shaposhnikov, On initial conditions for the Hot Big Bang. J. Cosmol. Astropart. Phys. **09**(06), 029 (2009)
18. F. Bezrukov, M.Y. Kalmykov, B.A. Kniehl, M. Shaposhnikov, Higgs boson mass and new physics. J. High Energy Phys. **10**, 140 (2012)
19. F.L. Bezrukov, D.S. Gorbunov, Distinguishing between R^2-inflation and Higgs-inflation. Phys. Lett. B **713**, 365–368 (2012)
20. F.L. Bezrukov, A. Magnin, M. Shaposhnikov, Standard Model Higgs boson mass from inflation. Phys. Lett. B **675**, 88–92 (2009)
21. F.L. Bezrukov, M. Shaposhnikov, The standard model higgs boson as the inflaton. Phys. Lett. B **659**, 703–706 (2008)
22. F. Bezrukov, A. Magnin, M. Shaposhnikov, S. Sibiryakov, Higgs inflation: consistency and generalisations. J. High Energy Phys. **01**, 016 (2011)
23. F. Bezrukov, M. Shaposhnikov, Standard model higgs boson mass from inflation: two loop analysis. J. High Energy Phys. **07**, 089 (2009)
24. F. Bezrukov, M. Shaposhnikov, Higgs inflation at the critical point. Phys. Lett. B **734**, 249–254 (2014)
25. K. Bhattacharya, B.R. Majhi, Fresh look at the scalar-tensor theory of gravity in Jordan and Einstein frames from undiscussed standpoints. Phys. Rev. D **95**, 064026 (2017)

26. M. Bounakis, I.G. Moss, Gravitational corrections to Higgs potentials. J. High Energy Phys. **04**, 071 (2018)
27. V. Branchina, E. Messina, Stability, higgs boson mass and new physics. Phys. Rev. Let. **111**, 241801 (2013)
28. V. Branchina, E. Messina, M. Sher, Lifetime of the electroweak vacuum and sensitivity to Planck scale physics. Phys. Rev. D **91**, 013003 (2015)
29. P. Burda, R. Gregory, I. Moss, Gravity and the stability of the Higgs vacuum. Phys. Rev. Lett. **115**, 071303 (2015)
30. P. Burda, R. Gregory, I. Moss, The fate of the Higgs vacuum. J. High Energy Phys. **06**, 025 (2016)
31. C.P. Burgess, H.M. Lee, M. Trott, Power-counting and the validity of the classical approximation during inflation. J. High Energy Phys. **09**, 103 (2009)
32. C.P. Burgess, H.M. Lee, M. Trott, Comment on higgs inflation and naturalness. J. High Energy Phys. **07**, 007 (2010)
33. D. Burns, S. Karamitsos, A. Pilaftsis, Frame-covariant formulation of inflation in scalar-curvature theories. Nucl. Phys. B **907**, 785–819 (2016)
34. D. Buttazzo, G. Degrassi, P.P. Giardino, G.F. Giudice, F. Sala, A. Salvio, A. Strumia, Investigating the near-criticality of the Higgs boson. J. High Energy Phys. **12**, 089 (2013)
35. X. Calmet, I. Kuntz, Higgs starobinsky inflation. Europ. Phys. J. C **76**, 289 (2016)
36. X. Calmet, T.-C. Yang, Frame transformations of gravitational theories. Int. J. Modern Phys. A **28**, 1350042 (2013)
37. S. Capozziello, R. de Ritis, A.A. Marino, Some aspects of the cosmological conformal equivalence between 'Jordan frame' and 'Einstein frame'. Class. Quant. Gravity **14**, 3243–3258 (1997)
38. T. Chiba, M. Yamaguchi, Conformal-frame (in)dependence of cosmological observations in scalar-tensor theory. J. Cosmol. Astropart. Phys. **13**(10), 040 (2013)
39. J.S.R. Chisholm, Change of variables in quantum field theories. Nucl. Phys. **26**(3), 469–479 (1961)
40. S.R. De Coleman, F. Luccia, Gravitational effects on and of vacuum decay. Phys. Rev. D **21**, 3305 (1980)
41. S.R. Coleman, J. Wess, B. Zumino, Structure of phenomenological Lagrangians. 1. Phys. Rev. **177**, 2239–2247 (1969)
42. A. De Simone, M.P. Hertzberg, F. Wilczek, Running inflation in the standard model. Phys. Lett. B **678**, 1–8 (2009)
43. G. Degrassi, S. Di Vita, J. Elias-Miro, J.R. Espinosa, G.F. Giudice, G. Isidori, A. Strumia, Higgs mass and vacuum stability in the standard model at NNLO. J. High Energy Phys. **08**, 098 (2012)
44. N. Deruelle, M. Sasaki, Conformal equivalence in classical gravity: the example of 'Veiled' general relativity. Springer Proc. Phys. **137**, 247–260 (2011)
45. B.S. DeWitt, *Dynamical Theory of Groups and Fields* (Blackie & Son, 1965)
46. G. Domènech, M. Sasaki, Conformal frame dependence of inflation. J. Cosmol. Astropart. Phys. **15**(04), 022 (2015)
47. A. Eichhorn, H. Gies, J. Jaeckel, T. Plehn, M.M. Scherer, R. Sondenheimer, The higgs mass and the scale of new physics. J. High Energy Phys. **04**, 022 (2015)
48. J. Elias-Miro, J.R. Espinosa, G.F. Giudice, G. Isidori, A. Riotto, A. Strumia, Higgs mass implications on the stability of the electroweak vacuum. Phys. Lett. B **709**, 222–228 (2012)
49. J. Elias-Miro, J.R. Espinosa, G.F. Giudice, H.M. Lee, A. Strumia, Stabilization of the electroweak vacuum by a scalar threshold effect. J. High Energy Phys. **06**, 031 (2012)
50. Y. Ema, Higgs scalaron mixed inflation. Phys. Lett. B **770**, 403–411 (2017)
51. Y. Ema, Dynamical emergence of scalaron in higgs inflation. J. Cosmol. Astropart. Phys. **19**(09), 027 (2019)
52. Y. Ema, M. Karciauskas, O. Lebedev, S. Rusak, M. Zatta, Higgs-inflaton mixing and vacuum stability. Phys. Lett. B **789**, 373–377 (2019)

53. V.-M. Enckell, K. Enqvist, S. Rasanen, E. Tomberg, Higgs inflation at the hilltop. J. Cosmol. Astropart. Phys. **18**(06), 005 (2018)
54. V.-M. Enckell, K. Enqvist, S. Rasanen, L.-P. Wahlman, Higgs-R^2 inflation—full slow-roll study at tree-level. J. Cosmol. Astropart. Phys. **20**(01), 041 (2020)
55. J.R. Espinosa, G.F. Giudice, E. Morgante, A. Riotto, L. Senatore, A. Strumia, N. Tetradis, The cosmological Higgstory of the vacuum instability. J. High Energy Phys. **09**, 174 (2015)
56. J.R. Espinosa, G.F. Giudice, A. Riotto, Cosmological implications of the Higgs mass measurement. J. Cosmol. Astropart. Phys. **08**(05), 002 (2008)
57. J.R. Espinosa, M. Quiros, Improved metastability bounds on the standard model Higgs mass. Phys. Lett. B **353**, 257–266 (1995)
58. R. Fakir, W.G. Unruh, Improvement on cosmological chaotic inflation through nonminimal coupling. Phys. Rev. D **41**, 1783–1791 (1990)
59. K. Falls, M. Herrero-Valea, Frame (In)equivalence in quantum field theory and cosmology. Europ. Phys. J. C **79**(7), 595 (2019)
60. V. Faraoni, E. Gunzig, Einstein frame or Jordan frame? Int. J. Theoret. Phys. **38**, 217–225 (1999)
61. K. Finn, S. Karamitsos, A. Pilaftsis, Grand Covariance in Quantum Gravity (2019) . arXiv:1910.06661v2 [hep-th]
62. E.E. Flanagan, The Conformal frame freedom in theories of gravitation. Class. Quant. Gravity **21**, 3817 (2004)
63. C.D. Froggatt, H.B. Nielsen, Standard model criticality prediction: top mass 173 +/- 5-GeV and Higgs mass 135 +/- 9-GeV. Phys. Lett. B **368**, 96–102 (1996)
64. J. Fumagalli, M. Postma, UV (in)sensitivity of Higgs inflation. J. High Energy Phys. **05**, 049 (2016)
65. D.M. Ghilencea, Two-loop corrections to Starobinsky-Higgs inflation. Phys. Rev. D **98**, 103524 (2018)
66. H. Gies, R. Sondenheimer, Higgs mass bounds from renormalization flow for a higgs-top-bottom model. Europ. Phys. J. C **75**(2), 68 (2015)
67. D. Gorbunov, A. Tokareva, Scalaron the healer: removing the strong-coupling in the Higgs- and Higgs-dilaton inflations. Phys. Lett. B **788**, 37–41 (2019)
68. A. Gundhi, C.F. Steinwachs, Scalaron-Higgs inflation. Nucl. Phys. B **954**, 114989 (2020)
69. A.H. Guth, The inflationary universe: a possible solution to the horizon and flatness problems. Phys. Rev D **23**, 347–356 (1981)
70. A.H. Guth, S.Y. Pi, Fluctuations in the new inflationary universe. Phys. Rev. Lett. **49**, 1110–1113 (1982)
71. Y. Hamada, H. Kawai, K.-Y. Oda, S.C. Park, Higgs inflation is still alive after the results from BICEP2. Phys. Rev. Lett. **112**, 241301 (2014)
72. C. Han, S. Pi, M. Sasaki, Quintessence saves higgs instability. Phys. Lett. B **791**, 314–318 (2019)
73. S.W. Hawking, The development of irregularities in a single bubble inflationary universe. Phys. Lett. **115B**, 295 (1982)
74. S.W. Hawking, I.G. Moss, Supercooled phase transitions in the very early universe. Phys. Lett. **110B**, 35–38 (1982)
75. M. He, R. Jinno, K. Kamada, S.C. Park, A.A. Starobinsky, J. Yokoyama, On the violent preheating in the mixed Higgs-R^2 inflationary model. Phys. Lett. B **791**, 36–42 (2019)
76. M. He, A.A. Starobinsky, J. Yokoyama, Inflation in the mixed Higgs-R^2 model. J. Cosmol. Astropart. Phys. **18**(05), 064 (2018)
77. L. Heisenberg, C.F. Steinwachs, Geometrized quantum Galileons. J. Cosmol. Astropart. Phys. **20**(02), 031 (2019)
78. M. Herranen, T. Markkanen, S. Nurmi, A. Rajantie, Spacetime curvature and Higgs stability after inflation. Phys. Rev. Lett. **115**, 241301 (2015)
79. M. Herrero-Valea, Anomalies, equivalence and renormalization of cosmological frames. Phys. Rev. D **93**, 105038 (2016)

80. A. Hook, J. Kearney, B. Shakya, K.M. Zurek, Probable or improbable universe? Correlating Electroweak Vacuum Instability with the Scale of Inflation. J. High Energy Phys. **01**, 061 (2015)
81. G. Isidori, V.S. Rychkov, A. Strumia, N. Tetradis, Gravitational corrections to standard model vacuum decay. Phys. Rev. D **77**, 025034 (2008)
82. L. Jarv, K. Kannike, L. Marzola, A. Racioppi, M. Raidal, M. Runkla, M. Saal, H. Veermae, Frame-independent classification of single-field inflationary models. Phys. Rev. Lett. **118**, 151302 (2017)
83. L. Jarv, P. Kuusk, M. Saal, O. Vilson, Invariant quantities in the scalar-tensor theories of gravitation. Phys. Rev. D **91**, 024041 (2015)
84. A. Joti, A. Katsis, D. Loupas, A. Salvio, A. Strumia, N. Tetradis, A. Urbano, (Higgs) vacuum decay during inflation. J. High Energy Phys. **07**, 058 (2017)
85. R.E. Kallosh, I.V. Tyutin, The equivalence theorem and gauge invariance in renormalizable theories. Yadernaya fizika **17**, 190–209 (1973)
86. S. Kamefuchi, L. O'Raifeartaigh, A. Salam, Change of variables and equivalence theorems in quantum field theories. Nucl. Phys. **28**, 529–549 (1961)
87. A.Y. Kamenshchik, E.O. Pozdeeva, S.Y. Vernov, A. Tronconi, G. Venturi, Transformations between Jordan and Einstein frames: Bounces, antigravity, and crossing singularities. Phys. Rev. D **94**, 063510 (2016)
88. A.Y. Kamenshchik, C.F. Steinwachs, Question of quantum equivalence between Jordan frame and Einstein frame. Phys. Rev. D **91**, 084033 (2015)
89. S. Kaneda, S.V. Ketov, Starobinsky-like two-field inflation. Europ. Phys. J. C **76**(1), 26 (2016)
90. A. Karam, T. Pappas, K. Tamvakis, Frame-dependence of higher-order inflationary observables in scalar-tensor theories. Phys. Rev. D **96**, 064036 (2017)
91. S. Karamitsos, A. Pilaftsis, Frame covariant nonminimal multifield inflation. Nucl. Phys. B **927**, 219–254 (2018)
92. C. Kiefer, M. Kraemer, Quantum gravitational contributions to the CMB anisotropy spectrum. Phys. Rev. Lett. **108**, 021301 (2012)
93. A. Kobakhidze, A. Spencer-Smith, Electroweak Vacuum (In)Stability in an Inflationary Universe. Phys. Lett. B **722**, 130–134 (2013)
94. K. Kohri, H. Matsui, Higgs vacuum metastability in primordial inflation, preheating, and reheating. Phys. Rev. D **94**, 103509 (2016)
95. O. Lebedev, A. Westphal, Metastable electroweak vacuum: implications for inflation. Phys. Let. B **719**, 415–418 (2013)
96. A.D. Linde, A new inflationary universe scenario: a possible solution of the horizon, flatness, homogeneity, isotropy and primordial monopole problems. Phys. Lett. **108B**, 389–393 (1982)
97. A.D. Linde, Chaotic inflation. Phys. Lett. **129B**, 177–181 (1983)
98. A.D. Linde, Quantum creation of an inflationary universe. Sov. Phys. JETP **60**, 211–213 (1984)
99. G. Magnano, L.M. Sokolowski, On physical equivalence between nonlinear gravity theories and a general relativistic selfgravitating scalar field. Phys. Rev. D **50**, 5039–5059 (1994)
100. J. Martin, C. Ringeval, V. Vennin, Encyclopædia Inflationaris. Phys. Dark Universe **5–6**, 75–235 (2014)
101. I. Masina, Higgs boson and top quark masses as tests of electroweak vacuum stability. Phys. Rev. D **87**, 053001 (2013)
102. K.A. Meissner, H. Nicolai, Conformal symmetry and the standard model. Phys. Lett. B **648**, 312–317 (2007)
103. S. Mooij, M. Shaposhnikov, T. Voumard, Hidden and explicit quantum scale invariance. Phys. Rev. D **99**, 085013 (2019)
104. I.G. Moss, Covariant one-loop quantum gravity and Higgs inflation (2014). arXiv:1409.2108v2 [hep-th]
105. V. Mukhanov, Quantum Cosmological perturbations: predictions and observations. Europ. Phys. J. C **73**, 2486 (2013)
106. V.F. Mukhanov, G.V. Chibisov, Quantum fluctuations and a nonsingular universe. JETP Lett. **33**, 532–535 (1981)

107. V.F. Mukhanov, G.V. Chibisov, The vacuum energy and large scale structure of the universe. Sov. Phys. JETP **56**, 258–265 (1982)
108. S. Nojiri, S.D. Odintsov, Quantum dilatonic gravity in $(D = 2)$-dimensions, $(D = 4)$-dimensions and $(D = 5)$-dimensions. Int. J. Modern Phys. A **16**, 1015–1108 (2001)
109. N. Ohta, Quantum equivalence of f(R) gravity and scalar-tensor theories in the Jordan and Einstein frames, Progr. Theoret. Exp. Phys. **2018**(3), 033B02 (2018)
110. M. Postma, M. Volponi, Equivalence of the Einstein and Jordan frames. Phys. Rev. D **90**, 103516 (2014)
111. T. Prokopec, J. Weenink, Frame independent cosmological perturbations. J. Cosmol. Astropart. Phys. **13**(09), 027 (2013)
112. V.A. Rubakov, Quantum mechanics in the tunneling Universe. Phys. Lett. **148B**, 280–286 (1984)
113. J. Rubio, Higgs inflation. Front. Astron. Space Sci. **5**, 50 (2019)
114. M.S. Ruf, C.F. Steinwachs, a), 'One-loop divergences for $f(R)$ gravity'. Phys. Rev. D **97**, 044049 (2018)
115. M.S. Ruf, C.F. Steinwachs, b), 'Quantum effective action for degenerate vector field theories'. Phys. Rev. D **98**, 085014 (2018)
116. M.S. Ruf, C.F. Steinwachs, c), 'Renormalization of generalized vector field models in curved spacetime'. Phys. Rev. D **98**, 025009 (2018)
117. D.S. Salopek, J.R. Bond, J.M. Bardeen, Designing density fluctuation spectra in inflation. Phys. Rev. D **40**, 1753 (1989)
118. A. Salvio, Initial conditions for critical higgs inflation. Phys. Lett. B **780**, 111–117 (2018)
119. A. Salvio, A. Mazumdar, Classical and quantum initial conditions for higgs inflation. Phys. Lett. B **750**, 194–200 (2015)
120. A. Salvio, A. Strumia, N. Tetradis, A. Urbano, On gravitational and thermal corrections to vacuum decay. J. High Energy Phys. **09**, 054 (2016)
121. I.L. Shapiro, H. Takata, One loop renormalization of the four-dimensional theory for quantum dilaton gravity. Phys. Rev. D **52**, 2162–2175 (1995)
122. M. Shaposhnikov, K. Shimada, Asymptotic Scale Invariance and its Consequences. Phys. Rev. D **99**, 103528 (2019)
123. M. Shaposhnikov, A. Shkerin, a), 'Conformal symmetry: towards the link between the Fermi and the Planck scales'. Phys. Lett. B **783**, 253–262 (2018)
124. M. Shaposhnikov, A. Shkerin, b), 'Gravity, Scale Invariance and the Hierarchy Problem'. J. High Energy Phys. **10**, 024 (2018)
125. M. Shaposhnikov, C. Wetterich, Asymptotic safety of gravity and the Higgs boson mass. Phys. Lett. B **683**, 196–200 (2010)
126. M. Shaposhnikov, D. Zenhausern, Quantum scale invariance, cosmological constant and hierarchy problem. Phys. Lett. B **671**, 162–166 (2009)
127. M. Sher, Electroweak higgs potentials and vacuum stability. Phys. Reports **179**, 273–418 (1989)
128. A.A. Starobinsky, Spectrum of relict gravitational radiation and the early state of the universe. JETP Lett. **30**, 682–685 (1979)
129. A.A. Starobinsky, A new type of isotropic cosmological models without singularity. Phys. Lett. B **91**, 99–102 (1980)
130. A.A. Starobinsky, Dynamics of phase transition in the new inflationary universe scenario and generation of perturbations. Phys. Lett. **117B**, 175–178 (1982)
131. A.A. Starobinsky, J. Yokoyama, Equilibrium state of a selfinteracting scalar field in the De Sitter background. Phys. Rev. D **50**, 6357–6368 (1994)
132. C.F. Steinwachs, *Non-minimal Higgs Inflation and Frame Dependence in Cosmology* (Springer international publishing, Switzerland, Springer Theses, 2014)
133. C.F. Steinwachs, A.Yu. Kamenshchik, One-loop divergences for gravity non-minimally coupled to a multiplet of scalar fields: calculation in the Jordan frame. I. The main results. Phys. Rev. D **84**, 024026 (2011)

134. C.F. Steinwachs, A.Y. Kamenshchik, Non-minimal Higgs inflation and frame dependence in cosmology. AIP Conf. Proc. **1514**, 161 (2013)
135. C.F. van der Steinwachs, M.L. Wild, Quantum gravitational corrections from the Wheeler-DeWitt equation for scalar-tensor theories. Class. Quant. Gravity **35**, 135010 (2018)
136. C.F. van der Steinwachs, M.L. Wild, Quantum gravitational corrections to the inflationary power spectra in scalar-tensor theories. Class. Quant. Gravity **36**, 245015 (2019)
137. K.S. Stelle, Renormalization of higher derivative quantum gravity. Phys. Rev. D **16**, 953–969 (1977)
138. M. Tanabashi et al., Review of particle physics. Phys. Rev. D **98**, 030001 (2018)
139. A. Vilenkin, Quantum Creation of Universes. Phys. Rev. D **30**, 509–511 (1984)
140. G.A. Vilkovisky, The unique effective action in quantum field theory. Nucl. Phys. B **234**, 125–137 (1984)
141. Y.-C. Wang, T. Wang, Primordial perturbations generated by Higgs field and R^2 operator. Phys. Rev. D **96**, 123506 (2017)
142. S. Weinberg, Ultraviolet divergences in quantum theories of gravitation, in S.W. Hawking W. Israel (eds.), General Relativity: An Einstein Centenary Survey (Cambridge University Press, 1980), pp. 790–831
143. C. Wetterich, Exact evolution equation for the effective potential. Phys. Lett. B **301**, 90–94 (1993)
144. C. Wetterich, Quantum scale symmetry (2019). arXiv:1901.04741v2 [hep-th]
145. R.P. Woodard, Avoiding dark energy with 1/r modifications of gravity. Lect. Notes Phys. **720**, 403–433 (2009)
146. Ya. Zeldovich, B. Starobinsky, A.A., Quantum creation of a universe in a nontrivial topology. Sov. Astron. Lett. **10**, 135 (1984)

The Gauge Theoretical Underpinnings of General Relativity

Thomas Schücker

Abstract The gauge theoretical formulation of general relativity is presented. We are only concerned with local intrinsic geometry, i.e. our space-time is an open subset of $\mathbb{R}^4$. Then the gauge group is the set of differentiable maps from this open subset into the general linear group or into the Lorentz group or into its spin cover.

To the memory of Christian Duval.

1 Introduction

One of the many satisfying features of general relativity is—at least in my view—that it allows for many different approaches, some of which are: a field theoretic approach, a chrono-metric approach, a geo-metric approach, gauge theoretic approaches, and causal, perturbative, numerical, discretized and quantum approaches.

An unpleasant corollary to the richness of having several approaches is that some colleagues get carried away by one of the approaches and ignore or dispraise other approaches. If mountaineer A in Fig. 1 thinks that his approach to his favorite summit is unique, then maybe he is sitting on a pre-summit. And he may even be tempted to build a chapel there and to throw stones at his colleague B, when she tries to climb higher.

The main subject of this contribution is a particular gauge theoretic approach to general relativity, another one being presented in the contribution by Hehl and Obukhov to this volume. But let us start by saying a few words about the field theoretic, the chrono- and the geo-metric approaches.

Supported by the OCEVU Labex (ANR-11-LABX-0060) funded by the "Investissements d'Avenir" French government program.

T. Schücker (✉)
Aix Marseille Univ, Université de Toulon, CNRS, CPT, Marseille, France
e-mail: thomas.schucker@gmail.com

S. De Bianchi and C. Kiefer (eds.), *One Hundred Years of Gauge Theory*, Fundamental Theories of Physics 199, https://doi.org/10.1007/978-3-030-51197-5_12

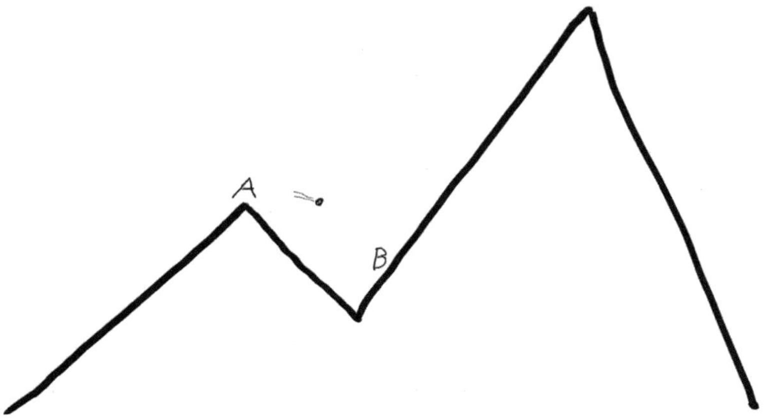

Fig. 1 Sitting on a pre-summit

2 Field Theoretic Approach

The field theoretic approach to general relativity is best set up in parallel to Maxwell's theory of electromagnetism.

Any electric charge Q is source of an electric field, while mass, or better energy E, is source of a gravitational field. The main difference is: charge is Lorentz invariant, while energy is only one component of a 4-vector, the other three components being the momentum $\mathbf{p}$. Therefore Einstein's theory has systematically one index more than Maxwell's.

$$\begin{array}{rcl}
Q & \text{sources} & p^\mu = (E/c, \mathbf{p}) \\
j^\nu = (c\rho, \mathbf{j}) & \text{densities} & T_{\mu\nu} \\
A^\mu = (V/c, \mathbf{A}) & \text{fundam. fields} & g_{\mu\nu} \\
\mathcal{D}_2 A = j & \text{field eq.} & \mathcal{D}_2 g = T \\
(\mathcal{D}_2 A)^\mu := \epsilon_0 c^2 \cdot & \text{diff. oper.} & (\mathcal{D}_2 g)_{\mu\nu} := \dfrac{c^4}{8\pi G} \cdot \\
\partial_\nu (\partial^\nu A^\mu - \partial^\mu A^\nu) & & (R_{\mu\nu} - \tfrac{1}{2} R\, g_{\mu\nu} - \Lambda\, g_{\mu\nu}) \\
{}^{\cdot} := d/d\tau, \quad m\ddot{x}^\mu - q F^\mu{}_\nu \dot{x}^\nu = 0 & \text{test part.} & \not{m}\ddot{x}^\lambda + \not{m}\Gamma^\lambda{}_{\mu\nu}\dot{x}^\mu \dot{x}^\nu = 0 \\
F_{\mu\nu} := \partial_\mu A_\nu - \partial_\nu A_\mu & \text{auxiliaries} & \Gamma^{\cdot}{}_{\cdot\cdot} = (g^{-1}\partial)^{\cdot} g. \\
\Delta\varphi = \dfrac{q}{\hbar} \displaystyle\oint A_\mu \dot{x}^\mu & \text{A-B} \mid \text{proper time} & c\Delta\tau = \displaystyle\oint \sqrt{g_{\mu\nu}\dot{x}^\mu \dot{x}^\nu}
\end{array}$$

We would like to use differential equations to compute the fields produced by *moving* sources. To this end we define charge- and current-densities j and the energy-momentum tensor T.

The question: "Is the electromagnetic force field ($\mathbf{E}/c$, $\mathbf{B}$) or the field of potentials ($V/c, \mathbf{A}$) the fundamental field?" has been source to a long controversy. Some suggested that the acceleration of a small, point-like test charge q of mass m is observable. This acceleration is given by the Lorentz force in terms of the electromagnetic force field ($\mathbf{E}/c$, $\mathbf{B}$). The latter is conveniently encoded in the "field strength" $F_{\mu\nu}$, an anti-symmetric 4×4 matrix and computed from the first partial derivatives of the 4-potential ($V/c, \mathbf{A}$). Later it has been recognized that distances are not observables and *a fortiori* accelerations are not observables. Also quantum mechanics, in particular the famous Aharonov-Bohm effect (A-B), tells us that the potential is fundamental, the force field is an auxiliary, derived field.

We write the field equations, the Maxwell and the Einstein equations, as second order partial differential equations with the potentials as unknowns and the source densities given. The potentials outside a lonely, static and spherically symmetric charge- or mass-distribution are the first exact solutions that one computes. For Maxwell we obtain a pure electric potential $V = Q/(4\pi\,\epsilon_0 r)$ where r is the distance to the center of the charge distribution. For Einstein, we obtain a gravitational potential with leading term $-MG/r$ plus a small term suppressed by $2MG/(c^2 r)$ and therefore falling off like $1/r^2$ plus a harmonic oscillator term $-\frac{1}{6}c^2\Lambda r^2$. The first correction to Newton's universal law is responsible for the perihelium advance of Mercury. The second correction, for positive cosmological constant Λ, induces a repulsive force which increases linearly with distance and which can explain the accelerated expansion of our universe.

Of course in the last two sentences, we have already used the law governing the motion of test particles, the Lorentz force for Maxwell and the "geodesic equation" for Einstein. Both laws give the acceleration in terms of first partial derivatives of the potentials A and g. The geodesic equation therefore identifies the "connection" Γ as gravitational force while the "metric" g encodes the gravitational potentials. Note the mismatch: the electromagnetic force field is the curvature F of the "gauge connection" or "gauge potential" A, while the gravitational force is the connection Γ of the metric g. The curvature R of the connection Γ describes the gradient of the gravitational force. Since this gradient is responsible for the tides, R is sometimes called tidal force.

Physically the geodesic equation expresses Newton's first axiom: force = mass × acceleration. Note that the mass of the test particle cancels. This is the equivalence principle visualized impressively in Newton's tube. Note that thanks to this cancelation Einstein's test particle can be massless. (This is only true for Maxwell's test particle in the trivial case $q = 0$.) Therefore general relativity predicts the bending of light by a massive object like the sun. (Warning: the geodesic equation is only valid for spinless test particles. Including the spin of the photon gives rise to more complicated phenomena: gravitational birefringence, studied in recent work [4] by Christian Duval who left us in September 2018.)

General relativity has three axioms, the Einstein equation, the geodesic equation (the latter is in fact a consequence of the former) and the axiom of time. This axiom will be discussed in the next section together with its electromagnetic analogue, the Aharonov-Bohm effect.

To close this section, we mention a few important features of Maxwell's and Einstein's theories.

8 additive terms	size of $\mathcal{D}_2$	$\sim 10^5$ add. terms in g & g^{-1}
Poincaré, gauge	symmetries	diffeomorphisms
$\partial_\mu j^\mu = 0$	(non-)conservation	$D_\mu T^{\mu\nu} = 0$
linear, 2nd order	uniqueness of $\mathcal{D}_2$	2nd order
$\epsilon_0 = 8.854187817 \cdot 10^{-12} \cdot$ $\text{N}^{-1}\text{C}^2\text{m}^{-2} \pm 0\,\%$	coupl. const.	$G = 6.674208 \cdot 10^{-11} \cdot$ $\text{N}\,\text{kg}^{-2}\text{m}^2 \pm 5 \cdot 10^{-5}$ $\Lambda = 1.11 \cdot 10^{-52}\,\text{m}^{-2} \pm 2\,\%$

Maxwell's operator $\mathcal{D}_2$ (written in inertial coordinates) has 8 additive terms in A. These terms were deduced from experimental facts and charge conservation. Einstein's operator $\mathcal{D}_2$ has some 10^5 additive terms in g and its inverse. These are hidden in the Ricci curvature $R_{\mu\nu}$ and the curvature scalar R and were deduced from invariance under general coordinate transformations and energy-momentum (non-)conservation.

Maxwell's theory (including the Aharonov-Bohm effect) is invariant under Poincaré transformations and under gauge transformations. Einstein's theory is invariant under general coordinate transformations (or "diffeomorphisms").

Maxwell's equations imply charge conservation, $\partial_\mu j^\mu = 0$. Einstein's equations would imply conservation of the energy-momentum of matter if there were no gravitational field. Otherwise there can be exchange of energy and momentum between matter and the gravitational field; one speaks about "covariant" (matter) energy-momentum conservation, $D_\mu T^{\mu\nu} = 0$. It should be noted that the linearity of Maxwell's equations implies that the electromagnetic field does not carry its own source = charge, while the non-linearity of Einstein's equations implies that a generic gravitational field does carry its own source = energy-momentum. (Warning: there is still no consensus on a definition of the energy density of a gravitational field.)

Maxwell's operator is the unique 1-parameter family of *linear* second order partial differential operators acting on the potentials A that are compatible with charge conservation and invariant under Poincaré and gauge transformations. This single parameter is the coupling constant ϵ_0.

Einstein's operator is the unique 2-parameter family of second order partial differential operators acting on the metric tensor g and that are compatible with covariant (matter) energy-momentum conservation and invariant under general coordinate transformations. The two parameters are the coupling constants G and Λ.

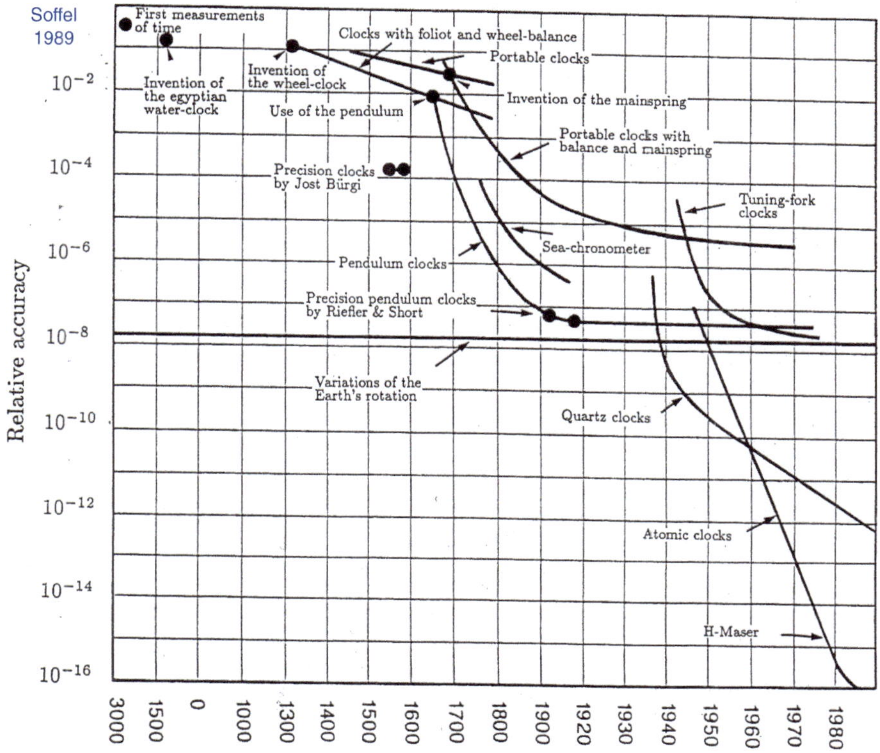

Fig. 2 The historical evolution of the accuracy in time keeping. Figure taken from Soffel's book [7]

3 Chrono- and Geo-Metric Approaches

The twentieth century has witnessed two revolutions concerning time, a technological one and a conceptual one.

Before the technological revolution, the official time keeping device was the rotating Earth with two draw-backs: its extended size and the inherent relative error of 10^{-8} due to the Earth not being rigid enough. The daily period of Earth was replaced by the period of light emitted by a caesium 133 atom. Atomic clocks are point-like enough for travelling and allowed for time keeping with relative accuracy of 10^{-16} in 1980, see Fig. 2, taken from Soffel's book [7]. Today the accuracy is reaching 10^{-18}.

Before the conceptual revolution, time was absolute. Einstein taught us already in 1905 to be more careful. He defined proper time τ for a point-like clock travelling with velocity less than the speed of light on a trajectory $x^{\mu}(p)$ in a gravitational potential $g_{\mu\nu}(x)$:

$$c\tau := \int_{p_0}^{p_1} \sqrt{g_{\mu\nu}(x(p))\,\frac{\mathrm{d}x^\mu}{\mathrm{d}p}\,\frac{\mathrm{d}x^\nu}{\mathrm{d}p}}\,\mathrm{d}p. \tag{1}$$

Of course the signature of our metric is $+---$.

By definition proper time is proper to each clock and depends on its entire history: how much time it spent in what gravitational potential and with which velocities it was cruising. Today we know with very high precision that point-like clocks do indicate proper time. To test the axiom of time we need two precise clocks and synchronize them when they are at the same location. Then we make them travel on separate paths, reunite them and read their proper time difference $\Delta\tau$. It coincides with the line integral over the closed path defined by the two paths between synchronisation and reunification.

The electromagnetic (quantum) analogue of proper time difference is the phase difference which the wave function of a charged particle accumulates when it passes through two holes in a screen and then follows two distinct paths in an electromagnetic potential. The two paths meet again and there the phase shift is measured. This phase shift can be computed from Schrödinger's equation coupled to the electromagnetic *potential* and is the line integral of this potential over the closed curve defined by the two paths between emission and reunification. The calculation was published in 1949 by Ehrenberg and Siday [5] and independently by Aharonov and Bohm [1] in 1959. It was measured for the first time in 1960 by Chambers [3].

The accuracy of length measurements evolved at a similar pace as the one of time measurements. The (French) definition of the metre as a part of the polar circumference of the Earth and materialized by a rigid rod of platinum was abandoned in favor of a multiple of the wave length of light emitted by a krypton 86 atom. However, as anticipated by Einstein, the precision of these new measurements killed the very notion of distance: rigid rods are incompatible with relativity as their speed of sound tends to infinity. But how useful can be a metre stick made of rubber? Now the wavelength emitted by an atom is not rigid either and changes when the atom is immersed in an electric field as shown experimentally by Stark [8] in 1913. Einstein's axiom of time implies that the wavelength emitted by an atom also changes in a gravitational potential (explaining that Stark should have become an admirer of Einstein).

Both, rods and waves, are by definition extended objects and cannot be described by a single trajectory.

In 1983 the official funeral of the metre took place when the 17th *Conférence Générale des Poids et Mesures* decided:

1. The metre is the length of the path travelled by light in vacuum during a time interval of 1/299 792 458 of a second.
2. The definition of the metre in force since 1960, based upon the transition between the levels 2p10 and 5d5 of the atom of krypton 86, is abrogated.

At the same time, the speed of light became an absolute quantity *defined* by 299 792 458 m per second and *without* error bars.

To see why this decision tolls the bell to distance measurements, let us take the distance between the Earth and the Moon. To simplify let us assume that they are both at rest. This distance is measured continuously since 1962 by the famous Lunar Laser Ranging experiment: A laser beam is sent to the Moon and reflected back to Earth. The time of the return flight, around 2.5 s, is measured and the distance Earth to Moon is obtained according to the definition in force: $\sim 3.8 \ldots \cdot 10^8$ m with a present error bar of ± 1 cm. This distance is found to grow by 3.8 cm per year. If on the other hand we wanted to measure the distance Moon to Earth with both, laser cannon and clock, on the Moon (a more expensive set-up), we would obtain a distance increased by 30 cm due to the weaker gravitational potential at the Moon's surface, $d(E, m) \neq d(m, E)$.

We close this section by briefly mentioning the geo-metric approach. Geometers like Gauß, Riemann, Christoffel, Levi-Civita, Ricci, Bianchi, ... worked out a definition of distance in curved spaces like the surface of the Earth. This definition, the arc-length of a geodesic, has the virtue of invariance under general coordinate transformations. This is precisely the virtue Einstein wanted for his theory of gravity. He simply replaced space by space-time, the arc length by proper time and kept the rest of the formalism. Therefore, although distances have been banned, the geometric language—metric, connection, parallel transport, geodesics and curvature – is still pertinent. The same language is pertinent to Maxwell's theory and to non-Abelian Yang-Mills theories describing the weak and strong nuclear forces. The next section tries to sketch out this common ground.

4 Gauge Theoretic Approach

To avoid subtleties of global geometry, let spacetime be a contractible open subset M of $\mathbb{R}^4$. Let G be a finite-dimensional Lie group. Define the gauge group ${}^M G$ as the set of all differentiable maps

$$\begin{aligned} M &\longrightarrow G \\ x &\longmapsto g(x) \end{aligned}$$

with pointwise multiplication $(g\tilde{g})(x) := g(x)\tilde{g}(x)$.

The elements $g \in G$ are sometimes referred to as *rigid* transformations in contrast to the spacetime dependent gauge transformations $g(x)$. The gauge group ${}^M G$ is obtained by 'gauging' or 'mollifying' or 'covariantizing' the rigid group G.

Both the gauge group ${}^M G$ and the diffeomorphism group $Diff(M)$, are infinite-dimensional groups, not Lie groups. Nevertheless they both admit infinite-dimensional Lie algebras: ${}^M \mathfrak{g}$ with $\mathfrak{g}$ the Lie algebra of G and commutators defined pointwise; and the Lie algebra of vector fields with the Lie bracket.

(Warning: some authors use the word gauge group in a wider sense including diffeomorphism groups and finite-dimensional groups.)

4.1 Linear Algebra

Fix a spacetime point $x \in M$ and consider its tangent space $T_x M =: V$. We have to parameterise the set of all Minkowski metrics on V. Choose a basis $b_i,\ i = 0,\ 1,\ 2,\ 3$ of V and define the metric tensor by

$$g_{ij} := g(b_i, b_j).$$

Parameterisation 1: For this fixed basis we have a one-to-one correspondence between the set of all metrics and the set of all symmetric matrices of signature $(+ - --)$.

The matrix g'_{ij} of the metric g with respect to a different basis b'_i,

$$b'_i = (\gamma^{-1})^j{}_i\, b_j, \quad \gamma \in GL_4,$$

is given by

$$g'_{ij} := g(b'_i, b'_j) = (\gamma^{-1T} g \gamma^{-1})_{ij}.$$

Attention, we use 4×4 matrices to describe a metric as well as a change of bases, two quite different mathematical objects.

Theorem (Gram & Schmidt): Any metric has an orthonormal basis e_a, i.e. a basis such that

$$g(e_a, e_b) = \eta_{ab} := \begin{pmatrix} +1 & 0 & 0 & 0 \\ 0 & -1 & 0 & 0 \\ 0 & 0 & -1 & 0 \\ 0 & 0 & 0 & -1 \end{pmatrix}. \tag{2}$$

Given an orthonormal basis e_a, any other basis e'_a with

$$e'_a = (\Lambda^{-1})^b{}_a\, e_b, \quad \Lambda \in GL_4 \tag{3}$$

is also orthonormal if and only if

$$\eta = \Lambda^{-1T} \eta \Lambda^{-1}, \quad \Lambda \in O(1,3) \subset GL_4. \tag{4}$$

Parameterisation 2: Choose a basis e_a of V and declare it orthornomal. This defines a metric. However two bases connected by a Lorentz transformation Λ define the same metric.

4.2 Connection and Curvature

Consider all tangent spaces together. They define a family of vector spaces indexed by the points x of spacetime M. By definition a (pseudo-Riemannian) metric on M is a differentiable family of metrics on each tangent space in the same sense as a vector field is a differentiable family of vectors. A frame is a differentiable family of bases $b_i(x)$ and an orthonormal frame is a differentiable family of orthonormal bases $e_a(x)$.

Choose coordinates x^μ, $\mu = 0, 1, 2, 3$ on spacetime. They define a particular frame

$$b_\mu := \frac{\partial}{\partial x^\mu}, \tag{5}$$

called holonomic.

Theorem: M is flat if and only if it admits an orthonormal and holonomic frame. The corresponding coordinates are called inertial.

Theorem: There is a unique metric and *torsion*-free connection Γ. It is a 1-form with values in the Lie algebra gl_4 and can be computed explicitly in terms of $b_i(x)$, $g_{ij}(x)$ and their first partial derivatives with respect to an arbitrary coordinate system. With respect to a change of frames $\gamma(x) \in {}^M GL_4$ the connection transforms as

$$\Gamma' = \gamma \Gamma \gamma^{-1} + \gamma \mathrm{d}\gamma^{-1}. \tag{6}$$

Theorem: The (Riemann) curvature

$$R := \mathrm{d}\Gamma + \tfrac{1}{2}[\Gamma, \Gamma] \tag{7}$$

is a 2-form with values in the Lie algebra gl_4 and transforms *homogeneously* under GL_4 gauge transformations:

$$R' = \gamma R \gamma^{-1}. \tag{8}$$

4.3 Three Formulations of General Relativity

For generic M there are *three parameterisations* of the metric:

- No name (no practical use): Choose an arbitrary frame $b_i(x)$ and a coordinate system x^μ. Compute the metric tensor g_{ij}, the connection Γ and the curvature R. Changes of frames are GL_4 gauge transformations and changes of coordinates are diffeomorphisms. They decouple in the sense of the semi-direct product: ${}^M GL_4 \rtimes Diff(M)$.

- Einstein: Choose a coordinate system x^μ and use the metric tensor $g_{\mu\nu}(x)$. Now a change of coordinates induces a change of frames with the GL_4 gauge transformation given by the Jacobian of the diffeomorphism:

$$\gamma^\mu{}_\nu = \frac{\partial x^\mu}{\partial x^\nu}. \tag{9}$$

"The GL_4 gauge group is lost due to gauge fixing".
- Elie Cartan: Choose an orthonormal frame $e_a(x)$ and a coordinate system x^μ. The metric tensor is constant, η_{ab}. The connection, traditionally denoted by ω, now takes values in the Lorentz algebra $so(1, 3)$ and is often called 'spin connection'.

$$\omega' = \Lambda\omega\Lambda^{-1} + \Lambda \mathrm{d}\Lambda^{-1}. \tag{10}$$

The symmetry group is ${}^M O(1, 3) \rtimes Diff(M)$.

Theorem: For a group G, denote by G^{id} the connected component of the identity. ${GL_4}^{id}$ is doubly connected. Its covering group has no faithful finite-dimensional represention.
$SO(1, 3)^{id}$ is doubly connected. Its covering group $Spin(1, 3)^{id}$ does have faithful finite-dimensional representions.

This theorem makes Cartan's parametrization mandatory if half-integer spin fields, e.g.. Dirac particles, are present.

Torsion can be added simply in Cartan's formulation by declaring the spin connection ω to be an additional fundamental field, independent of the metric described by the orthonormal frame e. Then, by Cartan's equation, the half-integer spin densities are the source of torsion in the same sense that by Einstein's equation the energy-momentum densities are the source of curvature. And we have the commuting diagram shown in Fig. 3.

For details the reader is referred to [6].

5 Concluding Remarks

Today, experimental evidence forces us to abandon the concepts of rigidity and staticity on scales ranging from atomic to cosmic. This was already foreseen by Einstein's theories of relativity. (However he had a hard time to accept the observational evidence that our universe is not static.) In this context it is natural to also abandon rigid transformations and to favor gauge theories.

All current models describing the four basic forces are of Euler-Lagrange type and are gauge theories in the sense of the definition at the beginning of Sect. 5. The main differences between general relativity and the standard model of electromagnetic, weak and strong forces are:

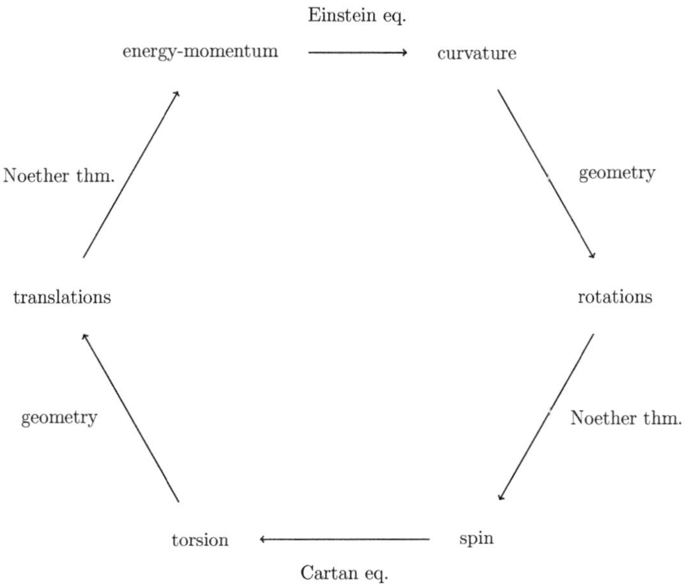

Fig. 3 A commuting diagram

- The fundamental field of general relativity (no torsion) is the gravitational potential = metric and the force is the connection = gauge potential; the fundamental field of the standard model is the gauge potential and the force is the curvature.
- The Lagrangian of general relativity is linear in curvature; the Lagrangian of the standard model is quadratic in curvature.
- The group $G = O(1,3)$ to be gauged in general relativity is simple but not compact;
  the group $G = [SU(2) \times U(1) \times SU(3)]/[\mathbb{Z}_2 \times \mathbb{Z}_3]$ to be gauged in the standard model is not simple but compact with finite-dimensional unitary representations.
- In general relativity, masses are put in by hand; in the standard model, masses are generated by spontaneous symmetry breaking.
- The standard model has a perturbatively renormalisable quantum theory. Though its perturbation series diverges, this theory is very well tested experimentally. Only recently such a quantum theory was discovered for general relativity by Damiano Anselmi [2].

After a century of efforts, we are still lacking a unified model of all basic forces and we are still lacking experimental evidence for general relativistic quantum effects.

Acknowledgements It is a pleasure to thank Silvia De Bianchi and Claus Kiefer for having organized a most stimulating conference and for their warm hospitality in Bad Honnef. This work has been carried out thanks to the support of the OCEVU Labex (ANR-11-LABX-0060) and the A*MIDEX project (ANR-11-IDEX-0001-02) funded by the "Investissements d'Avenir" French government program managed by the ANR.

References

1. Y. Aharonov, D. Bohm, Significance of electromagnetic potentials in quantum theory. Phys. Rev. **115**, 485 (1959)
2. D. Anselmi, Aspects of perturbative unitarity. arXiv:1606.06348 [hep-th], Phys. Rev. D **94**, 025028 (2016); D. Anselmi, On the quantum field theory of the gravitational interactions. arXiv:1704.07728 [hep-th], JHEP **1706**, 086 (2017); D. Anselmi, Fakeons, microcausality and the classical limit of quantum gravity. arXiv:1809.05037 [hep-th], Class. Quant. Grav. **36**, 065010 (2019)
3. R. Chambers, Shift of an electron interference pattern by enclosed magnetic flux. Phys. Rev. Lett. **5**, 3 (1960)
4. C. Duval, T. Schücker, Gravitational birefringence of light in Robertson-Walker cosmologies. arXiv:1610.00555 [gr-qc], Phys. Rev. D **96**, 043517 (2017); C. Duval, L. Marsot, T. Schücker, Gravitational birefringence of light in Schwarzschild spacetime, arXiv:1812.03014 [gr-qc], Phys. Rev. D **99**, 124037 (2019)
5. W. Ehrenberg, R. Siday, The refractive index in electron optics and the principles of dynamics. Proc. Phys. Soc. B **63**, 8 (1949)
6. M. Göckeler, T. Schücker, *Differential Geometry, Gauge Theories, and Gravity* Cambridge Monographs on Mathematical Physics (Cambridge University Press, 1987)
7. M. Soffel, *Relativity in Astrometry, Celestial Mechanics and Geodesy* (A& A Library, Springer, 1989)
8. J. Stark, Beobachtungen über den Effekt des elektrischen Feldes auf Spektrallinien I. Quereffekt. Annalen der Physik **43**, 965 (1914); also published in Sitzungsberichte der Königlichen Preuss. Akad. d. Wiss (1913)

Past and Future of Gauge Theory

Gerard 't Hooft

Abstract A brief account is sketched on how the doctrine based on local gauge invariance developed over the years, turning into a pivotal element in model building for elementary particles. This principle owes its success to being renormalizable order by order in the perturbation expansion for small coupling strengths. An important point is the requirement of unitarity and locality, which shows up in the details of the Feynman rules. After gauge fixing, one finds that the system displays an elegant new symmetry: BRST invariance. Recent experimental findings in the Large Hadron Collider may point the way to the future. To capture new clues for the future, we must bear in mind the fundamental successes of steps that were made in the past.

1 The Early Days

Shortly after Albert Einstein's publication of his theory of General Relativity, in which he succeeded to reconcile the gravitational force with his earlier ideas on relativity, Weyl [1] proposed to consider an additional class of transformations in Riemann spaces, of the form

$$\mathrm{d}x^{\mu} \to e^{\omega(x)}\mathrm{d}x^{\mu}, \tag{1.1}$$

which would transform the metric tensor $g_{\mu\nu}(x)$ as

$$g_{\mu\nu} \to e^{-2\omega(x)} g_{\mu\nu}. \tag{1.2}$$

Here, $\omega(x)$ may be taken as a space-time dependent, differentiable function. It is a transformation that affects the scale of things. Measuring things would require

G. Hooft (✉)
Institute for Theoretical Physics Utrecht University, Postbox 80.089, 3508, TB, Utrecht, The Netherlands
e-mail: g.thooft@uu.nl
URL: http://www.staff.science.uu.nl/~hooft101/

S. De Bianchi and C. Kiefer (eds.), *One Hundred Years of Gauge Theory*, Fundamental Theories of Physics 199, https://doi.org/10.1007/978-3-030-51197-5_13

the gauge of a weighing scale, so that now, also the sizes and time periods would be subjected to the principle of relativity.

If $\omega(x)$ is assumed to be x dependent, this would mean that the sizes of objects can no longer be compared to the sizes of neighbouring objects, unless one introduces a connection field $\omega_\mu(x)$. This enables one to define covariant derivatives for size-dependent fields of the form

$$D_\mu = \partial_\mu - \omega_\mu . \tag{1.3}$$

We are thus reminded of the electro-magnetic connection field $A_\mu(x)$, which only differs from this by a factor i: a charged field $\psi(x)$ can be rotated in the complex plane,

$$\psi \to e^{i\omega(x)}\psi , \tag{1.4}$$

and such fields can be connected at different space-time points by introducing as covariant derivative

$$D_\mu\psi = (\partial_\mu - iA_\mu)\psi . \tag{1.5}$$

This is why Eqs. (1.2) and (1.4) are now called 'gauge transformations', and the fields $\omega_\mu(x)$ and $A_\mu(x)$ are called 'local gauge fields'.

2 Quantum Electrodynamics, QED

Weyl may have hoped that electromagnetism could be accommodated for along these lines, by subjecting sizes to the general relativity principle, but, until this day, this approach has not shown to be very useful.[1] Nevertheless, progress was made in subjecting Maxwell's laws of electromagnetism to the principles of quantum mechanics. After a long period of small but crucial successes, it was found that this theory has to be handled as a perturbation expansion, which works well because it so happens that the primary quantity that must be used as a small perturbation parameter, is the finestructure constant,

$$\alpha = \frac{e^2}{4\pi\hbar c} \approx \tfrac{1}{137} , \tag{2.1}$$

which happens to be conveniently small.

The theory obtained its present shape through the works of Feynman [3], Schwinger [4], Tomonaga [5] and[2], Dyson [6], and many others but not before many essential and ingenious experiments had been performed that lead theoreticians in

[1] In this author's opinion, Weyl may have given up too soon.

[2] For a historical review see e.g. A. Pais [2]

the right direction. An important step to be taken was the realisation that the essential physical parameters such as the masses and the coupling strengths of the elementary particles have to be adjusted at various stages in the theory, a procedure called *renormalization* [3–6]. It was something of a shock to realise that renormalization can end up as an infinite correction to the theory—and yet, it works.

The infinities turned out neatly to cancel against one another if the calculations were done with sufficient care. Renormalizable theories were known to contain *scalar* fields $\varphi(x)$, *fermionic* fields $\psi(x)$, *electromagnetic* fields $F_{\mu\nu}(x)$ and their potential fields $A_\mu(x)$.

3 The Weak Interaction

The weak interaction also appeared to be mediated by a vector particle, just like the electromagnetic force is mediated by the photon. There should be at least two types of weak photons, W^+ and W^-, besides the photon γ itself. The $W^\pm$ should have large and equal masses, one being the antiparticle of the other. A massive neutral component, Z^0 was suspected to be needed as well, but this was not yet certain. Around the 1960s, Veltman [7] was convinced by the experimental evidence that *the weak interactions had to be some modification of a Yang–Mills theory*.

Yang and Mills [8] had constructed a magnificent theoretical edifice in which they generalised the local gauge transformations first thought of by Weyl and adapted to apply to electro-magnetism. The generalisation was obtained by replacing the local gauge symmetry by a *non-Abelian* local symmetry. This non-Abelian force would enable photons in turn to carry electric charges, unlike ordinary photons. A theory for charged photons, was just what Veltman needed.

However, the vector particles mediating the weak force must have mass, in order to match the exceedingly short range of the weak force, whereas the particles predicted by the Yang–Mills theory should be strictly massless. In addition, all Yang–Mills photons had to couple in the same way with the charged particles, while the Z^0 component, if it existed, would couple differently to charged and neutral currents. Veltman had some ideas on how to modify the Yang–Mills interactions so as to accommodate for these complications.

The problem now was the renormalization. There were good arguments—although not yet proven at that time—that the pure Yang–Mills theory should be renormalizable. The extra terms would not obey the local gauge symmetry that Yang and Mills had thought to be essential, but what does this symmetry have to do with renormalization? Veltman suspected that the only complication was the complexity of the theory he obtained. To handle this, he designed a computer program for *algebraic computations*.

This was pioneering work. Computers in those days were no match even to the simplest mobile phones of today. To speed up their performance, Veltman had to learn machine language. He taught his computer how to read off the Feynman diagrams from the Lagrangian of a theory, and to determine how the infinite renormalization effects should cancel out.

It was a disappointment when he found that effects that may safeguard renormalizability for a massless theory, do not work well as soon as one adds terms that violate local gauge invariance. His modified version of Yang–Mills theory did not seem to pass the tests. He found out that adding non-gauge invariant terms leads to substantial complications.

Veltman knew about spontaneous breaking of symmetries. But also for such theories there seemed to be problems. A well-known theorem by Nambu and Goldstone [9, 10] states that:

> One cannot have spontaneous symmetry breaking without generating one or more massless particles.

Indeed, chiral isospin symmetry, $SU(2)_{\text{left}} \otimes SU(2)_{\text{right}}$, generates a very light hadronic particle, the pion. Its mass can be attributed to a small explicit symmetry breaking due to the mass of the down quark. Thus we learned that gauge invariance and the Nambu Goldstone theorem were not to be messed with. And so the problem was: the weak force does not seem to be associated with any massless particle, yet exact gauge invariance would be needed to have it properly renormalized.

4 The Brout–Englert–Higgs Mechanism

The solution to this problem turned out to be what is now called the Brout–Englert–Higgs mechanism [11, 12]. Strictly speaking, this mechanism is not a symmetry breaking at all (neither explicit nor spontaneous), but a different realisation of a local gauge symmetry. This is why the Nambu–Goldstone theorem does not apply.

With today's understanding of perturbative quantum field theories, this is not difficult to understand; we see immediately that the BEH mechanism does not generate massless field degrees of freedom. But one must remember that, in the early days, quantum field theories were not considered trustworthy; one had to rely on abstract algebraical arguments, where mistakes were easily made. Not only did the BEH mechanism provide a mass term for the vector fields, thus making the weak force short-range, it also allowed us to add couplings between Higgs field and fermions in such a way that the masses of charged leptons and neutral leptons became different from one another. These couplings are now called the Yukawa couplings.

Actually, by attaching the appropriate quantum numbers to the Higgs field, we could generate the observed mixing between the electromagnetic force and the weak forces. Thus, the BEH mechanism solved three problems at once.

5 The Folklore of "the Origin of Mass"

To understand how the renormalization problem was solved by the BEH mechanism, one has to contemplate the *short distance structure of the theory.* At short distances, the effects of mass terms and the other interactions causing the BEH mechanism become negligible, so we re-obtain the unbroken theory. In the original Yang–Mills theory, the YM photons each only possess two helicity degrees of freedom, whereas massive spin 1 particles have three (at $s = 1,\ \ m_s = 1,\ 0,\ \text{or}\ -1$). The extra degrees of freedom (the longitudinal one, at $m_s = 0$), come from the scalar Higgs field.

To generate a mass term that happens to be manifestly gauge-invariant, no BEH mechanism is needed. Therefore, the BEH mechanism cannot be regarded as "the origin of mass"; it is the origin of mass terms that are not manifestly gauge invariant.

Let us clarify our use of the word 'manifestly' here. How is it that the BEH mechanism cannot be regarded as a 'spontaneous symmetry breaking'? The reason lies in the concept of spontaneous symmetry breaking. If the symmetry is a global one, then symmetry breaking at large distances suffices to break the symmetry everywhere else too. This is meant by the word 'spontaneous': if the symmetry is broken far from us, it will be broken here as well. In case of a global symmetry, there can be many different vacuum states that each break the symmetry. They all break the symmetry at the boundary of space-time at infinity. But if a symmetry is local, then breaking the symmetry at one point in space-time does not lead to symmetry breaking anywhere else. Most importantly, there still is only one single vacuum state.

Therefore, in the language of quantum field theory, the BEH mechanism should not be addressed as a 'spontaneous symmetry breaking'. The vacuum state is always invariant when a *local* gauge symmetry is considered—although it may be remarked that, in the zero coupling limit, the symmetry turns into a global one, in which case the vacuum state does break the symmetry. In short, the BEH mechanism only breaks the symmetry spontaneously in the limit of vanishing gauge couplings. At finite gauge couplings, it recovers a single vacuum state, and only its algebraic form reminds us of spontaneous breakdown.

In the case of local symmetries, one must impose a gauge condition such as the Landau gauge, $\partial_\mu A^a_\mu = 0$, or a unitary gauge constraint [13] such as

$$\varphi_a = (0,\ 0,\ F + \varphi'_3,\ 0),$$

where φ_a is the Higgs multiplet. It is a choice of a basis for the isovector field $\varphi_a(x)$, at all space-time points x^μ. The field φ'_3 is here invariant under local gauge transformations, and therefore it describes a neutral, massive particle, the Higgs particle.

6 Unitarity

Our original proofs of renormalizability were based on various techniques. One was the cutting rules for Feynman diagrams [14]. Consider the Feynman diagrams that have to be calculated to prove unitarity for the scattering matrix S,

$$S \cdot S^\dagger = \mathbb{I}, \tag{6.1}$$

and causality:

$$[\phi(x_1),\ \phi(x_2)] = 0 \quad \text{if} \quad (x_1 - x_2)^2 > 0, \tag{6.2}$$

that is, $(x_1 - x_2)$ is space-like.

The renormalization of the amplitudes in the theory must be carried out at the level of the Feynman diagrams. The rules for calculating the amplitudes associated to a diagram, the Feynman rules, consist of propagators, which, in momentum space, come with the factors

$$\Delta^F(k) = \frac{1}{2\pi i}\,\frac{1}{k^2 + m^2 - i\varepsilon}, \quad \text{when off mass shell, or} \tag{6.3}$$

$$\Delta^{\pm}(k) = \theta(\pm k_0)\,\delta(k^2 + m^2), \quad \text{when on the mass shell.} \tag{6.4}$$

Here, k is the 4-momentum, m is a mass, and ε is an infinitesimal, positive number. In position space, we work with their Fourier transforms

$$\Delta^F(x) = (2\pi)^{-3}\int \mathrm{d}^4k\, e^{ik\cdot x}\,\Delta^F(k); \qquad \Delta^{\pm}(x) = (2\pi)^{-3}\int \mathrm{d}^4k\, e^{ik\cdot x}\,\Delta^{\pm}(k). \tag{6.5}$$

These functions obey the very important *cutting rules*:

$$\begin{aligned}\Delta(x) &= \theta(x_0)\Delta^+(x) + \theta(-x_0)\Delta^-(z);\\ \Delta^*(x) &= \theta(x_0)\Delta^-(x) + \theta(-x_0)\Delta^+(x),\end{aligned} \tag{6.6}$$

where $\Delta^*(x)$ stands for the complex conjugate of $\Delta(x)$, and $\theta(z)$ stands for the Heaviside theta function (step function). Equations (6.6) are easily derived from

$$\theta(z) = \frac{1}{2\pi i}\int_{-\infty}^{\infty} \mathrm{d}\tau\, \frac{e^{i\tau z}}{\tau - i\varepsilon}. \tag{6.7}$$

The Feynman rules are formally defined by expanding functional integrals of the form

$$\int \mathcal{D}\varphi(x)\, e^{i\int(\mathcal{L}(\varphi, x) + J(x)\varphi(x))\mathrm{d}^4x} \tag{6.8}$$

in powers of the interaction parameters and the source fields $J(x)$.

These Feynman rules are subsequently complemented by the Feynman rules for the complex conjugates of amplitudes, for which we replace the propagators $\Delta(x)$ by $\Delta^*(x)$. A minus sign then emerges for every vertex of the diagram.

To find the expressions needed for verifying the unitarity condition (6.1), we underline the coordinates of the vertices in $S^\dagger$, and we use the on-shell propagators (6.4) for the particle trajectories connecting S with $S^\dagger$. The rules are then:

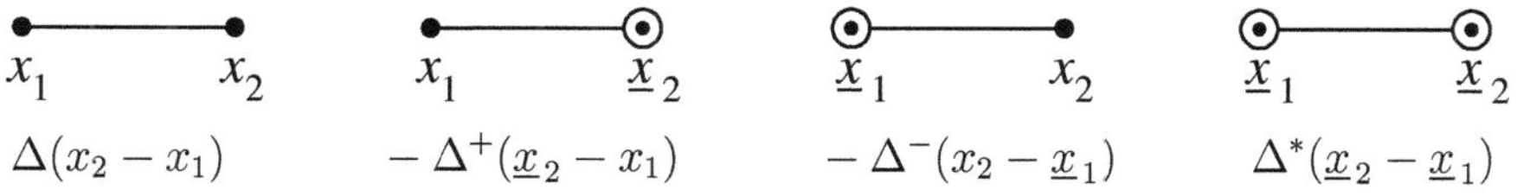

(6.9)

The underlined vertices of the Feynman diagram are here indicated by drawing small circles around them. The sum of all four expressions in (6.9) is zero, since, regardless the sign of the time component of $(x_2 - x_1)$, the propagators cancel pairwise.

Now consider a Feynman diagram with a given topology. Then consider the sum of the diagrams obtained by underlining any subset of the vertices. This is 2^V diagrams if V is the number of vertices. This sum vanishes, which can be verified in configuration space by applying the rules of Eq. (6.9) to the one vertex for which the time coordinate is the largest of all. This way, we see that all terms in the sum cancel pairwise. We write

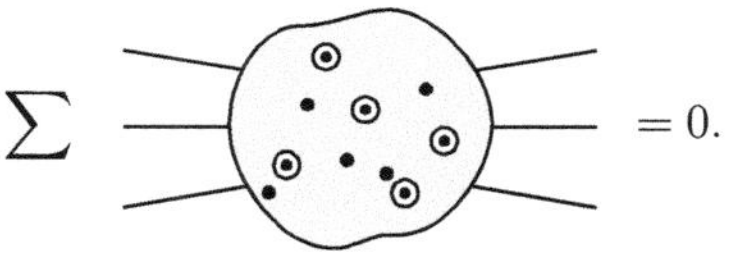

(6.10)

If we disentangle the diagrams by writing all underlined vertices at the right, and the others at the left, this reads as:

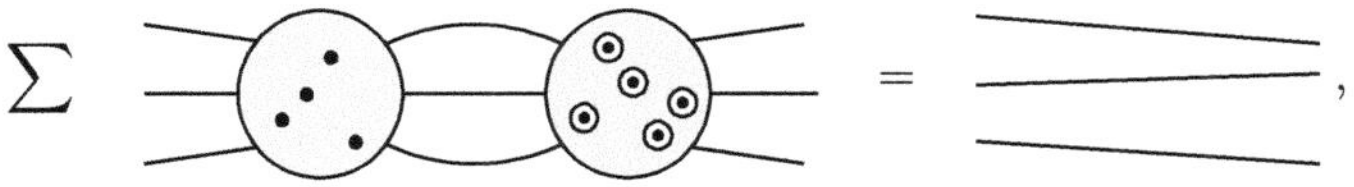

(6.11)

where only the diagram with zero vertices survives. This is the unitarity relation (6.1).

This exercise leads to a very important piece of insight: *Unitarity holds if all propagators of the theory take the form* (6.3), *where the mass terms m^2 must be non-negative,* and also any possible multiplicative factor in the propagator must be non-negative, otherwise the particle states associated to the external lines of the scattering matrix S cannot be normalized. Furthermore, all interaction terms in the Lagrangian must be real, otherwise the hermitian conjugate of the scattering matrix would not be described by $\Delta^*(k)$. It is these features that have to be carefully guarded when we renormalize a theory. How all ghost propagators cancel when all Feynman diagrams are added, is explained in [15].

7 BRST

In a local gauge theory, it is not always easy to see that the above conditions are met. The original Yang–Mills Lagrangian does not at all meet our requirement, exactly because of local gauge invariance: the bilinear terms cannot be inverted to give a propagator, because of degeneracies left by the symmetry structure. In principle, this bug is easy to repair: one must *fix the gauge*, by imposing a gauge condition:

$$C(A,\varphi)=0, \tag{7.1}$$

where C is a function of the fields A_μ and φ that can be tuned to zero by performing gauge transformations, which implies that C should *not* be gauge-invariant (A_μ stands short for all vector fields in the theory, and φ for all other matter fields).

In practice, this may seem to work, but it is more elegant to impose the identity by adding a term $-\frac{1}{2}C^2$ to the local Lagrange density $\mathcal{L}(x)$. For instance, if $C=\partial_\mu A_\mu$, one gets the Feynman propagator

$$\Delta^F_{\mu\nu}(k)=\frac{g_{\mu\nu}}{k^2-i\varepsilon}. \tag{7.2}$$

Here $g_{\mu\nu}=\mathrm{diag}(-1,1,1,1)$ is the metric of Minkowski space-time. The problem with this propagator is not that it is massless, as indeed photons are massless, but rather the minus sign in the metric. It would imply a component of the photon with spin in the time direction, contributing with a negative sign to the unitarity equation. In fact, the third component in the space direction would also be a problem since the photon is expected to possess only two helicities, not three. To prove that all these 'ghost particles' cancel out in the theory,

$\sum_{\text{ghosts}}$ 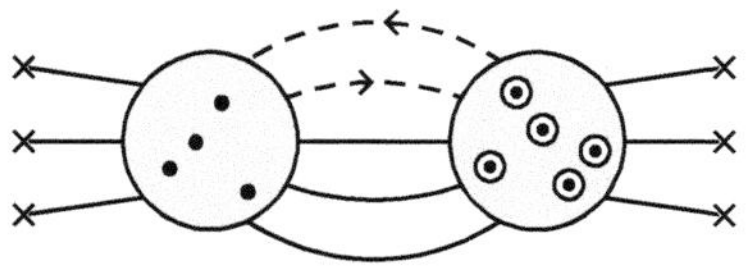 $=0,$

(7.3)

one must be able to transform them away. First however, one must realise that constraints such as Eq. (7.1) act as a Dirac delta in the functional integrals, and Dirac deltas, when transformed, lead to Jacobian factors. The Jacobian factor here turns out to behave as a new ghost field: *a scalar field with fermionic statistics*. It was found by Feynman, DeWitt [16], Faddeev and Popov [17]. The latter gave a very concise and elegant description of its origin. One writes the gauge fixing term and the Faddeev–Popov ghost in the Lagrangian $\mathcal{L}$ as

$$\mathcal{L}\to\mathcal{L}+\Delta\mathcal{L},\qquad \Delta\mathcal{L}=-\tfrac{1}{2}C(A,\varphi)^2+\overline{\eta}\,\frac{\partial C(A,\varphi)}{\partial\Lambda}\,\eta, \tag{7.4}$$

Here, Λ stands short for the generator of gauge transformations $\Lambda^{ab}(x)$, and η, $\overline{\eta}$ stand short for the (complex) fermionic ghost field $\eta^a(x)$ and its conjugate. The indices a, b, are the indices specifying gauge rotations in different directions (making these, in general, non-Abelian).

The fact that this system removes all unphysical particles (the time-like and the longitudinal gauge field components, as well as the Faddeev–Popov ghosts themselves), can be seen to be exactly as if a symmetry transformation is used to remove these fields, but this was not immediately recognised as such. This is why, in our original proofs [15], we meticulously considered all Feynman diagrams and their transformations, which lead to identities among diagrams that were first called Ward–Takahashi identities, after which they would be renamed Slavnov–Taylor identities [18, 19].

Today, however, we know that it is a new symmetry after all, a symmetry obeyed by the gauge-fixed Lagrangian (7.4). It was discovered by Becchi et al. [20], and independently by Tyutin [21]. But the reason why this was not found earlier is that this is a *supersymmetry*, that is, it transforms bosonic fields into fermionic ones and vice versa.

Writing the Lagrangian with a few more indices, we add a Lagrange multiplier field $\lambda^a(x)$:

$$\mathcal{L} = \mathcal{L}^{\text{inv}} + \lambda^a(x) C^a(A, x) + \overline{\eta}^a(x) \frac{\partial C^a(x)}{\partial \Lambda^b(x')} \eta^b(x') + f(\lambda^a), \tag{7.5}$$

where $\mathcal{L}^{\text{inv}}$ is the original gauge-invariant Lagrangian and the rest represents the gauge-fixing procedure. The function $f(\lambda^a)$ reproduces a more general function of C, so that this reproduces (7.4). One now considers the following infinitesimal transformation, generated by $\overline{\varepsilon}$:

$$\begin{aligned} \delta A^a(x) &= \overline{\varepsilon}\, \frac{\partial A^a(x)}{\partial \Lambda^b(x')}\, \eta^b(x'); \\ \delta \eta^a(x) &= \tfrac{1}{2} \overline{\varepsilon}\, f^{abc}\, \eta^b(x)\, \eta^c(x); \\ \delta \overline{\eta}^a(x) &= -\overline{\varepsilon}\, \lambda^a(x); \\ \delta \lambda^a(x) &= 0. \end{aligned} \tag{7.6}$$

With some algebra, one can establish that the action $S = \int \mathrm{d}^4 x\, \mathcal{L}(x)$, with $\mathcal{L}$ as in Eq. (7.5), is invariant:

$$\delta S = 0. \tag{7.7}$$

This is exactly what is needed to derive Eq. (7.3), with the ghosts transformed away.

Fig. 1 Quark confinement through vortex formation

8 Permanent Quark Confinement

In the early days, it was not realised that, actually, gauge theories *without* a BEH type of 'spontaneous symmetry breaking', are a lot harder to understand than the BEH theories. This is because these theories contain strongly interacting, massless particles: *gluons*. The prime example of such a theory is Quantum Chromo Dynamics (QCD). In this theory [22], it is now understood that the *magnetic dual* of the BEH mechanism takes place. This causes quarks to be confined by the formation of electric vortex configurations (Fig. 1).

We have the following general conjecture [23] for the vacuum structure of all local gauge theories in four space-time dimensions: the vacuum is one of three possible phase configurations:

(1) The standard BEH mechanism with a Higgs field whose vacuum orientation in the space gauge orientations can completely fix the local gauge invariance. In this case, there are no massless gauge bosons left, so that there is no infrared divergence anywhere in the system.
(2) The vacuum may be in the electric/magnetic dual of the BEH state, where we see massive glueball particles playing the role of gauge bosons, while all particles in non-trivial representations of the gauge group are confined, as we have in QCD, or
(3) A partial BEH mechanism. This may occur if the local gauge group is sufficiently large or complex. In this case, the Higgs field does not completely fix the gauge, leaving a non-trivial subgroup of the gauge group invariant. The complementary subset of gauge bosons acquires masses. An example of this is the actual electro-weak theory, in which a local $U(1)$ gauge group survives, generating the (long range) electromagnetic forces. In more general grand-unified theories, also an $SU(3)$ group survives, representing QCD.

In the sense of electric/magnetic duality, case (1) is dual to case (2), while case (3) is self-dual.

9 Glimpses of the Future

Superficially, it may seem that gauge theory has matured to become a quite well-established doctrine for describing particles of any sort. All we have to do is build more powerful particle accelerators to open up a world of new particles in mass regimes beyond what could be studied today. The Large Hadron Collider (LHC) was designed to do just that.

Yet, the indications found until now seem to point towards a new problem, called 'naturalness'. The Higgs particle has been found, and its mass was carefully measured. There were two surprises. First, the Higgs mass turns out to be very specially tuned: when we perform scale transformations, the Higgs field self coupling parameter, $\lambda_{\rm Higgs}$, will run towards an almost constant value, *as if we are approaching a domain of particle physics where we have almost exact scale invariance.* We may be running towards a fixed point, but it seems to be an unstable fixed point. Since scale invariance is not an exact symmetry, but a broken symmetry, it appears that such a system does not obey the conditions of naturalness. This may mean that our description of the physical world is incomplete. It is a difficulty that transcends our theoretical understanding of gauge theories, something new is to be suspected.

The other surprise may actually be related to this, in an unexpected way: the absence of new particle states that could turn the Standard Model into a supersymmetric system, or into a system with new confined building blocks called 'technicolour'. New heavy particles would strongly break scale invariance; is this why they are not there? If this aspect of our findings is not going to change soon, either in LHC or in its successors, it seems that scale symmetry will be a deeper and more fundamental symmetry than what was thought until now.

It is almost inevitable to extend scale transformations towards a group of *local* scale transformations. We then get local *conformal* symmetry. And this brings us towards quantum gravity. Gravity actually is a theory with local conformal invariance; no physical changes are necessary to establish this fact. This we see by re-writing the Einstein–Hilbert Lagrangian for gravity as follows. Entering, temporarily, a new scalar field $\varphi(x)$, we replace the metric tensor $g_{\mu\nu}(x)$ by $\varphi^2(x)\, g_{\mu\nu}(x)$, to obtain

$$\begin{aligned} \mathcal{L}_{\rm EH} &= \tfrac{\sqrt{-g}}{16\pi G}\,\big(g^{\mu\nu}\varphi^2\,R_{\mu\nu} + 6g^{\mu\nu}\partial_\mu\varphi\partial_\nu\varphi\big)\,; \\ \mathcal{L} &= \mathcal{L}_{\rm EH} + \sqrt{-g}\big(-\tfrac12 g^{\mu\nu}\partial_\mu\phi\partial_\nu\phi - \tfrac1{12}R\phi^2 - V(\phi) - \tfrac14 F_{\mu\nu}F^{\mu\nu} + \cdots\big). \end{aligned} \tag{9.1}$$

Here, $\phi(x)$ stands short for all scalar (Higgs) fields, while the fermionic fields have been momentarily ignored.

Returning to Weyl's transformations, Eqs. (1.1) and (1.2) in a slightly different notation, the *local conformal transformation*

$$
\begin{aligned}
g_{\mu\nu} &\to \omega^2(x)\, g_{\mu\nu}, \\
\sqrt{-g} &\to \omega^4(x)\sqrt{-g}, \\
\varphi(x) &\to \omega^{-1}(x)\, \varphi(x), \\
\phi(x) &\to \omega^{-1}(x)\, \phi(x), \\
A^a_\mu(x) &\to A^a_\mu(x), \quad \text{etc.}
\end{aligned}
\tag{9.2}
$$

is now a genuine local gauge symmetry. We see that the 'gauge condition' $\varphi(x) = 1$ restores ordinary gravity to its usual form. Note that, in many respects, the new scalar field $\varphi(x)$ acts as an ordinary scalar field (after scaling it in order to obtain a more familiar kinetic term), except for a sign switch.

This was just a short introduction to canonical conformal invariant gravity. It can be turned into an *almost* renormalizable theory [24, 25], except that a kinetic term for the metric field $g_{\mu\nu}$ does not have the appropriate form. Attempting to add a suitable kinetic term gives rise to sign difficulties and hence a breakdown of unitarity. These things are being disputed but no consensus has been reached as yet.

It is tempting to speculate further on conformal symmetries. Suppose we take the Standard Model for granted, possibly after adding a few more fields of whatever kind is needed. Then scale for some 20 orders of magnitude to higher energies. At such scales, all mass terms can be ignored. However, in the absence of all mass terms, *all presently known physical fields can be associated to Nambu–Goldstone bosons, and even Nambu–Goldstone fermions.* These are associated to an obvious global symmetry for the Lagrangian:

$$
\begin{aligned}
\varphi_a(x) &\to \varphi_a(x) + C^s_a, \\
A^a_\mu(x) &\to A^a_\mu(x) + C^{v\,a}_\mu, \\
\psi_k(x) &\to \psi_k(x) + \eta_k,
\end{aligned}
\tag{9.3}
$$

where $C^{s,\cdots}_a$ represent commuting constants and η_k are anticommuting. Thus, as many symmetries as particle types present. The commutators of these symmetry generators cannot be derived today, so we know little about the group structure of these symmetries.

When, at larger distance scales, the mass terms re-emerge, these all represent a relatively tiny explicit symmetry breaking.

The extremely high-energy behaviour of our theories cannot be detached from questions concerning the quantisation of the gravitational force, and this is why the problem becomes difficult. The appropriate mathematical instruments needed for incorporating gravity in our (gauge) theories are deficient.

References

1. H. Weyl, Raum, Zeit, Materie [Space, Time, Matter], in *Lectures on General Relativity* (Springer, Berlin, 1993 [1921]). ISBN 3-540-56978-2 (in German)
2. A. Pais, *Inward Bound: Of Matter and Forces in the Physical World* (Clarendon Press, 1988). ISBN: 9780198519973
3. R.P. Feynman, Relativistic cutoff for quantum electrodynamics. Phys. Rev. **74**, 1430 (1948)
4. J. Schwinger, On Quantum electrodynamics and the magnetic moment of the electron. Phys. Rev. **73**, 416 (1948)
5. Tomonaga, S. On a Relativistically Invariant Formulation of the Quantum Theory of Wave Fields. Prog. Theor. Phys. 1, 27–42 (1946)
6. F.J. Dyson, The S matrix in quantum electrodynamics. Phys. Rev. **75**, 1736 (1949)
7. M. Veltman, Perturbation theory of massive Yang-Mills fields. Nucl. Phys. B **7**, 637 (1968)
8. C.N. Yang, R.L. Mills, Conservation of isotopic spin and isotopic gauge invariance. Phys. Rev. **96** 191 (1954)
9. J. Goldstone, Field theories with superconductor solutions. Nuovo Cim. **19** 154 (1961)
10. Y. Nambu, G. Jona-Lasinio, Dynamical model of elementary particles based on an analogy with superconductivity. 1. Phys. Rev. **122**, 345 (1961)
11. P.W. Higgs, Broken symmetries, massless particles and gauge fields. Phys. Lett. **12**, 132 (1964); id., Broken symmetries and the masses of gauge bosons. Phys. Rev. Lett. **13**, 508 (1964); id., Spontaneous symmetry breakdown without massless bosons. Phys. Rev. **145**, 1156 (1966)
12. F. Englert, R. Brout, Broken symmetry and the mass of gauge vector mesons. Phys. Rev. Lett. **13**, 321 (1964)
13. G. 't Hooft, Renormalizable Lagrangians for massive Yang–Mills fields. Nucl. Phys. **B35**, 167 (1971)
14. G. 't Hooft, M. Veltman, Diagrammar. CERN Report 73/9 (1973), reprinted in Particle interactions at very high energies. Nato Adv. Study Inst. Series, Sect. B **4b**, 177
15. G. 't Hooft, Renormalization of massless Yang-Mills fields. Nucl. Phys. **B33**, 173 (1971)
16. B.S. DeWitt, Theory of radiative corrections for non-abelian gauge fields. Phys. Rev. Lett. **12**, 742 (1964); id., Quantum theory of gravity. 1. The Canonical theory. Phys. Rev. **160**, 1113 (1967); id., Quantum theory of gravity. 2. The Manifestly Covariant theory. Phys. Rev. **162**, 1195 (1967); id., Quantum theory of gravity. 3. Applications of the Covariant theory. Phys. Rev. **162**, 1239 (1967)
17. L.D. Faddeev, V.N. Popov, Perturbation theory for Gauge-invariant fields. Phys. Lett. **25B**, 29 (1967)
18. A. Slavnov, Ward identities in Gauge theories. Theor. Math. Phys. **10**, 153 (1972) (in Russian), Theor. Math. Phys. **10**, 99 (1972) (Engl. Transl.)
19. J.C. Taylor, Ward identities and charge renormalization of the Yang-Mills field. Nucl. Phys. B **33**, 436 (1971)
20. C. Becchi, A. Rouet, R. Stora, Renormalization of the Abelian Higgs-Kibble model. Commun. Math. Phys. **42**, 127 (1975); id., Renormalization of Gauge theories. Ann. Phys. (NY) **98**, 287 (1976)
21. I.V. Tyutin, Gauge invariance in field theory and statistical physics in operator formalism. Lebedev Prepr. FIAN **39** (1975) (In Russian). e-Print arxiv.org/0812.0580 [hep-th]
22. H. Fritzsch, M. Gell-Mann, H. Leutwyler, Advantages of the color Octet Gluon picture. Phys. Lett. **47B**, 365 (1973)
23. G. 't Hooft, *Confinement and topology in non-abelian Gauge theories.* Lectures given at the Schladming Winterschool, 20–29 February 1980, Acta Physica Austriaca. Suppl. XXII. **1980**, 531 (1980)
24. G. 't Hooft, Local conformal symmetry: the missing symmetry component for space and time. ITP-UU-14/25; SPIN-14/19, Int. J. Modern Phys. **D24**(12), 1543001 (2015), arxiv.org/1410.6675 v2 [gr-qc]
25. P.D. Mannheim, Solution to the ghost problem in fourth order derivative theories. Found. Phys. **37**, 532–571 (2007), arxiv.org/hep-th/0608154

MIX
Papier aus verantwortungsvollen Quellen
Paper from responsible sources
FSC® C105338

If you have any concerns about our products,
you can contact us on
ProductSafety@springernature.com

In case Publisher is established outside the EU,
the EU authorized representative is:
Springer Nature Customer Service Center GmbH
Europaplatz 3, 69115 Heidelberg, Germany

Printed by Libri Plureos GmbH
in Hamburg, Germany